MASTERPASONA
RUTGERS
SPRING 1973

DAVISON HALL
DOUGLASS COLLEGE
RM. 115
ASK FOR FRANK

The Growth

of Logical Thinking

from Childhood to Adolescence

BOOKS BY JEAN PIAGET
AVAILABLE IN ENGLISH TRANSLATION

The Language and Thought of the Child (1926)
Judgment and Reasoning in the Child (1928)
The Child's Conception of the World (1929)
The Child's Conception of Physical Causality (1930)
The Moral Judgment of the Child (1932)
The Psychology of Intelligence (1950)
Play, Dreams, and Imitation in Childhood (1951)
The Child's Conception of Number (1952)
The Origins of Intelligence in Children (1952)
Logic and Psychology (1953)
The Construction of Reality in the Child (1954)

[WITH BÄRBEL INHELDER]

The Child's Conception of Space (1956)
The Growth of Logical Thinking from Childhood to
    Adolescence (1958)

DATES GIVEN IN PARENTHESES ARE
THE FIRST PUBLICATION DATES OF THE ENGLISH TRANSLATIONS

The Growth of
Logical Thinking

FROM CHILDHOOD TO ADOLESCENCE

Bärbel Inhelder

and

Jean Piaget

AN ESSAY ON THE CONSTRUCTION OF FORMAL OPERATIONAL STRUCTURES

TRANSLATED BY ANNE PARSONS AND STANLEY MILGRAM

Basic Books, Inc., Publishers

Published in France
under the title *De La Logique de l'enfant à la logique de l'adolescent*
by Presses Universitaires de France

Translators' Introduction: A Guide For Psychologists

SOME of those who are drawn by the psychological title to choose this book from the display shelf may be tempted to put it down when they discover how many pages are filled with v's and ⊃'s or p's and q's. But their interest is not misplaced, for this is not a work on logic. It is a book which should be relevant both to the experimental psychologist interested in cognition and to the clinical psychologist or psychiatrist who deals with children or adolescents and who would like to know more about ego development. Logic does appear, both in that it is the more strictly logical aspects of the child's and the adolescent's thinking which make up the subject matter of the book and in that logical notation is used to provide a structural model of their thought processes. But this, we think, is not sufficient reason for putting it down, even for the person whose traumatic experiences with high-school mathematics have erected barriers around that part of the cognitive field labeled "abstract symbolism."

Nevertheless, the book poses a number of problems for the reader who is not familiar with the authors' methods and basic assumptions and who has no formal training in logic. On the one hand, it is a new installment in a long series of empirical works on the child's mental processes: it goes one step more up the genetic scale and covers the transition to adolescence. But in addition—

and in this respect it goes far beyond most of the earlier works —it is an attempt to isolate and describe the *mental structures* on which these reasoning processes are based. It is here that logic comes in. The empirical material is complemented by a structural analysis which uses symbolic logic as a tool. The set of mental structures which characterize the reasoning of the 7–11-year-old child is isolated and differentiated from the structures which characterize the reasoning of the 12–15-year-old adolescent.

As the authors state in the preface, this is a collaborative work based on an after-the-fact convergence. Professor Inhelder is primarily an experimental child psychologist and, at the time the work was conceived, was engaged in the study of adolescent reasoning. Professor Piaget, on the other hand, is an interdisciplinary thinker, and, besides the better known work on the thought of the child, he has also done independent work in logic. In comparing results of their respective recent work, it was discovered that Prof. Piaget's logical analysis provided the appropriate structural model for the data on adolescent reasoning collected by Prof. Inhelder.

Their collaboration rests on a view of the relationship between logic and psychology which Piaget has exposed elsewhere.[1] To sum it up briefly, logic and psychology are two independent disciplines: the first is concerned with the formalization and refinement of internally consistent systems by means of technically purified symbolism; the second deals with the mental structures that are actually found in all human beings, independent of formal training or the use of particular notational symbols and regardless of consistency or inconsistency, truth or falsehood.

But, although the formalization of systems as an activity in its own right belongs to logic alone, logic may be applied as a theoretical tool in the description of the mental structures that govern ordinary reasoning, as is done here. As an attempt to describe such structures, the present work is obviously of interest to the psychologist. But, since Piaget's work presupposes some understanding of the methods and concepts of two fields (and he does not attempt to translate concepts across academic boundaries for the uninitiate) his innovations are not easily assimilated. For this

[1] See Piaget, *Logic and Psychology* (New York: Basic Books, 1957).

reason we will try, in this "Guide," to furnish a few landmarks for the psychologist, experimental or clinical, by defining in terms more familiar to American psychology a few of the basic concepts used.

Although, over the last few decades, child psychology has on the whole been a more prominent focus of attention in the United States than in Europe, the work of the Piaget school has had little significant influence on this side of the Atlantic. The failure of the concepts to spread can be explained partially from the fact that, both theoretically and methodologically, Piaget occupies a sort of midway point between the main currents in American psychology. His direct plunge into complex human functioning and his neglect of tables of statistical significance or systematic response variation in favor of running commentary on selected protocols in the presentation of data have separated him from those groups which most emphasize methodological rigor. But his work is equidistant from that of the clinically-oriented psychologists and those currents which touch on sociology, social psychology, or anthropology, since he has no grounding in motivation theory and for the most part has chosen problems relative to cognitive functioning taken in isolation from any motivational variables. Moreover, since both sides of the American psychological world tend to divorce themselves from any philosophical tradition, his rationalist framework and ventures into philosophy have not been easily assimilated by either. Perhaps one could say that Piaget uses logic in a way analogous to the American use of theories of motivation (either reinforcement theory or psychodynamics) as an external frame of reference for study of the learning process.

But within its own framework, the Piaget method is both flexible and coherent and in a sense reconciles clinical and experimental approaches. Its basis is genetic—*i.e.*, intelligent behavior is analyzed with respect to the growth continuum. It is experimental in that constant problems or questions are presented to children of varying ages in samples large enough for general significance to be attributed to the differences found between age levels—*e.g.*, over 1500 subjects were tested for this work. Sometimes tests have been used; in these cases, success at a particular age is judged on the basis of a statistical norm.[2] Other quantitative devices, such as counting the occurrence of certain types of logical connectives relative to age in the spontaneous speech of children, have been used. For further problems, where response types may only partially correlate with age, medians have demonstrated that the age variable accounts for at least some of the systematic variation.[3] Thus some measurable growth in intelligence has been isolated.

But Piaget has not restricted his work to the gathering of measurable data. Like most clinicians, he has been very much concerned with some of the limitations and systematic biases inherent to quantifiable tests. Given the goal of describing the spontaneous intelligence of the child rather than his intelligence as seen through adult eyes, he usually chooses questions and evaluates answers in a way which the psychiatrist will find more sympathetic than will the experimental psychologist. Some works—in particular *The Child's Conception of the World* [4]—are based almost entirely on an intuitive attempt to explore an inner world and use no methodological paraphernalia beyond the skillful choosing of questions and evaluation of answers.

But, actually, the bulk of the research is neither experimental nor clinical in these two polar senses. It depends little on quantification of specific responses since intelligence is considered as a whole and often, as in the present study of propositional logic, in its most complex and highly integrated forms. Nevertheless, it is systematic and empirical in that various aspects of the child's intelligence have been taken up in turn and examined through the presentation of the same well-defined questions to large samples of subjects. And since a continual effort has been made to test hypotheses and reformulate the theoretical whole during more than twenty-five years of work, a solid body of knowledge

[2] A test is not considered passed until 75 per cent of the children tested succeed. See Piaget, *Judgment and Reasoning in the Child* (Humanities Press, 1952).

[3] See Piaget, *The Moral Judgment of the Child* (Free Press, 1948). It was discovered that children's expectations in regard to punishment differ significantly with age and at least to some extent independently of family socialization methods.

[4] Piaget, *The Child's Conception of the World* (Humanities Press, 1951). See the introduction to this work for a view of the relationship between testing and clinical method and a conception of the latter as used for the understanding of the child's mentality.

has been obtained. The criticism has been made that neglect of motivational factors detracts from the significance of the results. But, as Piaget states in the *Genetic Epistemology*,[5] the difference between science and philosophy is that the former tries to relate everything to everything else whereas the latter tries to delimit problems and find specific methods for dealing with them. The problems concerning validity of this schema across cultural lines, variation in function of motivational factors, etc., are still open.[6] In the last chapter of this work, the authors go beyond the purely cognitive and attempt to draw out the consequences of intellectual development for the affective and social psychology of the adolescent.

Over the series of works which attack intelligence at different points on the growth continuum and focus on different functions, the over-all aim has been to trace the development of intelligence as it comes to deal with increasingly complex problems or as it deals with simple problems in increasingly more efficient ways. The following are the four major stages of growth which have been delineated; the present work deals primarily with the transition from the third to the fourth.[7]

The first, covering the period from birth to about two years, is the sensori-motor stage. This is when the child learns to coordinate perceptual and motor functions and to utilize certain elementary *schemata* (in this context, a type of generalized behavior pattern or disposition) for dealing with external objects. He comes to know that objects exist even when outside his perceptual field and coordinates their parts into a whole recognizable from different perspectives. Elementary forms of symbolic behavior appear, as

[5] Piaget, *Introduction à l'épistémologie génétique* (Presses Universitaires de France, 1950), Vol. I, p. 7.

[6] Some attempts have already been made to combine the findings of Piaget with those of psychoanalysis. Piaget himself takes up problems relative to affectivity and discusses the relationship between his own theory and that of Freud in *Play, Dreams, and Imitation in Childhood* (Norton, 1952), and Charles Odier, in *Anxiety and Magic Thinking* (International Universities Press, 1956), attempts to relate the psychoanalytic theory of regression to Piaget's model of the early stages in ego formation. See also Rapaport, David, ed., *Organization and Pathology of Thought* (Austen Riggs Foundation, 1951), pp. 154–92.

[7] Concise definitions of the stages can also be found in Tanner and Inhelder, eds., *Discussions on Child Development* (Tavistock Publications, 1956), and Piaget, *The Psychology of Intelligence* (Routledge & Kegan Paul, 1956).

for example in the child who opens and shuts his own mouth while "thinking" about how he might extract a watch chain from a half-open matchbox. Expressive symbolism is also seen, as when Piaget's daughter at one year and three months lies down and pretends to go to sleep, laughing as she takes a corner of the tablecloth as a symbolic representation of a pillow.[8] From the behavioral standpoint, this period is covered in *The Origins of Intelligence in Children*[9] and from that of the organization of the perceptual field and the construction of the permanent object in *The Construction of Reality in the Child*.[10]

The preoperational or representational stage extends from the beginnings of organized symbolic behavior—language in particular—until about six years. The child comes to represent the external world through the medium of symbols, but he does so primarily by generalization from a motivational model—e.g., he believes that the sun moves because "God pushes it" and that the stars, like himself, have to go to bed. He is much less able to separate his own goals from the means for achieving them than the operational level child, and when he has to make corrections after his attempts to manipulate reality are met with frustration he does so by *intuitive regulations* rather than operations—roughly, regulations are after-the-fact corrections analogous to feedback mechanisms (*cf.* note p. 246). In the balance scale problem (Chap. 11), for example, we see that the preoperational subjects sometimes expect the scale to stay in position when they correct a disequilibrium by hand. They may, from an intuitive feeling for symmetry, add weight on the side where it lacks but may equally well add more on the overloaded side from a belief that more action leads automatically to success.

Protocols on this stage are found throughout the works, including this one. *The Child's Conception of the World* gives it the most attention from the standpoint of thought content. *The Child's Conception of Space*[11] takes up where *The Construction of Reality* leaves off in dealing with the perceptual aspect and the structuring of the spatial field.

[8] *Play, Dreams, and Imitation in Childhood*, p. 96.
[9] International Universities Press, 1952.
[10] Basic Books, Inc., 1954.
[11] Piaget and Inhelder, *The Child's Conception of Space* (Routledge & Kegan Paul, 1956).

Between seven and eleven years, the child acquires the ability to carry out *concrete operations*. These greatly enlarge his ability to organize means independently of the direct impetus toward goal achievement; they are instruments for dealing with the properties of the immediately present object world.

The stage of concrete operations has probably been more extensively studied than any other, but it is also that at which the greatest gaps are found in the list of English translations. *The Child's Conception of Physical Causality* [12] is especially interesting in that it covers many of the experiments used here, but it was written before the major phase of theoretical development. *The Child's Conception of Space* is theoretically closer to the present work and is in a sense its complement in covering the transition from the preoperational stage to the stage of concrete operations; it presents more fully some of the logical formulations used as the base line for discussing the transition to adolescence.

The fourth and final phase, preparatory to adult thinking, takes place between twelve and fifteen years and involves the appearance of *formal* as opposed to *concrete* operations. It is covered for the first time in detail in this work. Its most important features are the development of the ability to use hypothetical reasoning based on a logic of all possible combinations and to perform controlled experimentation.

Both the third and the fourth stages are *operational* as distinguished from the first two. The concept of operation has been elaborated gradually since Piaget's early work, partly in response to criticisms from Anglo-Saxon psychology [13] that the verbal aspects of intelligence had been overemphasized at the cost of actions. An operation is a type of action: it can be carried out either directly, in the manipulation of objects, or internally, when it is categories or (in the case of formal logic) propositions which are manipulated. Roughly, an operation is a means for mentally transforming data about the real world so that they can be organized and used selectively in the solution of problems. An operation differs from simple action or goal-directed behavior in that it is *internalized* and *reversible*. According to the authors: [14]

[12] Harcourt, Brace and Co., 1930.
[13] *Logic and Psychology*, p. 12.
[14] Definition given by the authors.

An operation is a reversible, internalizable action which is bound up with others in an *integrated structure*.

From the equilibrium standpoint [see Chap. 16], a transformation is reversible when it gives rise to complete compensation. From the structural standpoint [see Chap. 17], it is reversible when it can be canceled by an inverse transformation (as for example the direct and inverse operations comprising the "group" of transformations found in formal thinking).

The simplest definition of a reversible operation as it can be observed in concrete stage behavior is an action already performed which is symmetrically undone: *e.g.*, the child who puts a weight on the balance scale and realizes that it tips too far can take it off and search systematically for a lighter one, rather than add more weight simply for the sake of corrective action. With the advent of operations, the margin of trial-and-error is greatly decreased because the child selects means on the basis of an internal structure (in this example the structure is a *serial order* of weights). But even the most complex operations of propositional logic are seen as having their beginnings in actions which when internalized develop into highly differentiated mental structures.

From the theoretical standpoint, it is the *structural integration* of concrete and formal operations which is the principal concern of the present work. Although the number of intelligent acts of which a child is capable at any given age obviously depends on learning in a quantitative sense and on the situations which he happens to confront, the *range* of available operations can be described in terms of a limited number of interdependent structures. The structures found and the way in which they are integrated depends on the stage of development considered; each set of structures can be related to a particular group of logical forms. Thus the concrete structures depend on the logic of classes (for class-inclusion operations) and the logic of relations (for serial-ordering), whereas the adolescent's propositional logic depends on the integration of "lattice" and "group" structures in the *structured whole*. This obviously does not mean that the child or the adolescent acquires class logic or propositional logic in a formal sense. Rather, it means that in dealing with concrete problems he arrives at answers which imply the presence of certain logical forms although he does not isolate them from content as does the logician. The logical forms both are present in the child's reasoning and serve as a structural model for analysis of it.

The structures are integrated at each stage in the sense that each partial operation is used in relation to the totality of those which are available: [15]

Structural integration occurs when elements are brought together in a whole which has certain properties as a whole and when the properties of the elements depend partially or entirely on the characteristics of the whole. Some examples are: classifications, serial orders, correspondences, matrices, "groups," lattices, etc.

Each set of structures has its limitations in terms of the field which can be covered. The concrete stage protocols show the point at which the limits of the "grouping" structures are reached, necessitating the development of a new form of integration.

Below are some definitions of the operations based on the "grouping" structure which develop during the concrete stage (see Mays' introduction to *Logic and Psychology* for a more formal set of definitions):

1. *Class inclusion operations.* These relate to the child's ability to manipulate part–whole relationships within a set of categories. In order to define the operations for class inclusion, logicians use the terms addition, subtraction, and multiplication in a qualitative sense which is analogous to their use in defining arithmetical operations. Two classes can be added up so that they are included in a larger one: boys + girls = children; children + adults = people—*i.e.*, $A + A' = B$. By the same token, a part can be subtracted from the whole: people — adults = children. When the child can do this systematically, reversibility is present in that when the child needs to generalize or discriminate he can pass from the part to the whole and back again.

Likewise, classes can be multiplied. The child obtains four subclasses by discriminating between objects or properties according to two intersecting criteria: A (geometric figures) divided into B (squares) and B' (circles) and multiplied by ($\times$) A_1 (their colors) B (red) and B'_1 (green) gives $BB_1 + BB'_1 + B'B_1 + B'B'_1$ (red squares, red circles, green squares, and green circles). These are the double-entry tables referred to frequently in the text. The

[15] Definition given by the authors.

class multiplication operation is the means the 7–11-year-old child uses for discriminating between two (or more) independent variables.

Such relationships seem obvious, but experiments have shown that they are not made systematically when the preoperational child uses categories. Before the age of about seven, for example, children, given a box containing about eighteen brown and two white beads, all wooden, and asked whether there are more brown or more wooden beads, reply that there are more brown ones because only two are white.[16] That the categories are available and observations correct is shown by the fact that the younger children, when asked the questions separately, give correct answers as to the relative proportions of brown, white, and wooden beads. However, without class inclusion operations they cannot deal with the parts and the whole at the same time, and thus they make a false generalization.

2. *Serial ordering operations.*   These operations relate to ability to generalize along a linear dimension or to arrange objects (or their properties) in series. They are based on the logic of relations rather than class logic: the signs are $>$ and $<$ (greater than or less than). At about seven years, when the child is given a set of sticks to arrange in order of size, he proceeds by taking the smallest first (or the largest), then the smallest of those which are left, and so on, rather than beginning at random and rearranging when discrepancies are noticed. Mentally he is able to conclude, from $A > B$ and $B > C$, that $A > C$. Other empirical factors are ordered in the same way at different points during the concrete stage—*e.g.*, weights are ordered later than lengths, at about nine to ten years. When he has acquired serial ordering operations, the child is able to register in detail the changes in magnitude of a given variable—*e.g.*, in the angle problem (Chap. 1), he sees that the ball's course changes in direct relation to his angle of firing. Actually, in the mental operation he puts the angles into a series of increasing magnitude of which each member corresponds to a trial.

Secondly, given two independent series, the child learns to find *correspondences* between them (sign ↔). In other words, he begins to relate two variables accurately by observing concurrent

[16] *The Psychology of Intelligence*, p. 133.

changes. In the angle problem he sees that the series of angles of firing corresponds to the series of angles in the ball's course toward the goal, thus he is able to adjust his firing correctly. As opposed to the preoperational child, he comes to know *which means* goes with *which end*.

In sum, the concrete operations are based on the logic of classes and the logic of relations; they are means for structuring immediately present reality. During the formal stage, on the other hand, the adolescent comes to control formal logic. Rather than reasoning with directly given data alone, he begins to reason with propositions and with hypotheses. The concept of the "concrete operation" was developed as a means of applying logical analysis to the child's operations when he is dealing directly with objects and thus of avoiding the fallacy of judging the child's intelligence in terms of that of the logician. But in the study of adolescent reasoning, which is much closer to that of the logician than that of the seven-year-old child, the concept of operation has by no means been abandoned. Rather, formal logic is also conceived of as a set of mental operations, although they are based on a different structure. From one standpoint the formal operations differ from the concrete in that they are performed on propositions rather than directly on reality: [17] they are a set of transformations which can be made as a way of generalizing from the data at hand. As opposed to concrete operations which are limited to the empirically given, they make it possible for the subject to isolate variables and to deduce potential relationships which can later be verified by experiment.

The *propositions* on which formal operations are performed refer both to variables hypothesized as causal and to the effects they produce in the experimental situation. In the flexibility problem, for example (Chap. 3), where the adolescent is able to isolate and combine variations in flexibility (which depends on a number of factors), what the subject does from the behavioral standpoint is to ascertain a number of facts and formulate them as propositions—*e.g.*, "This rod is steel; it is also long," "That rod is steel, but it is shorter," etc. These refer to the potentially causal properties of the rods he has chosen to test. He also ascertains the

[17] This does not mean that formal logic is verbal and concrete logic is nonverbal. (See pp. 252–254.)

results of his experiments—*e.g.*, a long steel rod bends, a short brass one does not, a short steel one does, etc. The formal operations enable him to combine these propositions mentally and to isolate those which confirm his hypotheses on the determinants of flexibility. The *combinatorial system* is the structural mechanism which enables him to make these combinations of facts.

In other words, formal operations are ways of transforming propositions about reality so that the relevant variables can be isolated and relations between them deduced. The operations described (see p. 103, the sixteen binary operations) are *different kinds of combination,* any one of which may be appropriate depending on the particular situation observed. The frequently recurring term "association" refers to an observed conjunction of facts—*e.g.*, "this rod is steel" and "this rod (the same) bends enough to touch the water." The kind of relationship formulated depends on the particular association of facts observed: *e.g., implication* means that every time one variable appears in the experimental situation, a particular result is present—every time the string of a pendulum is lengthened, the amplitude of the swing increases—and if this variable is not present, that result is never present—the amplitude cannot be increased without lengthening the string. The formula for implication is $p \supset q$. Another type of combination is *disjunction*—*e.g.*, the situation in the flexibility problem when the subject sees that sometimes short rods bend but at other times they do not $(p.q \vee \bar{p}.q \vee p.\bar{q} \vee \bar{p}.\bar{q})$—you can have short rods that bend, or short rods that do not bend, or long rods that bend, or long rods that do not, with $\vee$ symbolizing "or." The p's and q's with their negations $\bar{p}$ and $\bar{q}$ stand for the observation that a given variable or its result is or is not found in the experimental situation. When the disjunctive relationship is observed, the subject concludes that length alone does not determine flexibility. Since, unlike the concrete level child, he does not have to limit his consideration to a single relationship at a time, he can then proceed to the consideration of other variables which might determine it. He "feeds" his information into a general mechanism—the combinatorial system or structured whole—which assimilates the facts in the form of propositions and arranges them according to all possible combinations (the logical term is *composition*). He can move around among these possibilities (there is reversibility and complete compensation) so as to select a situation that would tell him which other variables are involved and which of a number of potential explanations in fact explain what he saw: for example, that length alone does not determine flexibility so that another factor was involved for the short rod that bent; length does partially determine flexibility so that, in the situation in which the long rod did not bend, there was a counteracting factor, etc. Thus the propositional operations always operate as a whole and as a whole which is structured internally. The adolescent both discriminates between parts (variables or specific events which occur, such as the rod's bending beyond the degree required for flexibility, etc.) and generalizes to an over-all explanation of the results and to other potential situations. As in Gestalt psychology, the development of thought is seen as moving toward the construction of wholes, but, as is emphasized to a greater extent, it also moves toward a finer discrimination of elements within the whole. The *structured whole* is structured precisely in the sense that the relationships between its parts are separable as well as integrated.[18]

In reading the book, various aspects of the structural model are best seen in individual experimental problems. Thus, the *combinatorial system* is presented in its purest form in the coloring liquids problem (Chap. 7) where the experiment itself calls for the systematic combining of a number of variables given as discrete; the differences between the adolescent method, which goes around the full circle of possibilities each time, and the child's method of one-by-one combination, which always leaves some

[18] In the translation "structured whole" we have had to sacrifice some of the connotations of the original in the interest of securing a meaningful and communicable equivalent. The French term is *ensemble des parties*, where ensemble means both "whole" (with the implication of integration as used by Gestalt psychology) and "set" as used by mathematical set theory. For logicians the term should be translated "the set of all sub-sets." Readers of Mays' translation of *Logic and Psychology* will notice a terminological difference in that he has used "structured whole" for the more general *structures d'ensemble*. There are *structures d'ensemble*, which we have translated as "structural integration" or "integrated structures," at both concrete and formal stages, but the type of structure found differs between the two. The combinatorial system or "structured whole," on the other hand, is the particular type of structural integration which characterizes formal thinking. Moreover, the term *structure d'ensemble* is employed by a school of mathematicians, in particular the Bourbaki, whose structural research lies at the foundation of this work. Here also, *ensemble* means "set" as used in set theory as well as "whole" or "integrated."

steps untouched, become obvious in the protocols. The contrast-
ing process in which discrete variables have to be extricated from
a situation where they appear in combination is described in
Chap. 3, the multivariate flexibility problem. There one sees that
at the concrete stage the variables somehow tend to stick together
once given combinations have been observed in sharp contrast to
the easy detachability from the given which they acquire at the
formal level. In sum, the structured whole, by virtue of which the
subject is able both to combine parts into a whole and to separate
them from it, might be impressionistically characterized as a sort
of mental scaffolding held up by a number of girders joined to
each other in such a way that an agile subject can always get from
any point—vertically or horizontally—to any other without trap-
ping himself in a dead end. The other structural forms—the "lattice" and the "group"—we had better leave to the logicians.

ANNE PARSONS

Preface

THE DOUBLE TITLE of this work [on the original French edition, it was *De La Logique de l'enfant à la logique de l'adolescent: Essai sur la construction des structures opératoires formelles*] does not indicate simply that the authors collaborated in a new way, or simply a desire to distinguish between their respective contributions to a common task. Actually it reflects the twofold nature of the questions which they have asked, and it in no way impairs the ultimate unity of the conclusions. With reference to the choice of this title, there is an anecdote worth citing, particularly because it is a good example of how experimental and deductive methods can at the present time converge in the area of the operational analysis of intelligence, given a deductive analysis based on precise logical techniques. It happened that the second author left his experimental work for a while in order to complete his *Genetic Epistemology*,[1] *Treatise on Logic*, and an *Essay on the Transformations of Logical Operations*. The aim of the two latter works is to furnish a possible symbolic model of the actual processes of thinking. During this period the first author and her assistants undertook a systematic empirical study of the induction of physical laws in children and adolescents. But this genetic study of experimental induction led to two unexpected results.

[1] Piaget, *Introduction à l'épistémologie génétique* (Presses Universitaires de France). Volume I will appear shortly in English translation.

First, in our earlier works we had repeatedly stressed the importance of the stage of development beginning at 11–12 years. In our new study it became increasingly clear that this stage was not simply the culmination point of the 7–11-year stage (when concrete operations are worked out by the child) but also involved a period of new structuring leading to another level of equilibrium at about 14–15 years. So it seemed possible to describe the adolescent's thought in terms of the structuring of certain methods of experimental induction, and above all in terms of those methods of systematic verification not found in the child.

As to the second result, the methods of discovery and experimental proof found in the adolescent but not in the child were shown to be bound up with an entirely new set of operational structures. These are based on propositional logic and a "formal" mode of thought as distinguished from the "concrete" operational thought found between 7 and 11 years. (The latter requires only a limited number of operations taken from the logic of classes and relations.)

The second author intervened at this point, and for the following reasons, which well illustrate the convergence referred to above between the results of experiments and those of deduction. It became clear that the well-known techniques of propositional logic were inadequate to analyze the integrated structures of operations found in the adolescent's formal thinking. For we also had to make use of the group of four transformations (inversions and reciprocities) which one of us has described as necessary to the functioning of the mechanisms of formal thought.[2]

Now, the outstanding feature of the data of the empirical investigation was that they showed that formal thought is more than verbal reasoning (propositional logic). It also entails a series of operational schemata which appear along with it; these include combinatorial operations, propositions, double systems of reference, a schema of mechanical equilibrium (equality between action and reaction), multiplicative probabilities, correlations, etc. But in trying to explain how these operational schemata and propositional logic develop together, we found it was not enough to

[2] See Piaget, *Traité de Logique* (Colin, 1949), pp. 264–286, and especially *Essai sur les transformations des operations logiques* (Presses Universitaires de France, 1952), Chap. II. Neither of these works is available in English.

refer only to the specific operations of propositional logic. For in addition—and this is most important—we have to refer to the "integrated structures" on which they are based—*i.e.*, to the dual structure of the lattice and the group of four transformations (Klein group or *Vierergruppe*) analyzed by the second author in his work on the transformation of propositional operations.

In other words, while one of us was engaged in an empirical study of the transition in thinking from childhood to adolescence, the other worked out the analytic tools needed to interpret the results. It was only after we had compared notes and were making final interpretations that we saw the striking convergence between the empirical and the analytic results. This prompted us to collaborate again, but on a new basis. The result is the present work.

But this is not all. The operational structures of adolescent logic are not only interesting in themselves; they also cast a backward light on an earlier set of structures, those of the child's concrete logic. Actually, the only logical operations the child can handle at the concrete level are the "elementary groupings" of classes and relations; the class groupings are based on a form of reversibility which can be called *inversion* (negation), and the groupings of relations on another such form, called *reciprocity*. At this stage there is no general structure to integrate transformations by inversion and transformations by reciprocity into a single system. But analysis of the set of four transformations found in the propositional logic of the adolescent shows how the two forms of concrete operational reversibility finally do come to be coordinated into a single system.[3] Meanwhile, the combinatorial system of propositional lattices develops as a result of a generalization of classification. In other words, it seems clear that the twofold structure found in formal thought is the end product of a series of coordinations as they attain a final level of equilibrium. (This is no bar to new integrations and continual growth in adult thinking.) Therefore, an analysis of the mechanisms of formal thought is indispensable if we are to draw up an operational theory of intelligence which aims at a step-by-step explanation of the successive and hierarchical organization of thinking as it develops.

This book has two aims: to set forth a description of changes

[3] The four transformations: inversion; reciprocity; inversion of the reciprocal, or reciprocation of the inverse; and identical transformation.

in logical operations between childhood and adolescence and to describe the formal structures that mark the completion of the operational development of intelligence. To tie these together the authors have tried to present the material in a way that would stress the close relationship between the two. Each of the first fifteen chapters (Parts I and II) includes an experimental part by the first author and a brief final analysis by the second author. This analysis aims to isolate the formal or propositional structures found in each case.[4] Chapters 16 and 17 (beginning of Part III) are the work of the second author, whereas Chapter 18 is a joint production. In addition, the specific problems of experimental induction analyzed from a functional standpoint (as distinguished from the present structural analysis) will be the subject of a special work by the first author.[5]

BÄRBEL INHELDER

JEAN PIAGET

[4] For a more detailed presentation of the symbolism of propositional operations employed in the chapter conclusions and in Chap. 17, see *Traité de Logique*, Chap. V.

[5] *Translators' note:* The experimental diagrams from this work have been included in this translation.

Contents

PART III

The Structural Integration of Formal Thought

Part I

THE DEVELOPMENT

OF PROPOSITIONAL LOGIC

IF WE are to explain the transition from the concrete thought of the child to the formal thought of the adolescent, we must first describe the development of propositional logic, which the child at the concrete level (stage II: from 7–8 to 11–12 years) cannot yet handle. Experimentation shows that after a long period during which only operations appropriate to class and relational groupings and to the numerical and spatiotemporal structures which resulted from them are used, the beginnings of stage III (substage III-A, from 11–12 years to 14–15 years; and substage III-B, from 14–15 years onward) are distinguished by the organization of new operations performed on the propositions themselves and no longer only on the classes and relations that make up their content.

To study the questions raised by this development, we must analyze how children or adolescents at stage III go about solving problems which appear purely concrete but which experiments indicate can be resolved only at stage III and which actually presuppose the use of interpropositional operations. Part I of the present work will be devoted to this analysis.

1

The Equality of Angles
of Incidence and Reflection
and the Operations
of Reciprocal Implication[1]

OUR AIM in this chapter, and in the remainder of Part I, is not a systematic study of the concept of the equality of two angles. Actually, we already know how the concept is constructed: that it is first acquired at the level of concrete operations.[2] But it is precisely the fact that the concept is already so well known by the time the formal level (stage III) is reached that makes the reasoning process involved in the discovery of the equality between the angles of incidence and reflection so instructive. One of the aims of this study, then, is to isolate the operational mechanisms involved in the formal reasoning process itself, when this reasoning rests on notions already constructed at the concrete level.

The experimental apparatus consists of a kind of billiard game. Balls are launched with a tubular spring device that can be pivoted and aimed in various directions around a fixed point. The

[1] With the collaboration of H. Aebli, former research assistant, Laboratory of Psychology, Science Faculty, University of Geneva, professor, École normale supérieure, Zurich; L. Müller, former research assistant, Institut des Sciences de l'Éducation, University of Geneva; and M. Golay-Barraud, student, Institut des Sciences de l'Éducation.

[2] See Piaget and Inhelder, *The Child's Conception of Space* (Routledge & Kegan Paul, 1956), Chap. XII, and Piaget, Inhelder, and Szeminska, *La Géométrie spontanée de l'enfant,* Chap. VIII. (Not transl.)

ball is shot against a projection wall [3] and rebounds to the interior of the apparatus. A target is placed successively at different points, and subjects are simply asked to aim at it. Afterwards, they report what they observed.

But the equality between the angles of incidence and reflection is discovered only at stage III-A (11–12 to 14 years) and is often not formulated until stage III-B (14–15 years). Our problem is to understand why a concept as familiar after 7–9 years as that of the equality of two angles is utilized in the induction of an elementary law only at this late date and, especially, why formal operations are necessary for its use. We shall try to answer this question by retracing briefly the ground covered by the child before his arrival at the formal level, then by examining the latter more closely.

§ Stage I

In the course of stage I (up to about 7–8 years) subjects are most concerned with their practical success or failure, without consideration of means; often even the role of rebounds is overlooked. The result is that, except toward the end of the stage, the trajectories are not generally conceived of as formed of rectilinear segments but rather as describing a sort of curve:

DAN (5 ; 2) * succeeds at first: *"I think it works because it's in the same direction."* He adjusts the plunger by himself, but proceeds by empirical trial-and-error. Then he asks spontaneously: *"Why do you have to turn the plunger sometimes? . . . No, you have to put it there* [he fails]. *If it could be pushed a little further"* [he does this and succeeds]. But, although he knows how to control the rebounds successfully, DAN has no idea that they are made up of angles: the curve he describes with his finger is not tangent to the wall; he takes into account the starting point and the goal but not the rebound points.

WIRT (5 ; 5): *"It came out here and it went over there. . . . I'm sure to make it,"* etc. He succeeds occasionally but describes the trajectories

---

[3] With a rubber buffer.

* Figures within parentheses indicate age in years and months—*i.e.*, five years; two months.

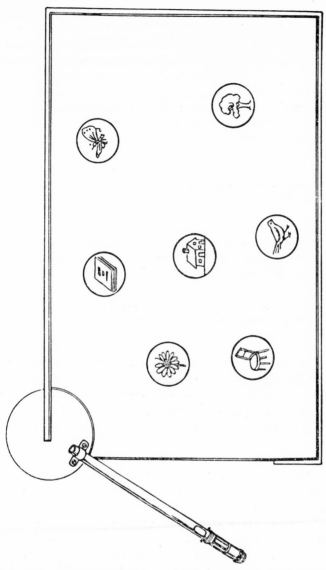

FIG. 1. The principle of the billiard game is used to demonstrate the angles of incidence and reflection. The tubular spring plunger can be pivoted and aimed. Balls are launched from this plunger against the projection wall and rebound to the interior of the apparatus. The circled drawings represent targets which are placed successively at different points.

with his finger only in the form of curves not touching the walls of the apparatus; he considers only the goal as if there were no rebounds.

NAN (5 ; 5), on the other hand, is astonished by the detour made by the ball which first touches the walls. *"It always goes over there."* But he does not succeed in adjusting his aim: *"Oh, it always goes there. . . . it will work later."*

PIT (5 ; 5) notes about one of his tries [a failure]: *"It was straight* [as if this were an exception].*—Why didn't it hit it?—I thought I hit it"* [no comprehension].

ANT (6 ; 6) becomes aware of the existence of rebounds at the same time that he notices the rectilinear character of the trajectory segments: *"It* [the ball] *hits there, then goes over there"* [his gesture indicates straight lines].

PER (6 ; 6), in contrast, in spite of his age, resorts to the curvilinear model: *"It goes there and it turns the other way"* [gesture indicating a curve].

The reactions of this stage are extremely interesting, for although the children demonstrate by their behavior that they know how to act in the experimental situation, sometimes successfully, they never *internalize their actions as operations,* even as concrete operations. In a general sense, by *concrete operations* we mean actions which are not only *internalized* but are also *integrated* with other actions to form general *reversible systems.* Secondly, as a result of their internalized and integrated nature, concrete operations are actions accompanied by an awareness on the part of the subject of the techniques and coordinations of his own behavior. These characteristics distinguish operations from simple goal-directed behavior, *and* they are precisely those characteristics not found at this first stage: the subject acts *only* with a view toward achieving the goal; he does not ask himself why he succeeds. In the experiment under consideration he is not aware of either the rectilinear nature of the trajectory segments or the existence of rebounds except toward the end of the stage (toward 6 or 6–7 years); consequently he cannot take note of the presence of angles at the rebound point.

§ Stage II (Substages II-A and II-B)

Substage II-A is distinguished by the appearance of concrete operations in the sense just defined:

VIR (7 ; 7) succeeds after several attempts. He points out and then draws trajectories with two distinct rectilinear segments, saying: *"To aim more to the left, you have to turn* [the plunger] *to the left."*

TRUF (7 ; 10): *"I know about where it will go";* in fact, he shows by his gestures that he realizes that the angle of rebound is extremely acute when the plunger is raised and extremely obtuse when it is lowered. Thus, he shows us that he has a vague global intuition of the equality between the angles of incidence and reflection. But he does not make it explicit, since he fails to divide the total angle indicated by his gesture into two equal angles.

BEND (8 ; 0): *"It's the corner* [the angle of rebound] *that makes it turn; you change the contour* [the size of the angle] *when you change the plunger"* [inclination of the plunger]. He demonstrates as did the preceding subject that the angle is extremely acute when the plunger is slightly inclined and extremely obtuse when it is sharply inclined. We ask him what he means by the contour, and he points to the opening of the angle with a gesture indicating that he is thinking of the very generation of the angle by the progressive rotation of the plunger and of the rebound of increasing amplitude which results.

DESI (8 ; 2): *"The ball always goes higher when the plunger is higher."* Then: *"The ball will go there* [further] *because the plunger is tilted more; I put my eyes high up* [= I pinpoint the rebound point] *and from the rubber* [= the rubber band attached to the wall on which the ball rebounds] *I look at the round pieces"* [= the disks serving as goals].

At substage II-B the preceding operations, which give rise to a model that includes straight lines and angles, are complemented by an increasingly more accurate formulation of the relations between the inclination of the plunger and that of the line of reflection:

NIC (9 ; 4): *"You have to move the plunger according to the location of the target; the ball has to make a slanting line with the target."*

KAR (9 ; 6): *"The more I move the plunger this way* [to the left—*i.e.,* oriented upwards], *the more the ball will go like that* [extremely acute angle], *and the more I put it like this* [inclined to the right], *the more the ball will go like that"* [increasingly obtuse angle]. KAR reaches the point of discovering that the ball returns to the starting point when the plunger is *"straight"—i.e.,* perpendicular to the rebound wall.

BAER (9 ; 6): "How do you explain it?"—*"It has to be at the same distance as the target"* [he points out the angle increasing with the withdrawal of the target and not the length of the line between plunger and rebound point or between the latter and the target].

ULM (9 ; 8): *"As you push the plunger up, the ball goes more and more like that* [acute angle], *and the more I put it like that* [inclined to the right], *the more the ball will go like that"* [obtuse angle].—"But, tell us more about what you are looking at."—*"I am still looking at that* [the goal], *and that's all, because it turns with the plunger"* [—because the direction of the path between the rebound and the goal changes with the inclination of the plunger].

DOM (9 ; 9): *"It hits there, then it goes there"* [he points out the equal angles, repeating his phrase for different inclinations of the plunger].

Thus we see that the subjects succeed in isolating all of the elements needed to discover the law of the equality of the angles of incidence and reflection, yet they can neither construct the law *a fortiori* nor formulate it verbally. They proceed with simple concrete operations of serial ordering and correspondences between the inclinations of two trajectory segments (before and after the rebound), but they do not look for the reasons for the relationships they have discovered. And they do not consider the segments except from the standpoint of the directions taken; thus the idea of dividing the total angle made up of the two segments into two equal angles (incidence and reflection) fails to occur to them.

However, in contrast to the stage I subjects, substage II-A and II-B subjects no longer limit themselves to overt performance but internalize their actions in the form of operations of placing or displacing: thus they have become aware of the facts that the plunger can be adjusted to specific slopes, that the trajectory of the ball is composed of two rectilinear segments, and, above all, of the fact that these two segments form an angle (whose peak coincides with the rebound point) whose size varies according to these slopes. They manage to order serially these latter inclinations, distinguishing between *"sharper"* or *"more to the left,"* *"higher,"* etc., and *"less sharp,"* etc., which amounts simply to a translation of the more or less well-ordered operations that they know how to execute beginning with stage I into coordinated operations of serial ordering. Similarly, they succeed in ranking the degrees of incline or the directions of the trajectory segments included between the rebound point and the goal ("the ball keeps going higher" or lower, it "will go here" or there, etc.). Finally, and particularly important, they establish a correspondence between the slope or direction of the plunger (and consequently of the first segment of the ball's trajectory) and the inclinations or directions of the second segment: "The more the plunger is (inclined, etc.), the more the ball will go (downwards, etc.)."

If the increasing inclinations of the plunger (and of the first segment of the trajectory for the ball leaving the plunger) are symbolized by the letters a, β, γ, etc. and the inclinations of the second segment (between the rebound point and the goal) by the signs a', β', γ', etc., the serial ordering and correspondence operations which subjects of this second stage can perform are as follows:

$$a < \beta < \ldots \text{ or } a' < \beta' < \gamma' \ldots \text{ and} \qquad (1)$$
$$a \leftrightarrow a', \ \beta \leftrightarrow \beta', \ \gamma \leftrightarrow \gamma', \text{ etc.} \qquad (2)$$

(where the sign $\leftrightarrow$ stands for the correspondence).

Why, then, does the correspondence between the two rank orderings fail to lead to the discovery of the law of the equality of the angles of incidence and reflection? It is because the subjects stick to the concrete rank ordering and correspondences without looking for the reasons behind this correspondence, just as the subjects at stage I knew what to do to attain their goal but did not look for the reasons behind their reactions (displacement of the plunger, etc.). The stage II subjects stick to dealing with facts whose accuracy is due to serial ordering and correspondence operations, but they do not seek to explain these facts further in terms of the formal operations of implication, etc., which are the conditions of hypothetico-deductive thought.

Since they do not seek an explanation of the observed facts, they must remain at the level of rough, global observation, certainly a great advance over that of stage I but still too global to lead to an analytic breakdown of the observed angles. Thus, because they are content to point out slopes of directions and to deal with the total angle composed of the two segments of the trajectory (BEND: the "contour"; ULMS "It turns"; etc.), they do not divide this total angle into the two equal parts that would give us the angle of incidence and the angle of reflection. That is why, although the subjects are very close to the discovery of the law and already possess all of its elements, it is not yet discovered; the formal operations needed for the quest for an explanatory hypothesis are lacking.

§ *The Formal Stage (Substages III-A and III-B)*

At this last stage the subjects finally discover the law of the equality of angles. At first the discovery is slow and partial, including verification or rejection of several specific hypotheses, then complete and rapid because subjects are oriented by the hypothesis that there is a necessary equivalence between two successive segments of the trajectory.

First, let us look at a case typical of substage III-A:

BON (14 ; 8) first invokes the launching force, then realizes that the trajectories are the same whether the balls are shot hard or soft. Next he invokes the role of the *"distances, how you have to place the rod."* Then he establishes concrete correspondences in the same way as the stage II subjects: *"It's the position of the lever* [of the plunger]: *the more you raise the target, the more you raise it here"* [the lever]. He uses a ruler to mark the trajectory of the ball between the rebound point and the target in such a way as to verify its correspondence with the orientation of the plunger. Then he hypothesizes that the angle is always a right angle: *"It has to make a right angle with the lever."* But after several trials he concludes: *"No, above* [= when the plunger is straightened] *it won't work."*—"It isn't ever a right angle?"—*"Yes, that's correct for one position."*—"And without that?"—*"When you turn, one should be smaller, the other larger. Ah! They are equal"* [he points out the angles of incidence and reflection].

Thus, in the later stages there is a search for a general hypothesis which can account for the concrete correspondences between the inclinations as soon as they are found. Subject BON thinks first of the right angle, then ascertains that the total angle is sometimes acute, sometimes obtuse; then he breaks it down to form two equal angles (incidence and reflection).

But the hypotheses found at substage III-A are still very close to concrete correspondences in that they attempt only to express the general factor which the correspondences contain. Substage III-B, on the other hand, is distinguished by a new exigency which is absent at substage II-B and still implicit at III-A: the need to find a factor which is not only general but also necessary —*i.e.*, which will serve to express beyond the constant relations the very reason for these relations.

In other words, the subjects at substage III-B are not completely satisfied with the establishment of a correspondence between the inclinations of the plunger and the line included between the buffer and the target, as are those of II-B. Nor are they satisfied with the search for a single constant factor which translates these correspondences, as are those of III-A. Initially they ask themselves why a certain difference in inclination x_1 of the plunger necessarily corresponds to a difference x_2 in the buffer-target line. This pursuit of a necessary reason, in certain cases going as far as an immediate appeal to the concept of necessity, is what distinguishes formal thinking, with its operations of implication or equivalence (= reciprocal implication), from concrete thinking, with its simple statements of constancy. This is demonstrated by the following subject, who begins, like BON, with the hypothesis of the right angle but soon afterwards turns to a search for necessity:

DEF (14 ; 8) imagines at first that the two trajectory segments always form a right angle. But after three trials he says: "*The more the target approaches the plunger, the more the plunger must* [necessity] *also approach the target*" [which signifies evidently that the two inclinations of the plunger and of the line between the target and the buffer imply each other reciprocally].—"What do you mean by 'must also approach the target?' "—"*For example, if there were a line here* [he indicates a line perpendicular to the buffer], *the ball would come back exactly the same way.*" Then he puts the plunger at 45°: "*That makes a right angle here and you have about the same distance as there*" [= the two angle openings]. Then he continues with several angles chosen at random and again verifies his law of equality. We object that the law may not be a very general one: "*It depends on the buffer too; it has to be good and straight—and also on the plane—it has to be completely horizontal. But if the buffer were oblique, you would have to trace a perpendicular to the buffer and you would still have to take the same distance from the plunger* [to the line and from it] *up to the target: the law would be the same.*" The buffer is turned around, replacing the rubber by wood: "*Perhaps the wood is less elastic: the ball would be sent back with less force.*"—"Then what about your law?"—"*The law doesn't vary.*"

Even in this first case we see several new factors appear which psychologically distinguish formal from concrete thinking: the requirement of necessity ("the plunger *must* also, . . ." "it would *have to be* the same width here and there," "you still would *have to* take the same distance again," etc.); the ability to formulate hypotheses or hypothetical constructions not given by direct observation (trace an ideal perpendicular from the buffer, etc.); confidence in the generality of the law because it is conceived of as necessary, thus as holding true even if conditions are modified ("the law doesn't vary," etc.).

It is clear that new operations appear at this level (after a preparatory period beginning with substage III-A) which are superimposed on the concrete operations. Specifically, of what do these new operations consist? This should be made clear in the course of the examination of the following protocols, to be considered jointly with the preceding one.

GUG (14 ; 4), after several trials, says: "*The more you go toward the right angle* [*i.e.*, the more the plunger approaches a position perpendicular to the buffer] *the closer to the starting point the ball comes back.*"—"Is that always true?"—"*Yes, or at least I think so: you'd have to check.*" He continues his trials and, when there is chance dispersion due to deficiencies of the apparatus, he concludes: "*There must be something wrong.*" After several new trials he concludes: "*You have to find the angle,*" and he looks first for equality in the complementary angles included between the walls of the apparatus and the plunger or the line buffer-goal. At last he discovers that: "*You have to trace the perpendicular"* [in relation to the buffer]. He then realizes the constant equality between the angles of incidence and reflection.

MUL (14 ; 3) begins with a series of correspondences: *"I was here and it went in this direction,"* etc.; *"You change the angle to see how it goes."* By systematically diminishing the total angle, he discovers the fundamental proposition: *"If I shoot it straight, at a right angle* [*i.e.*, when the plunger is perpendicular to the buffer], *it will come right back."* Then he inclines the plunger progressively, according to the angles a_1, β_1, γ_1, etc., and ascertains that, as these angles increase, their complementaries a_1', β_1', γ_1' decrease [a', β' standing for the angles included between the plunger and the buffer]: *"The smaller you make the angle here* [a_1', β_1', etc.], *the larger the angle there"* [a_1, β_1, γ_1]. Then he perceives the equality which he had been seeking from the time he understood that, in the case where the plunger is perpendicular, the ball returns to its starting point. *"This angle* [a_1'] *is the same as that one* [a_2']; *you have to make it parallel to that one* [a_2']. *I am going to see* [he checks for several different angles]. *Yes, I think that's it. You have to carry over exactly that angle"* [the complementaries a_1', and a_2', etc.].

POM (15 ; 5) also begins by noting the correspondences between the angles: *"I look a bit at an angle. . . . The higher up you want to aim, the wider the angle has to be"* [he calculates on the complementary as did MUL]. In order to verify this hypothesis, he spontaneously places the plunger perpendicular to the buffer: *"If the lever is straight, the ball returns exactly."* Afterwards he adjusts the plunger in three different positions, but without moving the target and without firing, and concludes immediately: *"You have to have two angles: the inclination of the lever equals the angle that the trajectory of the ball makes"* [from the buffer to target].

LAM (15 ; 2): *"The rebound depends on the inclination* [of the plunger], . . . *Yes, it depends on the angle. I traced an imaginary line perpendicular* [to the buffer]: *the angle formed by the target and the angle formed by the plunger with the imaginary line will be the same."*

REV (15 ; 4): *"It's a right angle* [several trials]. *No, this slant has to be the same as that one."* When there are chance misses due to the apparatus, he says, *"I didn't move; the gadget isn't fair."*

GOD (15 ; 9), after several fruitless trials: *"You would have to find the rebound angle."* First he indicates $a_1' = a_2'$ as did MUL. Then he traces the perpendicular and points out the angles of incidence and reflection: *"The two must be equal."*

FORT (16 ; 0) begins with several trials: *"You have to move the lever according to the target position and vice versa* [reciprocity]. *You must have an angle there, but it isn't always the same* [he continues his trials]. *It's obvious that everything changes."* Then: *"You have to think in straight lines. To the extent that the lever is displaced, you find the same distance in the other direction: you have to displace it according to the mean* [= the perpendicular from the rebound point to the buffer which he has spontaneously designed]. *The two distances* [the angle openings], *the two sides, always indicate the angles"* [of incidence and reflection].

JAN (16 ; 4): *"You have to find the corresponding angle: the more acute the target angle, the more the lever goes towards the middle and vice versa."*—"Can you measure it?"—*"It's more or less a right angle. No, it varies here the same way as there"* [same design as FORT].

BERG (16 ; 6), after analogous explanations, is shown a wooden buffer rather than the rubber one: *"I think that it's the same law. Yes, I'm sure of it. I take the perpendicular and I focus on the distances. Yes, now the angles have to be equal."*

Although they differ from each other in a number of respects, these examples of reasoning have in common several essential elements which must be differentiated before we can describe the differences between formal and concrete thinking as they relate to the experimental problem under consideration.

In the discovery of the law, the general starting point for these subjects seems to be the fact that the establishment of the concrete correspondences between the inclines of the plunger and the path of the ball after it strikes the buffer seems to lead automatically to the idea of a necessary reciprocity—*i.e.,* each incline implies the other and vice versa. For example, this is expressed by FORT: "You have to move the lever according to the target and vice versa." But this reciprocity, which adds the idea of mutual implication to that of one-to-one reciprocal correspondence, does not in itself entail the realization that the two angles are equal (as FORT, who is at first struck by the variation of the angles, demonstrates).

The bridge from the idea of reciprocity to that of equality—and this is the second point common to all of the answers—is actually furnished by the assertion (explicit, or in certain cases purely mental) that the ball returns to the starting point when the plunger is perpendicular to the buffer. Then it follows that if the null incline of the plunger implies the null incline of the ball's return course, any inclination of either implies an equal inclination of the other.

Once in possession of this double assertion (mutual implication of inclines and return of the ball to its starting point in the case of null incline), the subject will either imagine a perpendicular to the buffer from the rebound point, which leads him to discover the equality of the angles of incidence and reflection; or he will look for the complementary angles (located between the plunger and the buffer or between the former and the trajectory of the ball after the rebound), which step also leads him to the idea of equality.

In either case, the construction of the law is due to the quest for a necessary explanation of the observed inclinations; the serial orders and correspondences established prior to this point are not in themselves sufficient for the subject to discover the relationship between the angles, or even for him to break up the total angle included between the two successive segments of the trajectory into two partial angles.

§ *Conclusion: The Transition from (Concrete) Correspondence to (Formal) Reciprocal Implication*

In spite of what we have just said about the discoveries of our stage II subjects, we have yet to understand just what formal thought adds to concrete operations in the specific case, since subjects at stage II seem *a posteriori* so close to the formulation of the law. What is the contribution of formal operations to the solution of a problem that at first glance seems to require nothing more than correspondences and equalization? Actually, the context of stage III reactions is quite different from that of preceding stages: reasoning by hypothesis and a need for demonstration have replaced the simple stating of relations. In other words, henceforth thought proceeds from a combination of *possibility, hypothesis,* and *deductive reasoning,* instead of being limited to deductions from the actual immediate situation.

The distinction between the one-to-one correspondence of the angles of incline (at stage II) and the reciprocity leading to the idea of the equality of angles (discovered at stage III) is extremely fine as long as we are not in a position to state exactly what the differences are between the operations used at these two stages. Nevertheless, there is a difference. And though it is slight in this first case, it does give us an example of the general opposition of concrete and formal operations that we shall encounter again in increasingly clearer form in the following chapters.

The difference can be stated as follows: Although concrete operations consist of organized systems (classifications, serial ordering, correspondences, etc.), they proceed from one partial link to the next in step-by-step fashion, without relating each partial link to all the others. Formal operations differ in that all of the possible combinations are considered in each case. Consequently, each partial link is grouped in relation to the whole; in other words, reasoning moves continually as a function of a "structured whole."

Stated in symbolic terms, when two classes, A_1 and A_2, with their complementaries, A'_1 and A'_2, are taken, concrete class logic furnishes only four elementary products $(A_1A_2 + A_1A'_2 + A'_1A_2 + A'_1A'_2)$. On the other hand, formal logic, taking the two propositions p and q with their negations $\bar{p}$ and $\bar{q}$, furnishes sixteen possible combinations derived from the four elementary propositional conjunctions $(p.q) \vee (p.\bar{q}) \vee (\bar{p}.q) \vee (\bar{p}.\bar{q})$, which define respectively relations of implication, disjunction, etc., depending on whether the conjunctions are taken one-by-one, two-by-two, three-by-three, the four together, or none at all. The implication of q by p, for example, corresponds to the sum of three conjunctions, $(p.q) \vee (\bar{p}.q) \vee (\bar{p}.\bar{q})$; the implication of p by q corresponds to the sum of $(p.q) \vee (p.\bar{q}) \vee (\bar{p}.\bar{q})$; and the equivalence of p and q (or reciprocal implication) corresponds to the sum of the two conjunctions $(p.q) \vee (\bar{p}.\bar{q})$. But, in order to affirm the truth of one of these three links, $p \supset q$ or $q \supset p$ or $p = q$, one also has to establish the respective falsehood of $(p.\bar{q})$ for $p \supset q$, of $(\bar{p}.q)$ for $q \supset p$, and of $(p.\bar{q})$ as well as of $(\bar{p}.q)$ for $p = q$.

In other words, the difference between the concrete level subjects (who do not go beyond the formulation of term-by-term correspondences between the inclinations of the plunger and the
course of the ball from the buffer to the target) and the formal
level subjects (who look for necessary reciprocity immediately)
can be wholly accounted for by distinguishing the step-by-step
operations based on simple correlations found in class and relational groupings from the combinatorial operations based on the
"structured whole" which constitute propositional logic. Thus,
subjects at stage II are limited to stating successively the correspondences in question and to constructing from the resulting
table that the more the plunger is inclined, the more the course
of the ball between buffer and target is inclined. Certainly this
could be called a law, but it is a law which is a simple summary
of formulations made one by one.

In contrast, stage III subjects view the experiment from the
start both in terms of the total number of possibilities and in terms
of necessary relations, since they possess operations which both
are combinatorial and contain the potential assurance of deductive
necessity. In their first correspondence operations they do not
merely take note of the empirical relationships but immediately
proceed to search for an explanation—*i.e.*, they consider the correspondences as implications. Of course, in a sense the implication $p \supset q$ is still a statement of fact, equivalent to establishing
that the case $(p.\bar{q})$ never occurs. Still, in order to establish this it
is necessary to consider the four possibilities $(p.q)$ v $(p.\bar{q})$ v $(\bar{p}.q)$ v
$(\bar{p}.\bar{q})$; in any case, the implication is nothing more than the addition of three possibilities (the first, the third, and the fourth) combined by the operation (v) which signifies "or"—*i.e.*, it is an addition of what is possible and not of "realities."

Actually, when faced with a correspondence $p.q$ (let p be the
term for a certain angle of incline of the plunger and q the term
for the corresponding angle of incline of the course of the ball
between the buffer and the target), the stage III subjects are not
restricted to pointing out the existence of the conjunction, as are
those of stage II, who are satisfied at this point. They exclude the
possibility $(p.\bar{q})$—*i.e.*, they introduce by hypothesis an implied link
between p and q; but they also exclude $(\bar{p}.q)$—*i.e.*, they also introduce by hypothesis an implied link between q and p. Thus

they proceed immediately from stating the conjunction $p.q.$ to stating the hypothesis of a reciprocal implication $p \supsetneqq q$, with the assumption that this reciprocity $p \supsetneqq q$ or $p = q$ (which is not in itself an equality of content but a simple equivalence from the point of view of the truth of the propositions) covers the equality of some real factor.

At this point, the reasoning process of the 14–16-year-old subjects, based from the start on the twofold consideration of possible combinations and necessary links, is elaborated into a true hypothetico-deductive construction. Unlike stage II subjects, who are limited to noting the occurrence of various correspondences,[4] the adolescents at stage III sooner or later (and often very early) try to uncover the general principle underlying the special case of null inclination. Having established that the ball returns to its starting point, they immediately draw the conclusion that the corresponding inclinations must be equal and consequently the angles which determine them must also be equal; after verification with one or two they generalize the conclusion to all cases.[5]

In symbolic terms, the subject's reasoning at substage III-B is approximately the following (see as an example the extremely clear case of DEF):

$$p \supsetneqq q, \text{ because } (p.q) \vee (\bar{p}.\bar{q}) \text{ are true and} \tag{1}$$

$(p.\bar{q}) \vee (\bar{p}.q)$ false where p and q state corresponding inclinations having the respective values x and y. But

$$(x = 0) \supsetneqq (y = 0), \text{ and} \tag{2}$$
$$(x = a) \supsetneqq (y = a) \tag{3}$$

where a is a determinate inclination > 0. Therefore,

$$x \supsetneqq y, \text{ and} \tag{4}$$
$$\wedge x \supsetneqq \wedge y \tag{5}$$

where $\wedge x$ and $\wedge y$ are the angles of incidence and reflection (or their complementaries).

---

[4] Which may include the case in which the plunger is not inclined and the ball returns to the starting point, but from which they do not abstract the general principle.

[5] Note that the elementary reasoning by recurrence is itself accessible at the concrete level (see *La Géométrie spontanée de l'enfant*, Chap. IX, no. 4). It appears so late in this case because all of the subject's deduction is directed by preliminary reciprocal implications.

In sum, the discovery of the equality of the angles is the result of the reciprocal implication between the corresponding inclinations postulated from the start and not the inverse; this reciprocal implication differs from simple concrete correspondence by the fact that it results from a calculation of possibilities and not merely from an account of the empirical situation.

2

The Law of Floating Bodies
and the Elimination
of Contradictions[1]

A GIVEN NUMBER of disparate objects are presented to the subject, who is asked to classify them according to whether or not they float on water. Then (the classification completed) he is asked to explain the basis of his classification in each case. Next, the subject himself experiments, having been given one or several buckets of water; finally, he is asked to summarize his observations, this latter request suggesting that he is to look for a law, if this has not already spontaneously occurred to him.[2]

Unlike the law considered in the problem presented in the first chapter, the law of floating bodies cannot be derived from concepts which are entirely accessible at the level of concrete operations. Neither the conservation of volume nor, consequently, of density, is worked out in systematic fashion before substage III-A (11–12 years); however, the conservation of weight and certain schemata preparatory to the concept of density are acquired at substage II-B.

---

[1] With the collaboration of J. Nicolas, former research assistant, Laboratory of Psychology, and M. Meyer-Gantenbein, former research assistant, Laboratory of Psychology.

[2] With the older subjects, in addition to the objects to be classified we present three cubes of equal volume having different densities and an empty cube with "plexiglass" or plastic walls (with a density of about one) to facilitate accurate comparisons with the density of water.

But given that the law to be found is that objects float if their density or specific gravity is less than that of water, two relationships are essential to the solution of the problem: *density—i.e.,* the relation of weight to volume—and *specific gravity—i.e.,* the relation between the weight of the object (its density if it is solid, or the weight of its matter plus that of the air which it contains) and an equivalent volume of water.

In addition, the problem requires the construction of a classification including both the class of bodies which float on water and the class of bodies which do not float plus two other eventual classes—that of bodies which may float in certain situations and not in others (such as empty bodies which can be either full of air or full of water) and that of bodies which remain suspended. The law ultimately to be discovered states a relation between only two large classes; that of bodies whose density is less than the density of water and that of bodies whose density is greater.

Thus *the law* states a single and noncontradictory relationship. But in order to construct it empirically the subject first has to eliminate a series of contradictions that frequently characterize the early stages. For example, at first the explanation may be formulated in terms of weight alone, although in fact it is sometimes the heavier, sometimes the lighter bodies which will float. Secondly, the element common to several different explanations (weight, volume, air, etc.) must be isolated. Although the simplest contradictions can be overcome by means of concrete operations alone, the elimination of the more subtle ones, and particularly the formulation of a unified explanation, requires the use of implications—*i.e.,* the intervention of formal propositional operations. In the light of these considerations we feel that the problem of floating bodies, like that of the equality of angles, is an appropriate choice for a preliminary analysis of the transition from concrete to formal thinking.

§ *Stage I (Substages I-A and I-B)*

The stage I reactions (until about 7–8 years) are very interesting, for they are far from demonstrating that a search for a single and noncontradictory explanation is primitive. We find instead that the youngest children are satisfied with multiple and often contradictory formulations. Furthermore, although the problem calls for classification of objects into two groups (floating and non-floating bodies), a substage I-A must be distinguished during which the subject does not even formulate this elementary dichotomy because successive judgments in time relative to a single object are still contradictory.

Although the children of this first substage, once they have determined whether or not a particular object floats, may come to predict that it conserves its properties, because they lack a general frame of reference they do not extend the same properties to other analogous objects. Furthermore, they do not always conceive of these properties as constant even for an identical object:

IEA (4 years) says, for example, of a piece of wood that *"it stays on top. The other day I threw one in the water and it stayed on top."* But a moment later: *"Wood? It will swim anywhere."*—"And this one?" [a smaller piece].—*"The little wood will sink."*—"But you told me that the wood would swim."—*"No, I didn't say so."* On the first presentation of a wire, he says, *"The wire goes to the bottom"* [he has not done the experiment].—"And this weight?" [metal].—*"It will swim."*—"The wood?"—*"It will swim anywhere."*—"The wire?" [third presentation].— *"It will swim."* Finally, for two metal needles of identical appearance he says the opposite: "This one?"—*"It will float."*—"And that one?"— *"It will sink."* We must add that although IEA generalizes little, his explanations can be reduced to the format: "The pebble?"—*"It will sink."*—"Why?"—*"Because it stays on the bottom."*

MIC (5 years) predicts that a plank will sink. The experiment which follows does not induce him to change his mind: [He leans on the plank with all his strength to keep it under the water.] *"You want to stay down, silly!"*—"Will it always stay on the water?"—*"Don't know."* —"Can it stay at the bottom another time?"—*"Yes."*

Classification cannot follow from such responses. First, there is no basis for sorting the objects into floating and nonfloating classes. One way to construct these two classes would be to invoke a constant quality which would in itself furnish the explanation of the fact that a given object floats or fails to float. However, at this stage the subjects do not yet use such explanations and are restricted to looking for the cause in the description of a particular case.

One could maintain from another standpoint that the subjects could classify the objects *a posteriori* after having observed their properties in the experiment. However, in this case as well the properties would have to be perceived as constant. Thus classification is no less impossible, since for these subjects: (1) The same object does not necessarily conserve its properties over time (*cf.* the wire for IEA and the plank for MIC); (2) Different properties may be attributed to two identical objects (*cf.* the needles for IEA); (3) Analogous objects may also be given different properties (*cf.* the small piece of wood which sinks and the large one which floats—IEA).

The reader could object that the child's reasoning is not actually contradictory in that it is analogous to the reasoning of the meteorologist who knows that the same cloud may send forth rain at a given moment and not at another, or that of two similar clouds one might produce rain and the other not. Here the fluctuations or lack of constancy appear in reality itself and not in the thinking of the observer. Nevertheless, the meteorologist seeks to generalize. In spite of the risks inherent in his profession, he goes on to assume that one can fit deductions to the empirical world; he ascribes discrepancies to the operation of chance factors and concludes that under constant conditions his predictions would be accurate. In contrast, the substage I-A child does not try to fit deductions to the situation and does not yet know how to distinguish the deductible from the random and he does not assume that results will be similar under equivalent conditions. Rather, he assumes invariance and deductibility only in certain cases (*cf.* the wood which floats because it floated "the other day"—IEA). But that is exactly the point; since his assumptions vary from case to case he is not able to discern either the reasons for invariance or the reasons for variation.[3]

Thus, to attribute the probabilistic reasoning of the meteorologist to the subjects at substage I-A would be to commit the "fallacy of the implicit." As additional evidence we can say that, since from substage II-A on the subjects do seek invariances and do

---

[3] *Translators' note:* For further experimental evidence concerning the failure of the substage I-A child to understand probabilistic reasoning, see Piaget and Inhelder, *La Gènese de l'idée de hasard chez l'enfant* (Presses Universitaires de France, 1951).

construct classes having general qualities, there is no reason why we should not interpret the successive reactions as they appear genetically. Thus, beginning with substage II-A an attempt to eliminate contradictions can be assumed, whereas for the substage I-A subjects, to assume either an attempt to discover contradictions or a radical impermeability to them would be equally misleading; rather, we must speak of indifference before contradictions when the causal problem cannot be resolved. (The problem is not resolved at substage II-A either, but the child does assume the possibility of a coherent solution and this expectation is sufficient to urge him to try to resolve contradictions.)

At substage I-B the child tries to classify the objects in a stable way into floating and nonfloating, but he does not achieve a coherent classification for the following three reasons (the first of these is logically legitimate, but the other two relate to preoperational thinking): (1) Since the law is not discovered (although he begins to look for it), the subject is satisfied with multiple explanations or a series of subclasses difficult to arrange hierarchically; (2) In the experimental situation, he finds new explanations and thus adds new divisions to his classification but does not reformulate the whole; (3) There are contradictions between some of these classes.

TOSC (5 ; 6) divides the objects presented into two classes prior to the experiment: class B (objects remaining above water) and class B' (objects which sink). Class B includes seven subclasses: (A_1)—Objects which "swim" or float because it is their nature: boats and ducks ("*My little duck that swims like the real ones*"). (A_2)—Small objects ("*little tiny pebbles*," tokens, needles). (A_3)—Light objects (small pebbles float "*because they aren't heavy*," and thus belong simultaneously to A_2 and A_3, but an aluminum plate floats because it is light although it is not small). (A_4)—Flat objects (example: "*This pebble, because it's so flat*"). (A_5)—Thin objects (a wooden blade). (A_6)—Objects which are the same color as the receptacle ("Why will this plank stay on top?"—"*Because they are both the same color*" [the plank and the bucket]). (A_7)—Objects which have already floated (example: a piece of wood "*because it stayed up before*").

Class B' includes the following subclasses: (A_1)—Objects "*that don't belong on the water*" by nature (for example, a piece of a candle: "Where will it go?"—*"To the bottom."*—"Why?"—*"Because the candle doesn't belong on the water."* We put it in the water: "It floats. Why?"—*"Because it swims on the water."* Thus the candle is classified parallel to subclass A_1 of class B.). (A_2)—Large objects. (A_3)—Heavy ones (with the same difficulty in identification and the same interference as for A_2 and A_3 of class B). (A_7)—Those that *"went to the bottom before."* (A_8)—Long objects (a copper wire sinks *"because it's long"*). (A_9)—Those which have been shoved (example: a metal cover).

We notice in this classification some effort at assimilation (small = light, and sometimes thin = flat and flat = small), but it fails because the criteria adopted are inadequate. Initially the child assumes that class B is composed of only two subclasses which, moreover, are heterogenous; there are the objects which float by function or nature (A_1) and the "small ones" (A_2). The subject does not seek the common quality which defines the first category. For the second, he thinks that the quality "small" involves other properties, such as light, etc. However, before the experiment, when he enumerated the objects thus collected, the subject felt he had to specify (without either order or hierarchy) the connotations of the concept "small": light, flat, and thin. In addition, new criteria unrelated to the preceding were brought in at particular points—*i.e.*, the color. Finally, a global category analogous to that of A_1 was constructed (but after the fact): that of objects which have already stayed above water.

As for the objects in B', those which sink, we find three subclasses which correspond in the negative to A_1, A_2, and A_3. But the categories derived from flat, thin, or color have no negatives; reciprocally, two new subclasses (A_8 and A_9) have no corresponding class in B.

Summing up, this type of classification (of which there are other analogous examples) can be defined by the following characteristics: (1) The subclasses are not all disjunctive; (2) They do not all have negatives; (3) They do not allow for grouping either by simple hierarchy (inclusions and complementarities) or by multiple hierarchy (double- and triple-entry tables). Thus, for the child, the experiment complicates rather than simplifies matters. For TOSC, for example, it is responsible for the *post facto* introduction of subclasses A_8 and A_9 in class B' without correspondence in *B*. It is true that "long" could be the opposite of "small," but that is the point; the subject predicted that the wire would float because it was "thin" (A_5) and afterwards that it sinks because it was "long" (A_8), although long is not conceived as contrary to "thin" for "thin" derives from "small" by specification.

Thus the diversification of subclasses without hierarchy must sooner or later introduce a contradiction. Actually, in TOSC's reasoning the contradiction was present from the start, for two interfering classes, "small" and "light," were implicitly regarded as identical or included within each other; classification by simple class inclusion was not differentiated from the double-entry table system. This type of confusion of two potential sorts of classification is accentuated as each new addition is made until the contradiction becomes explicit—*e.g.*, "length" makes an object sink and "thinness" makes it float, but the copper wire is both long and thin; by the same token, "smallness" makes an object float and "heaviness" makes it sink, but several objects are both small and heavy.

Several typical examples of the types of contradiction found at substage I-B are presented below:

TOSC (5 ; 6, same subject), after having said in reference to the plank: *"It goes to the bottom."*—"Why?"—*"Because it is heavy,"* adds a little while later *"because it is big."* Then he sees that the plank floats and explains the fact as follows: *"It's too big and then there's too much water"* [to touch the bottom]. A moment later he tries to hold it at the bottom with another plank and a wooden ball; the two come back up *"because this plank is bigger and it came back up."*—"And why does the ball come up?"—*"Because it's smaller."*—"And this cover?"—*"It will come back up."*—"Why?"—*"Because it is smaller than this piece of wood, than the plank."*—"Try."—*"It stayed down because I pushed too high up."*

BEZ (5 ; 9) explains the floating by the weight [inversely to TOSC]: "Why do these things [previously classified] go to the bottom?"—*"They are little things."*—"Why do the little ones go to the bottom?"—*"Because they aren't heavy, they don't swim on top because it's too light."* —"And these?" [class of floating objects].—*"Because they are heavy, they swim on the water."* We go on to the experiment: the key sinks *"because it's too heavy to stay on top"* whereas the cover sinks *"because it's light."* Comparing two keys: the larger does not stay above water *"because it's light"*—"And the little one?"—*"It will go to the bottom too."*—"Why?"—*"Because it's too light."*

GIO (6 ; 0) "These things [previously classified] go to the bottom?"— *"Yes, that one"* [the wooden ball].—"Why?"—*"Because it's heavy."*— "And these?" [the class of floating objects].—*"That one swims because it's light."* We do the experiment with the cover. It floats *"because it's light."*—"And if you push it?"—[It sinks.] *"It's because it's light, and light things never stay on top."*—"And that plank?"—*"It will stay on top."*—"Why?"—*"Because it's heavy."*—"Why?"—*"Because it's big."*— "And if you lean on it?"—[He does.] *"It comes back up because it's light."*—"And this?" [large needle].—*"It goes to the bottom because it's big."*—"And that [metal plate] if you push?"—*"It will stay at the bottom."*—"Why?"—*"Because it's light."*

ELI (6 ; 10): "That?" [candle].—*"It goes to the bottom."*—"Why?"— *"Because it's round."*—"And that?" [ball].—*"It stays on top."*—"Why?" —*"Because it's round too."* Thus the contradiction does not relate only to the weight. "And that needle?" [placed on the water].—*"It floats because it's light."*—"And if you push?"—*"It will go under."*—"Why?"— *"Because it will be heavy."* Here contradiction goes with nonconservation.

In reference to analogous observations, a logician once maintained that such assertions are not contradictory, just because the same result can be due to either of two opposite causes; for example, persons who pay taxes that are low proportionate to their wealth may be either those who are very rich or those who have hardly anything. But children at the present stage are far from such a subtle schema, and it seems to us that three kinds of considerations demonstrate that they remain indifferent to contradictions or, more accurately, that they do not perceive their continual contradictions:

1. From the beginning, the subjects predict a simple distribution but according to two contrary explanations (which already reveal the ambiguity of the concepts used). The bodies that float are either those that are light because they are small, or those that are heavy because they are large. However, each of these two explanations already includes a number of implicit contradictions, for light and heavy do not coincide with small and large and the floating is due to relative and not to absolute weight.

2. The experiment does not set the child right, but he tries to reconcile the whole by adopting either explanation alternately without perceiving the incompatibility; bodies float or sink equally well if they are large, small, heavy, or light (or even, by association, because they are round, long, etc.). Thus the contradiction is made explicit, but it is not any closer to being noticed, doubtless because of the initial ambiguity of the pairs small × light and heavy × large.

3. The same assertions judged mutually compatible by the subjects at substage I-B would appear irreconcilable beginning with substage II-A. Here we find the best proof that they do not constitute the reflection of an implicit coherence but rather of a thinking process in a state of disequilibrium for lack of instruments of coordination (operational classification, etc.) which will attain equilibrium only at the point when concrete operations are structured.

§ Stage II (Substages II-A and II-B)

The behavior of the 7–9-year-old subjects is marked by an effort to remove the main contradiction to which they have submitted previously without reaction: that certain large objects can float and certain small ones sink without, however, barring the possibility that in general the light ones float and not the heavy. The contradiction tends at this point to be surmounted by a revision of the concept of weight, now seen in relation to that of volume— i.e., the child begins to renounce the notion of absolute weight in order to look toward density and, above all, toward specific gravity.

Specific gravity refers to the relationship between the weight of a given volume of a body and that of an equal volume of water, and density refers to the weight of a cubic centimeter of the body considered. But we will take these two concepts in a more elementary sense. We will speak of density when the subject explicitly relates the weight and the volume—i.e., when the concept is understood as a relationship—and of specific gravity when the subject understands that for the same volume each substance has a characteristic weight. (In the latter case the subject does not refer explicitly to the volume.) Thus, substage II-A children acquire the beginnings of the conception of specific gravity and try to resolve the earlier contradictions by invoking it.

The two problems that arise at this stage in the development of logical operations are: (1) Do the stage I contradictions tend to disappear of themselves because the subject grasps the notion of specific gravity, or is it in trying to surmount these contradictions that he constructs the concept in question? (2) If the latter is true, how does the child come to perceive the contradictions with the aid of concrete operations alone?

KER (7 ; 6) classifies as floating objects wood, matches, corks, a cover, metal clamps, an eraser, small nails, and a small cylinder of hollow metal; as nonfloating a key, some stones, a metal disk, a needle, and a heavy wooden ball. After the experiment he constructs a third class, that of objects which float or sink depending on whether they are empty or filled with water—the cover and the needle [whose eye may permit the water to pass]. The first two classes are defined by "light" and "heavy," but notice that KER wavers between two possible meanings of these notions: the earlier or absolute sense [the small nails for the "light" and the large ball of wood for the "heavy"] and the new or relative sense—*i.e.*, the specific gravity. "The little pebble goes to the bottom?"—"*Yes.*"—"But isn't it light?"—"*No, it's stone.*"—"And the nail?"—[He does the experiment.] "*It's because it's iron.*"

BAR (7 ; 11) first classifies the bodies into three categories: those which float because they are light [wood, matches, paper, and the aluminum cover]; those which sink because they are heavy [large and small keys, pebbles of all sizes, ring clamps, needles and nails, metal cylinder, eraser]; and those which remain suspended at a midway point [fish]. "The needle?"—"*It goes down because it's iron.*"—"And the key?"— "*It sinks too.*"—"And the small things?" [nails, ring clamps].—"*They are iron too.*"—"And this little pebble?"—"*It's heavy because it's stone.*" —"And the little nail?"—"*It's just a little heavy.*"—"And the cover, why does it stay up?"—"*It has edges and sinks if it's filled with water.*"— "Why?"—"*Because it's iron.*"

DUF (7 ; 6): "That ball?"—"*It stays on top. It's wood; it's light.*"—"And this key?"—"*Goes down. It's iron; it's heavy.*"—"Which is heavier, the key or the ball?"—"*The ball.*"—"Why does the key sink?"—"*Because it is heavy.*"—"And then the nail?"—"*It's light but it sinks anyway. It's iron, and iron always goes under.*"

The principal difference between these reactions and those of substage I-B lies in the real effort made to resolve the contradictions. This is done by improving the classification system by the utilization of class inclusion operations. These permit the subject to distinguish systematically between "all" and "some" by means of the reversible addition $A + A' = B$ and $B - A' = A$, from which it follows that $A < B$. As soon as this operation permits him to determine the accurate inclusion of the part in the whole, the subject is led to the most significant discovery of the first operational level; that small objects do not always weigh less than large ones or, in other words, that it is false to consider all small objects as light and all large ones as heavy. In the case of floating, in particular, all the objects which float are not small and all those which sink are not large. Thus the child succeeds in making a double-entry classification with reference to weight and volume which gives four possibilities: the small light objects, the small nonlight objects, the large light, and the large heavy. As a result of the operation of class inclusion, the subject becomes sensitive to contradiction and, by coordinating two classes now perceived as distinct from each other, can separately formulate a double-entry table. There would be contradiction if weight and volume were identified in the presence of these four subclasses.

Thus the child is led to revise his notion of weight and to place the concept of absolute weight—*i.e.*, of weight equal to the volume or to the quantity of matter—in opposition to a new concept of weight perceived as relative to the matter under consideration—*i.e.*, of weight as a quality of distinct types of matter—which is a rough approximation of specific gravity. But we must insist that the way they achieve this rough restructuring of the concept of weight—one which avoids the inconsistencies of earlier formulations—cannot be understood without considering the new logical apparatus composed of concrete class and relational operations. Actually, even reporting on the experimental data relative to weight and to the quantity of matter presupposes that the parts are distinguished from the whole for a given class ("all" and "some")—*i.e.*, the presence of a coherent structure is indispensable in order to avoid contradictions.

However, the notion of weight approximated in this way remains insufficient. As yet it is no more than a quality inherent in various types of matter, not a relation between the weight and the volume. The reason for this is simple. As we have seen elsewhere, at this level the child cannot yet conceptualize either the conservation of weight or the conservation of volume, and the only invariant available to him is the quantity of matter; thus he is not able to make any accurate composition of the relationship between weight and volume from the standpoint of the relations between bodies or their internal configurations. It would serve no purpose to refer back to earlier experiments, which are completely confirmed by the present results.[4] It is sufficient to note that, without conservation or composition of the relationships referring either to weight or to volume, specific gravity could not be conceived as other than a simple quantity inhering in each respective substance.

Moreover, given the incompleteness of the concept of specific gravity, the failure to distinguish between the concepts of absolute weight and of specific gravity naturally persists at this level. This is a residual source of contradictions in spite of the visible effort of the child to overcome them. (See, for example, in KER's report, the large wooden ball which he sometimes conceives of as heavy, sometimes as light; likewise the nail, etc.) Furthermore, the subject vacillates between the two concepts he applies to weight because he is not entirely aware of the fact that he is dealing with two concepts, though he can distinguish them to some extent. Actually, in order to distinguish the two explicitly, he would have to possess the operational means for such a distinction. But we have just seen that he does not possess them. Even the serial ordering of weights between objects of the same volume is not acquired until substage II-B.[5] Thus, the nascent notion of specific gravity marks only the beginning and not the completion of the separation of the variables of weight and volume. It is the expression of the discovery that not all the small objects are light nor all the large ones heavy; but the concept remains at this stage of preliminary classification and does not yet reach a higher level of organization.

[4] Piaget and Inhelder, *Le Développement des quantités chez l'enfant* (Delachaux and Niestle, 1940). See Chaps. I–III and especially VIII–IX on the composition of relationships between weight and volume.

[5] *Développement des quantités*, p. 233.

At about 9 years or 9 ; 6, weight begins to be conserved by the child.[6] Thus, from then on he knows how to apply to weight the concrete operations of serial ordering and equalization and even, up to a certain point, of measuring. As for specific gravity and density, he is no longer limited to qualifying the various materials in terms of simple weight: iron is heavy, wood light, etc. Instead, he introduces a new general explanatory scheme: the objects with high specific gravity are more "full" than the others.[7] But, since volume is not automatically conserved at this stage, we do not yet see the formulation of an operational relationship between the two.

For this reason, in comparing the weights of specific bodies to the weight of water (which begins at this stage), the child does not relate the object's weight to that of an equal volume of water but rather to the water contained in the entire receptacle.[8]

It is on this point that a new residual core of contradiction can be observed during this stage. Although several of the contradictions of the preceding level are eliminated as a result of the progress which we have just described, in contrast the subjects still assimilate the weight of the body, compared with that of the total volume of water, to a substantial force or to a motor activity, giving rise to a new group of dynamic explanations which are mutually contradictory. In addition, the notion of "filled," in spite of the fact that it permits the unification of the explanations relating to solid homogenous matter, gives rise, in the case of hollow objects (boats, covers, etc.), to the hypothesis that the latter float because they are filled with air. However, without being wrong, this explanation provides another source of possible difficulties.

The following examples illustrate these various types of reaction:

[6] The term "conservation" is used in a sense specific to the authors' meaning—a particular empirical factor (weight, volume, etc.) remains an invariant in the child's mind throughout observed changes of state. The timing of the appearance of conservation for various factors differs, but those discussed here all appear during the concrete stage.

[7] *Développement des quantités*, p. 173.

[8] *Cf.* Piaget, *The Child's Conception of Physical Causality* (Harcourt Brace and Co., 1930), Chap. VI: at this stage a boat which can float on a lake would be too heavy for the Rhone, etc.

BAR (9 years): [class 1]—Floating objects: ball, pieces of wood, corks, and an aluminum plate. [class 2]—Sinking objects: keys, metal weights, needles, stones, large block of wood, and a piece of wax. [class 3]— Objects which may either float or sink: covers. Later BAR sees a needle at the bottom of the water and says: *"Ah! They are too heavy for the water, so the water can't carry them."*—"And the tokens?"—*"I don't know; they are more likely to go under."*—"Why do these things float?" [class 1].—*"Because they are quite light."*—"And the covers?"—*"They can go to the bottom because the water can come up over the top."*— "And why do these things sink?" [class 2].—*"Because they are heavy."* —"The big block of wood?"—*"It will go under."*—"Why?"—*"There is too much water for it to stay up."*—"And the needles?"—*"They are lighter."*—"So?"—*"If the wood were the same size as the needle, it would be lighter."*—"Put the candle in the water. Why does it stay up?" —*"I don't know."*—"And the cover?"—*"It's iron, that's not too heavy and there is enough water to carry it."*—"And now?" [it sinks].—*"That's because the water got inside."*—"And put the wooden block in."— *"Ah! Because it's wood that is wide enough not to sink."*—"If it were a cube?"—*"I think that it would go under."*—"And if you push it under?" —*"I think it would come back up."*—"And if you push this plate?" [aluminum].—*"It would stay at the bottom."*—"Why?"—*"Because the water weighs on the plate."*—"Which is heavier, the plate or the wood?"—*"The piece of wood."*—"Then why does the plate stay at the bottom?"—*"Because it's a little lighter than the wood, when there is water on top there is less resistance and it can stay down. The wood has resistance and it comes back up."*—"And this little piece of wood?"—*"No, it will come back up because it is even lighter than the plate."*—"And if we begin again with this large piece of wood in the smallest bucket, will the same thing happen?"—*"No, it will come back up because the water isn't strong enough: there is not enough weight from the water."*

BRU (9 years): *"The water can't carry the pebbles. The wood can be carried."*—"And if it is pushed under?"—*"It will come back up because the water isn't strong enough: it doesn't have enough weight"* [= this time the weight operates to maintain it at the bottom and no longer to carry it!]. And a moment later, *"The wood comes up when you let go because it springs up."*

The case of BAR clearly illustrates most of the characteristics of this stage. In the first place, he classifies the objects according to specific gravity and not absolute weight; there were three exceptions, two of which are due to ignorance (wax and aluminum), and one (the large piece of wood) to the fact that it is related to the total volume of water in the bucket. At one point the subject even gets at an explicit relation between the weight and the volume of the body; the needle is heavier than the piece of wood because "if the wood were the same size as the needle, it would be lighter." Why, in this case, after such a favorable beginning, is the subject unable to find the law, at the end losing himself in an increasing number of contradictions? His failure results from the fact that in relating weight and volume, he has not yet found a general operational form (logical multiplication for equal volumes and different weights or for equal weights and different volumes) and has stayed within the limits of the particular case of the comparison of iron and wood. In addition, whenever the principal relationship relevant to the formulation of the law appeared— i.e., that between the weight of the body under consideration and that of the water—he did not compare weights with equal volumes (body and water) but the weight of the object and that of the total quantity of water; "heavier than water" signified "too heavy for the water to be able to carry them." But once he began to conceptualize the relationship between weights in terms of active forces, all explanations became possible as his observations proceeded and sooner or later he was bound to entangle himself in a contradiction. This is brought out at the end of our questioning of bar (as well as the text cited from bru) up to the point where bar returns to the explanation in absolute weight, which is easier to reconcile with his dynamic imagination.

These initial efforts at unification and internal consistency which, for lack of adequate operational instruments, are not crowned with final success, reappear in the cases which are most difficult from the standpoint of an integrated explanation; the case of hollow objects, where the air plays a part, and that of the needle, which floats in certain cases because of the surface tension. Thus, certain subjects who explain the specific gravity by the notion of more or less filled generalize the case of covers or boats, which float when they are empty (but supported by the air) and sink when they are filled with water, up to the point of using it as a prototype for the specific gravity of all sorts of objects.

RAY (9 years): *"The wood isn't the same as iron. It's lighter: there are holes in between."*—"And steel?"—*"It stays under because there aren't any holes in between."*

DUM (9 ; 6): The wood floats *"because there is air inside"*; the key does not *"because there isn't any air inside."*

But the analogy cannot be considered valid except on condition that the "holes" in the wood stay closed. This leads to the following type of explanation given in the case of needles poised delicately on the surface of the water:

RAY (9 years, same subject): *"The needle pricks and goes in the water because it is thin and heavy."*—"Look" [it floats].—*"Ah! It's because there was a hole in the other needle that went under."*—"But this one has a hole too."— . . . —"And that ring clamp?"—*"It will go to the bottom because there are holes; the water comes in."*

AND (10 ; 0): *"The needle floats because there is a little hole."*—"And if it were big?"—*"It falls."*—"How can you tell beforehand?"—*"It depends on whether it is big or small. If the water doesn't come over the top of it, it stays up."*

On the whole, substage II-B shows significant progress in the direction of internal consistency and in the search for a single explanation based on the preliminary relating of the weight to the volume presupposed by the schema of more or less "filled." However, since the volume of water envisaged is not that of the displaced water but rather of the total quantity of water contained in the receptacle, the relationship between the weight of the body and that of the water remains one between active forces, thus reintroducing a complexity rich in contradictions. The probability that they will appear is greater when the air is seen as intervening and holes, open or closed, are assigned a role. In short, for lack of operational relations sufficiently worked out to dominate the sum of the relationships between weight and volume, the explanation, although vaguely intuited, is not clearly discovered, and a coherent system is not as yet formulated.

§ *Stage III*

We have put a great deal of emphasis on the preoperational levels and the concrete operational stages with two purposes in mind;

in order to point out, first, what a long road thought processes must tread before even the attempt to find a single noncontradictory explanation appears, and second, why the completion of a model for such explanations cannot be achieved without the aid of formal operations even in the present case, in which the law to be found can be stated by using purely concrete operations. We must now try to analyze the role of the formal operations needed to discover the law. But the problem is somewhat more complex than that in the case of the law of equality of angles studied in the preceding chapter. Actually, in the latter case, in themselves the correspondences between the inclinations found at substages II-A and II-B gave a first approximation of the law; only the reason for these correspondences remained beyond the subject, and formal thought introduced nothing more than an element of necessary implication to a set of relations which were already exactly formulated. In the present example, on the other hand, the law is not completely discovered at substage II-B, and formal thought is indispensable to its formulation in a complete form. This difference can be given two explanations, which are as follows:

In the first place, even if the relation between densities, once it is found, can be expressed in a purely concrete form, formal schematization is still needed to work out the relevant concepts. The concept of density in fact presupposes that of volume. However, we have stated before that the conservation of volume is not worked out conceptually before the beginnings of the formal level—i.e., toward 11–12 years.[9] Without a doubt the reason for this is that, in contrast to simple forms of conservation, which the subject masters by simple additive compensations, the conservation of volume throughout changes of form presupposes the ability to handle proportions.[10] However, we shall see in the course of chapters 11–14 of the present work why the concept of proportion does not itself appear before the formal level, when it arises in connection with certain general properties of the group structures characteristic of propositional operations.

In the second place, formal operations are particularly impor-

[9] *Le Développement des quantités chez l'enfant,* Chap. III and *La Géométrie spontanée de l'enfant,* Chap. XIV.

[10] If the three dimensions of a volume, x, y, and z, are transformed into x', y', and z', there is conservation when $xy/x'y' = z'/z$, from the formula $xyz = x'y'z'$. These are multiplicative compensations, thus proportions.

tant in the case of the law of floating bodies in order to make possible both the exclusion of the too-simple interpretations of stage II and the purely imaginative construction of a hypothesis which does not correspond to any of the directly observable concrete data. The stage II explanations are not actually absurd and do not directly contradict the facts, and, if they are to be excluded, the fact that they are not coherent enough must be felt. But this can be done only by a thought process able to deduce the consequences of simple hypotheses with necessity. On the other hand, to relate the weight of the body under consideration to the weight of an equal volume of water is to invent a situation which has no empirical correlate, because only the total volume of the water in the receptacle is actually observed, whereas the conceptualization of a volume of water equal to that of the object to be compared is the product of a subtler separation of variables which once more requires hypothetico-deductive thought.

Doubtlessly the subjects we are going to examine now are much more likely to use acquired knowledge, for they are approaching the academic level where they deal with such questions. But when this acquired knowledge does not correspond to the mental structures indispensable to their assimilation this is immediately recognized in the questioning, and we have not used the cases prematurely influenced in this way. In addition, we have seen how, as early as 9 years of age, subject BAR compares wood and iron at equal volumes. The generalization of the same mental operation to the water itself is made so naturally in the course of stage III that it is hard not to allow for the role of spontaneity in the progressive structuring of the data, even if it is hastened by the surrounding social environment. The following examples, beginning with two intermediate cases, illustrate this stage:

FRAN (12 ; 1) does not manage to discover the law, but neither does he accept any of the earlier hypotheses. He classifies correctly the objects presented but hesitates before the aluminum wire. "Why are you hesitating?"—"Because of the lightness, but no, that has no effect."—"Why?" —"The lightness has no effect. It depends on the sort of matter: for example, the wood can be heavy and it floats." And for the cover: "I thought of the surface."—"The surface plays a role?"—"Maybe, the surface that touches the water, but that doesn't mean anything." Thus he discards all of his hypotheses without finding a solution.

FIS (12 ; 6) also, in the transition phase between stages II and III, comes close to solution, saying in reference to a penny that it sinks *"because it is small, it isn't stretched enough. . . . You would have to have something larger to stay at the surface, something of the same weight and which would have a greater extension."*

ALA (11 ; 9): "Why do you say that this key will sink?"—*"Because it is heavier than the water."*—"This little key is heavier than that water?" [the bucket is pointed out].—*"I mean the same capacity of water would be less heavy than the key."*—"What do you mean?"—*"You would put them* [metal or water] *in containers which contain the same amount and weigh them."*

JIM (12 ; 8) classifies floating or sinking objects according to whether they are *"lighter or heavier than water."*—"What do you mean?"—*"You would have to have much more water than metal to make up the same weight."*—"And this cover?"—*"When you put up the edges, there is air inside; when you put them down, it goes down because the water comes inside and that makes more weight."*—"Why does the wood float?"—*"Because it is light."*—"And that little key?"—*"No, this piece of wood is heavier."*—"So?"—*"If you measure with a key* [= with the weight of a key], *you need more wood than lead for the weight of the key."*—"What do you mean?"—*"If you take metal, you need much more wood to make the same weight than metal."*

MAL (12 ; 2): *"The silver is heavy, that's why it sinks."*—"And if you take a tree?"—*"The tree is much heavier, but it is made of wood."*—"The silver is heavier than that water?" [bucket].—*"No, you take the quantity of water for the size of the object; you take the same amount of water."*—"Can you prove that?"—*"Yes, with that bottle of water. If it were the same quantity of cork, it would float because the cork is less heavy than the same quantity of water."* And again: *"A bottle full of water goes to the bottom if it is full because it's completely filled without air, and that bottle stays at the surface if you only fill it halfway."*

We see how, rejecting any suggestion that they relate the weight of the objects in question to the weight of all the water in the receptacle, these subjects reach the point of comparing the first weight to that of an equivalent volume of water. FRAN begins by assigning a possible role to the contact surface; FIS believes that a piece of metal would float if, without adding to its weight, its "extension" could increase; then ALA, JIM, and MAL are able to reason about the amount of water equal to the volume of the object. "You take the quantity of water [equal to] the size of the object," says MAL, for example. Thus the "more or less full" schema used at substage II-B is transformed into a relationship between the weight and the correct volume (for FIS, the "stretch" becomes the relationship of weight to "extension") and finally between the weight and volume of the object in question and the corresponding weight and volume of water displaced by that object.

These facts raise three related problems: (A) How does the subject start to discard the hypotheses he has had up to this point? (B) How does he go about constructing the new hypothesis? (C) How does he go about verifying it?

A. On the first question, it is worth noting that from this point on the subject discards only crude hypotheses without verification, whereas he is more and more likely to verify the superior hypotheses. He even discards the first almost without explicit reasoning, as when FRAN says, "That has no effect," or "That doesn't mean anything." In other words, he finds that in order to refute an explanation it is sufficient to invoke verbally or mentally a case where the purported factor is associated with the opposite effect. Thus FRAN eliminates absolute weight in saying, "For example, wood can be heavy and it floats." Likewise, ALA and MAL discard all comparison with the total volume of water in the receptacle, knowing well that the variations of this volume leave the floating or nonfloating of the bodies in question unchanged. In comparison with stage II, the innovation is the same in the case of floating bodies as in the problem of the equality of angles studied in Chap. 1; the subject views the problem in terms of all possible combinations in such a way as to draw out their implications or nonimplications instead of noting the empirical links simply in order to draw tables of correspondences or classifications from them. We have seen in the first chapter how implication is substituted for simple correspondence, and in a moment we will return to the subject of implication in reference to the construction of the new hypothesis characteristic of this stage. But the elimination of the hypotheses deemed inadequate is accomplished by the following three procedures, each of which supposes comprehension of nonimplication.

In the first place, if we call p the assertion that the bodies will float and let q be any factor associated with p—for example, lightness (absolute)—the subject will find that to state the occurrence of the association $p.\bar{q}$ is all that is needed to discard the factor q; for example, the large block of wood is heavy, nevertheless it floats. It is known, in effect, that $p.\bar{q}$ is the negation of the implication $p \supset q$:

$$p.\bar{q} = \overline{p \supset q}. \tag{1}$$

In the second place, the subject may note the two possibilities combined $(p.q) \vee (p.\bar{q})$—i.e., of $p.(q \vee \bar{q})$—which constitutes the operation we may speak of as the affirmation of p independently of the truth or falsity of q. But this operation contains $p.\bar{q}$ and amounts also to discarding $p \supset q$. This is what FRAN, for example, says when he declares that "the wood can be heavy (or light) and it floats":

$$p.(q \vee \bar{q}) = (\bar{p}.\bar{q}) \vee (p.\bar{q}). \tag{2}$$

Finally, the subject may not retain a factor because he knows that all the possible combinations are true. For example, if p states that the bodies float and q that there is a large quantity of water in the receptacle at the present level, the subjects do not attribute any more importance to statement q because they know well that one can observe the occurrence of all four combinations, $(p.q) \vee (p.\bar{q}) \vee (\bar{p}.q) \vee (\bar{p}.\bar{q})$—i.e., one object may float equally well on much or little water, another may sink in the same two situations. But this operation with four conjunctions, which is called "tautology" or "complete affirmation," again contains the non-implication $(p.\bar{q})$:

$$(p * q) = (p.q) \vee (p.\bar{q}) \vee (\bar{p}.q) \vee (\bar{p}.\bar{q}). \tag{3}$$

B. As for the explanation of how the child gets to the hypothesis found at stage III according to which "lighter (or heavier) than the water" signifies also "at equal volumes," it can be explained in terms of what we have said for both the operations and the concepts themselves.

As for the concepts, the child at stage II has already learned that one body may be heavier than another with equal volume (see BAR's comments on the wood and on the metal of which the needle is made), but he believes that the weight of the body in question is to be compared to the total volume of the water in the receptacle. The stage III child, on the other hand, rejects the latter hypothesis. All that remains for him to do is to relate the weight of the body to that of a quantity of water—no longer any quantity whatsoever but a quantity equal to the volume of the body itself. In other words, the discovery distinctive of stage III is nothing more than the generalization of the mode of comparison roughly formulated at stage II for two solid bodies, but henceforth it is applied to the water itself as well as to the object judged heavier or lighter than it. That this comparison should be more difficult when it involves a solid body and water than when it involves two solid bodies should be obvious, since the volume of water equal to the volume of the immersed solid has no visible contours and can be conceptualized only after a preliminary abstraction. But this discovery is nothing less than the resultant of all the previous conceptualization of relative weight or specific gravity.

Thus, from the standpoint of the relevant operations this comparison with a hypothetical equal volume touches on a reasoning process of which we will find numerous examples later and which consists of considering the variation of a single factor "all other things being equal." If we let p be the assertion that a given object floats and $\bar{p}$ the assertion that it does not, q the assertion that its volume is equal to that of a certain quantity of water, r the assertion that it is lighter than that quantity of water, and $\bar{r}$ the assertion that it is heavier, the relationship which the subject establishes is the following:

$$p.q.r. \vee \bar{p}.q.\bar{r}, \tag{4a}$$

which is in fact the schema of proof based on the assumption "all other things being equal." But this expression is itself equivalent to the product of two formal operations, $(p \gtreqless r)$ and $(q.r) \vee (q.\bar{r})$ —i.e., the reciprocal implication (or equivalence) between the floating of the body and its weight (relative to the same volume of water) and the assertion that weight and volume vary independently. The operation $(p.q) \vee (p.\bar{q})$—i.e., proposition (2)—by means of which the subject shows that a given factor does not play a causal role, can also be recognized in the operation $(q.r) \vee (q.\bar{r})$.

Thus the explanation discovered at stage III covers all possible cases including hollow objects without requiring the assignment of a causal role to the opposing forces of water and air or to any sort of hole. In the case in which the weight of a body equals that of the water (at the same volume) we have, of course (if $p_0 =$ neither floats nor sinks and $r_0 =$ neither heavier nor lighter than water):

$$p_0.q.r_0 \vee \bar{p}_0.q.\bar{r}_0 \,. \qquad (4b)$$

For example RAY (12 ; 7) says, in reference to a very thin cube of plastic (whose density is approximately equal to that of water), that if it were filled with water *"It would stay in the middle, in the liquid, because the weight is the same."*

After the multiple attempts at unification seen throughout stage II, the subjects finally attain a unified noncontradictory explanation; the two principal previous sources of contradiction (absolute weight and active forces) are eliminated by the single hypothesis of density or of the relation between weight and volume.

C. If we study the verification processes used by the subjects, we find they confirm completely what has been said earlier and, in particular, allow us to verify the fact that the subjects' reasoning no longer operates in a simple formulation of relationships or concrete correspondences but requires a formal combinatorial system. Whereas at the first preoperational level the subject is not capable of any proof, at the level of concrete operations (substages II-A and especially II-B) he does not feel spontaneously the need for it, but he can furnish it if asked. However, in keeping with the entire logic of concrete operations, which is simply a matter of organizing the reading of the raw experimental results (by classification, setting up of relationships, etc.), at this point the only method of verification of which he conceives is to accumulate facts until more or less complete certainty is reached but without going beyond the general—*i.e.*, without introducing necessary links by isolating these facts from their contextual interdependence and deducing the relations thus isolated.

BON (11 ; 0) wants to prove that *"all wooden objects float."* Therefore he puts two in the water [wooden cube and the ball]: *"I only have to put the two things in the jar. They both float. All wooden objects float."*

But how valid is this jump from "some" particular cases to "all," which reminds one of the amplifying induction that classical logic wanted to regard as a fundamental reasoning process? In the absence of probabilistic reasoning (excluded at substage II-B), it is only a worthless extrapolation, for $p.q \supset (p \supset q)$ gives $(p \supset q) \lor (p.\bar{q})$, therefore $p.q \supset (p * q)$—i.e., "some wooden objects float" could imply any one of several results in a particular case.

At the level of formal thought (from substage III-A on), on the other hand, proof consists in demonstrating the truth or falsehood of a particular or general assertion which takes into account (or tries to take into account) the total number of possible combinations, thus permitting the subject to group combinations in a demonstrative fashion. However, grouping these combinations is exactly the same as selecting the cases where a single factor varies (the others being held constant) so as to isolate universal relationships from simple contingent conjunctions and above all so as to be able to discover necessary relationships between variables. Such a composition of relations requires that we must resort to what we have called in Chap. 1 the "structured whole" (the combinations having 0, 1, 2, 3, and 4 conjunctions); thus it contrasts to the simple additive and multiplicative class inclusion—i.e., "some" and "all"—characteristic of concrete operations.

In other words, the verification process found at stage III makes explicit use of the schema "all other things being equal" to which we have just compared the explicative hypothesis of which the subject conceives [4a, 4b]. Two further remarks should be made with regard to both the difference between substages III-A and III-B and the relationship between the present problem and problems which will be taken up in the following chapters.

As a general rule and in its authentic form, the schema "all other things being equal" appears only at substage III-B, as we shall see later in reference to flexibility: but then the problem is, given n factors $A, B, C, D, \ldots$ independent of each other, to vary A leaving B, C, D, etc., unchanged. But in the present case the two factors, weight and volume, are not independent in this sense, since the subject is trying to determine the relation between them and to link them in a new concept—i.e., density. Thus, in comparing an object to the water it floats on, it is easier to vary weight and leave volume constant than to hold several independent factors constant (such as temperature, pressure, etc.), as one would in studying the role of weight. This is why the schema "all other things being equal," in the elementary form it takes for the present problem, appears as early as substage III-A:

GER (12 ; 7) to prove that the coin has a higher density than water says: *"If there were a jar filled with water and another just like it filled with pennies . . .* [the latter would be] *heavier and would go to the bottom of the lake."*

AL (12 ; 8): *"With the same volume, the water is lighter than that key."* To prove it: *"I would take some modeling clay, then I would make an exact pattern of the key and I would put water inside: it would have the same volume of water as the key . . . and it would be lighter."*

But it is still true that it is only at substage III-B that this schema acquires its general value. Actually, for the present question, it is only toward 13–14 years of age that it gives rise to the search for a common metric unit. Some of the objects we used in our experiments were (among others) a cube of wood, a cube of iron, and an empty plastic cube (density about 1), all three of the same volume. But it is striking to see the subjects at substage III-B turn sooner or later to these units; they are the only subjects who do so spontaneously:

LAMB (13 ; 3) correctly classifies the objects that sink: *"I sort of felt that they are all heavier than the water. I compared for the same weight, not for the same volume of water."*—"Can you give a proof?"— *"Yes, I take these two bottles, I weigh them. . . . Oh!* [he notices the cubes] *I weigh this plastic cube with water inside and I compare this volume of water to the wooden cube. You always have to compare a volume to the same volume of water."*—"And with this wooden ball?" —*"By calculation."*—"But otherwise?"—*"Oh, yes, you set the water level* [in the bucket]; *you put the ball in and let out enough water to maintain the original level."*—"Then what do you compare?"—*"The weight of the water let out and the weight of the ball."*

WUR (14 ; 4): *"I take a wooden cube and a plastic cube that I fill with water. I weigh them, and the difference can be seen on the scale according to whether an object is heavier or lighter than water."*

Here we notice that the factor left invariant, as well as the common unit sought, is always seen in terms of volume, although theoretically it would be equally possible to say that for equal weights of water and of the object in question, the latter would float if it had a greater volume (*cf.* the case of FIS and his piece of money which would float if it had greater extension). But experimental verification would be more difficult in this case.

Thus, generally, verification at stage III consists of two procedures: (1) separating out variables according to combinations not given by direct observation, and (2) the composition of these relationships according to operations of conjunction and implication such as those of proposition (4). It is in this respect that, in the present problem as in the case of the equality of the angles of incidence and reflection, in the end the required law must be worked out formally even though the discovery of this law has been prepared by a long process of concrete structuring. But, without a doubt, in neither case are the possible combinations numerous enough for the role of formal operations to be clearly distinguished from that of concrete operations and particularly for the schema "all other things being equal" to acquire all its general significance. For this reason we need to pass on to the analysis of more complex problems.

3

Flexibility and the Operations Mediating the Separation of Variables[1]

THE FLEXIBILITY of a rod depends on the material it is made of, its length, its thickness, and the form of its cross-sections. All other things being equal, the degree to which it bends varies as a function of the weight that is placed at its tip. To study the reasoning processes mediating the separation of variables and the verification of their respective roles, it seemed worth while to give our subjects a problem involving much greater empirical difficulty than the earlier ones, though not requiring for its solution concepts essentially more complex. In the case of floating bodies, we have just had a glimpse of the importance which the schema "all other things being equal" plays in hypothetico-deductive thinking. But the interference of five distinct variables, as in the flexibility problem, furnishes a situation particularly favorable for the study of the formation of this experimental schema and of the formal operations which it presupposes, for if a complete solution is to be attained each factor must be varied independently and the others held constant.[2]

[1] With the collaboration of A. M. Weil, former research assistant, Institut des Sciences de l'Éducation, and J. Rutschmann, research assistant, Laboratory of Psychology.

[2] The experimental technique is as follows: The experimenter presents the subject with a large basin of water and a set of rods differing in composition (steel, brass, etc.), length, thickness, and cross-section form (round, square,

§ Stage I

If we are to understand in what way formal operations complement concrete operations at stage III, we must first find out what
the latter contribute to the separation of variables; but in order

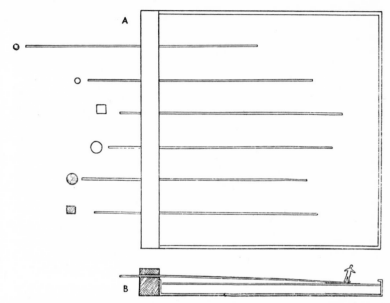

FIG. 2. Diagram A illustrates the variables used in the flexibility experiment. The rods can be shortened or lengthened by varying the
point at which they are clamped (see B for apparatus used). Cross-
section forms are shown at the left of each rod; shaded forms represent
brass rods, unshaded forms represent non-brass rods. Dolls are used
for the weight variable (see B). These are placed at the end of the rod.
Maximum flexibility is indicated when the end of the rod touches the
water.

rectangular). Three different weights can be screwed to the ends of the rods.
In addition, the rods can be attached to the edge of the basin in a horizontal
position, in which case the weights exert a force perpendicular to the surface
of the water. The subject is asked to determine whether or not the rod is
flexible enough to reach the water level. His methods are observed and his
comments on the variables he believes influence flexibility are noted; and
finally, proof is demanded for the assertions he makes.

to do that we must start by describing responses at the preoperational level (until about 7 years). The reactions of this stage are simple; in all his explanations the child is limited to describing what he sees. As neither classifications nor organized operations of serial ordering are yet available, he fills in his observations with precausal linking (finalism, animism, moral causality, etc.):

RIC (5 ; 0) puts 200 grams on the 40 cm. square steel bar: *"It doesn't touch the water"* [the rods to be compared represent mobile "bridges" attached to a plank which in turn represents a "road"]. At the end of these bridges are found small dolls, or "fishermen," which reach the edge of the water if the "bridges" bend enough. Next he takes up the round brass rod 7 mm.^{2} in diameter which, unlike the steel bars above, touches the water. "Why?"—*"Because it is lower down."* For the round steel rod, 22 cm. long, diameter 16 mm.^{2}: "Why doesn't it touch the water?"—*"Because the bar is too high."*—"Why does it stay too high?"—*"Because it's on a plank* [= attached to a plank, but so are all of the rods!].—"But why with that one [brass] and not with this one?"—*"Because it's too small"* [= too short].—"And why didn't it work with the first one?" [40 cm.].—*"It didn't work because there is wood* [the attachment plank] . . . [attached] *to the second there is wood too. I am going to try again* [he begins again]. *No, that doesn't work."*—"Why?"—*"Because it's heavier and it goes down in the water."*—"And this one?" [new rod].—*"It doesn't work because it's too high . . . ,"* etc.

HUC (5 ; 5) after a number of trials puts 100 grams on a rod and waits as if it were going to descend in a moment. "Why don't all the sticks go down the same way?"—*"Because the weight has to go in the water."* —Then he places 200 grams on a thick rod and 100 grams on a fine one: "Which one bends the most?"—*"That one"* [the fine rod].—"Why?" —*"The weight is bigger here* [he points out 200 grams on the other one]; *it ought to go into the water."*—[We put 200 grams on the thin one, which then touches the water. He laughs.] "Why does it touch now?"—*"Because it has to."*

We see that these subjects are generally limited to a simple report of what they perceive; the rod does not touch the water because it remains too high or it touches the water because it descends too low, etc. Finalism and moral causality ("It has to"), etc., are added. They also start to formulate relations, which process has a certain logical interest for us in that the child is satisfied with undifferentiated, overly-general classes. Just as, when making the transition from definitions which correspond in form to the finalism and moral causality of primitive explanations, he defines by generic classes which lack internal differentiation (as when he states that "a mama" is "a lady" without referring to her children), so at the present level (which immediately follows the level of precausal explanations) objective relational processes do appear but in terms of generic rather than specific inclusions. Thus RIC declares that a certain bar does not touch the water because it is attached to the plank, although those that do touch the water are similarly attached.[3] An instant later he takes up the same explanation again; *"Because there is wood"* (= the plank), but adds spontaneously *"There is wood* (attached) *to the second too."* In explaining why a thin rod bends more than a heavy one, HUC limits himself to noting that the heaviest weight is on the rod that bends the least, as if to imply that both "should" (in the moral rather than the logical sense) touch the water. There is still a great gap between this kind of inclusion, exclusively generic because its form is even more primitive than that of concrete operations, and the formal type of implication that will eventually succeed the latter.

§ Stage II (Substages II-A and II-B)

With the appearance of concrete class and relational operations, it becomes possible to report on raw empirical data through the use of classifications—coherent and differentiated serial ordering and correspondences—but this is not in itself sufficient to assure the separation of variables—*i.e.*, to assure the organization of a valid experiment.

MOR (7; 10), after having put the weight on a narrow rod which reaches the water, says: *"It won't fall the same way with this one* [thick] *because the other one is thinner."* Then he changes the weight: *"This one isn't so heavy as the other one"*; he places the heavy weight on a short rod and the light one on a long rod, predicting that the

---

[3] *Cf.* Piaget, *The Moral Judgment of the Child* (Free Press, 1948): The subject SCHMA (6 ; 6) thinks that a little liar fell in the water because he lied, but that if he had not lied he would nevertheless have fallen in "because the bridge was old."

curve will be sharper *"because the other weight is lighter than this one."* The experiment does not confirm his prediction, and then he lengthens the short rod: *"Oh! with the* [thick] *one you have to do that . . . ,"* etc.—The subject is asked to summarize what he has discovered up to that point by ordering the rods serially according to flexibility: "Which one bends most?"—*"This one because it is the thinnest."*—"Next?"—*"That one"* [long and thin, metal].—"Next?"—*"That one"* [short, wood].—"Next?"—*"This one* [thicker]; *it goes with the weight"* [heavy].—"Next?"—*"That one* [heavy, metal]; *it didn't go in the water because I had to do that"* [lengthen it].

BAU (9 ; 2): *"Some of them bend more than others because they are lighter* [he points out the thinnest] *and the others are heavier."*— "Show me that a light one can bend more than a heavy one [he is given a short thick rod, a long thin one, and a short thin one].—[He places 200 grams on the long thin rod and 200 grams on the short thick one without noticing the fact that the thin rod that he has chosen is also the longest.] *"You see."*—"Show me that the long one bends more than the short."—Again he puts 200 grams on the same two rods and this time the result is supposed to demonstrate the role of length.—"If I take away the long one, can you compare again to find out whether it's the lightest rod that bends more?"—*"Yes, this one and that one"* [the two short rods, one thick, one thin].—"Which is better, to compare these two or to compare the way you did before?"—*"These two"* [long and thin, short and thick].—"Why?"—*"They are more different."*

These two cases are sufficient to show us both the progress made over stage I and the inability of the subjects at substage II-A to separate out the experimentally relevant variables.

As before, the advance over stage I lies in the fact that the subject becomes capable of systematically registering the raw data —*i.e.*, the facts as directly observed—though not as they might be selected with the question in mind of the verification of a hypothesis or the separation of variables. The registration of data is systematic in that, instead of depending on the formulation of a simple global relationship (such as the unspecified generic class inclusion found at stage I), the subject is capable of differentiated classification, serial ordering or equalizations, correspondences, etc., which are all accurate when considered independently. For example, MOR manages to compare lengths, thicknesses, weights, etc., by serial ordering and even to set up a series of five rods arranged in order of observed inclinations. Furthermore, each one of these operations is correct including the last one, which is the most complicated. But taken together they prove nothing when the subject is left to his own initiative. When the experimenter chooses two terms of comparison with respect to a certain factor, all other things being equal, the operation of comparison that the subject accomplishes seems meaningful. But when the subject is left to himself, everything is mixed up. Thus, the series of five inclinations that MOR arranges to summarize what he has observed by himself is a multifactor confusion from which nothing can be deduced. Likewise, in order to demonstrate the role of the width (which moreover he confounds with the weight), BAU compares two rods, one of which is the narrowest but also the longest. Afterwards he chooses the same elements to demonstrate the role of the length, and when we try to encourage him to separate the two factors he answers that it is best to compare the terms which differ most widely.

In such cases the difference is clear between (1) the formal operations that would enable the subject to separate out the variables by use of the indispensable combinatorial system and (2) the concrete operations needed to report the facts but insufficient to structure an experiment which could utilize this separation. Less clear are the reasons why concrete operations are insufficient. Before analyzing this problem further, let us reexamine the reactions of substage II-B; these add explicit multiplicative schemas to the operations used at substage II-A, which appeal only to implicit logical multiplication (BAU knows that his rod is "more different" because "at the same time" thinner and longer, but he does not say so and proceeds by simple addition of relations).

The only change found at substage II-B is the successful use of multiplications between asymmetrical relations. While the subjects at substage II-A do not use logical multiplications except under the elementary form of one-by-one correspondences, at substage II-B subjects use double-entry tables with orders oriented serially in different directions [4] as well as multifactor groupings (several links for the same result):

[4] *Cf.* the coordinate axes for space which also appear towards 9–10 years (*The Child's Conception of Space*, Chaps. XIII–XIV).

OT (9 ; 3) begins by referring to length: *"You see that because the bar is longer it can go down better."*—"And if you take two bars of the same length?" [he is given a thin and a thick one].—*"There it goes down further, because it is thinner than the other which is fat, and that one isn't."* Next he determines the influence of weight and predicts for a short rod: *"That won't work: the rod is too short and the weight is too light for the rod."*

HAE (10 ; 9) discovers the roles of the material the rods are made of, thickness, and length. "Can you tell me without trying whether that [weight] will reach the water with this rod?"—*"It could, but by pulling it in* [only] *a little; it's made of the same metal as the other one but it is thicker, so you wouldn't have to pull it in as much as the other."* Thus [A same metal as B] × [thicker] × [longer] = [same inclination]. In addition there is understanding of the compensation between two relations oriented in opposite directions: [less thin] × [longer] = [thinner] × [shorter].

This last example indicates the appearance of both double or triple-entry tables (condition of the multiplication of transitive asymmetrical relations, with or without compensations) and multifactor multiplication (several causes are possible for the same effect).

Still, subjects at this level are unable to verify the action of one factor by leaving all of the other known factors constant. Likewise, although they understand the compensation of length and thickness for identical matter, they do not know how to generalize the concept of compensation to the mutual compensation of all known factors. Why this should be so raises a problem. The subjects' failure to generalize is even more difficult to explain when we consider the fact that they seem to be in possession of all the requisite operational instruments. But although the subject OT, for example, when given two bars of the same length, well understands that the thinner one will bend more, when asked to demonstrate the role of thickness he compares bars of unequal width without assuring equivalence among the other factors and does not realize that his verification is worthless. Likewise HAE does not proceed any more skillfully, in spite of his discovery of a potential compensation between two specific factors. Everything seems to indicate that we have come across two different systems of thought: one, the concrete, permitting simply the composition of relations and of classes which depend on the immediate data, and the other, the formal, permitting a restructuring of necessary links.

The concrete system consists of tables of associations or correspondences either of classes or of relations. For purposes of simplification we can express this in the language of classes. Let us call A_1 the class of rods which are 50 mm. or more long and A'_1 that of rods < 50 mm.; A_2 the class of weights of 300 grams or more and A'_2 that of weights < 300 grams; A_3 will be the class of brass rods and A'_3 that of non-brass rods; etc. Finally, X will be the class of rods touching the water and X' that of rods which do not. Remember that when joined together the two classes A and A' give the total class B.

In this case we have, for the two couples of classes (a single factor and its result X or X'), the double-entry table:

$$(B_1) \times (X + X') = A_1X' + A_1X + A'_1 + A'_1X'. \qquad (1)$$

For two factors and their result there are eight combinations (triple-entry table):

$$(B_1) \times (B_2) \times (X + X') = A_1A_2X + A_1A_2X' + A_1A'_2X + \\ A_1A'_2X' + A'_1A_2X + A'_1A_2X' + A'_1A'_2X + A'_1A'_2X'. \qquad (2)$$

Similarly, from three factors and their results, sixteen combinations can be derived, from four factors, thirty-two combinations, and finally from five factors and their results X or X' there are sixty-four combinations. In the course of the experiments, in a more or less empirical way (depending on his level) the child executes these sixty-four combinations fully or partially; they allow him to correlate the variation of factors with the result X or X' (the combinations are in fact even more numerous, since the factors of length, thickness, weight, and inclination themselves give rise to more variations than A or A' and since the class of A'_3 of non-brass rods is in fact subdivided).

First, we must find out whether or not the subject at substage II-B can construct such tables. It is likely that he can when one or two factors are involved, because he uses reasoning based on congruent structures and we know how easy it is for subjects of 7–10 years to structure serial correspondences. For three to five factors they can proceed by addition of new elements in succession (doubling the preceding table each time). But it is obvious that a complete calculation would not be possible, and, furthermore, it is not needed as long as the subject works with immediate correspondences from one element to the next.

Furthermore, in the presence of a single factor and its result, X or X', it is generally sufficient for the child to find an immediate correspondence in order to establish the relation between A_1 and X: if $A_1X + A'_1X'$ occurs and the combinations A'_1X and A_1X' are null (= not given in observation because nonoccurring), it is clear the subject will conclude that A_1 influences X. But if all four combinations or the three combinations $A_1X + A'_1X + A'_1X'$ occur empirically, the method of formulating simple correspondences between A_1 and X (or between their negations) will no longer suffice; the subject must hypothesize a second factor which operates in the case A'_1X (*i.e.*, a factor which produces X for causes other than A_1). However, although he knows how to do this for the simple cases (when the width compensates the length, etc.), his efforts are less and less systematic when the number of factors increases and the experimenter does not simplify the experiment in his successive presentations of the factors to the child. In other words, at the stage of concrete operations (II-A and II-B) the subject knows how to observe the experiment in terms of the various correspondences which actually occur, which means that he can construct increasingly more complex tables from the empirical associations (positive and negative). But he does not know how to interpret his tables except when immediate correspondences are sufficient. And he does not know how to separate variables when they are too thoroughly mixed.

The reason for this failure is that in order to separate variables one needs to vary each in turn while holding the others constant ("all other things being equal"). To do so, it no longer is enough to consider the table as a whole in which all the correspondences are simultaneously given; the associations A_1X, etc., must be analyzed situation by situation so that one may see which are linked and which are mutually exclusive. But to arrive at this analysis the subject would have to use a complete combinatorial system, one which is no longer the mere construction of a table of associations such as tables (1) with its 4 associations or (2) with its 8 associations. This complete combinatorial system involves considering the associations one-by-one, two-by-two, etc., so that 16 combinations can be derived from table (1) or 256 combinations from table (2) instead of the 4 or 8 derived from the double- or triple-entry tables. In other words, while tables (1) and (2) constitute simple wholes composed of 4 or 8 parts (or associations) brought together, the combinatorial system necessary for the formal analysis of the associations is based on what can be called a "structured whole," which here is composed of 16 combinations in case (1) and of 256 in case (2).

Moreover, we have already seen in chapters 1 and 2 that this complete combinatorial system is precisely the mark of formal thought, for its structure goes beyond additive or multiplicative groupings of classes and relations (with their simple concrete inferences founded on the transitivity of class inclusions (or of relational linkings) and engenders the structuring of propositional logic. Actually, for two propositions p and q the 16 possible operations (conjunction, disjunction, implication, incompatibility, etc.) correspond exactly to these 16 combinations which can be derived from table (1), and for three propositions p, q, and r the 256 possible operations (which, moreover, are all reducible to compositions of binary operations) correspond to the 256 combinations that can be derived from table (2).

In other words, if the substage II-B subjects do not yet isolate the variables but simply establish the empirically given correspondences, it is because they have not acquired the combinatorial system which constitutes propositional logic. The result is, on the one hand, that they do not know how to combine empirical results in such a way as to demonstrate which among the possible associations of variables actually occurs and, on the other, that they do not know how to reason by implication, etc., in such a way as to combine the various factual data that they observe in a form that is both necessary and conclusive. However, these two failures can actually be reduced to a single one, since the same combinatorial system will permit the stage III subjects to devise experiments for separating variables and to deduce the results of these experiments by the means of interpropositional operations.

Moreover, we must emphasize that it is because this same propositional logic is not available that the reasoning process which would be used to prove the empirical hypotheses remains inaccessible at substages II-A and B. Assuming $A \to B$ and $B \to C$ (where $\to$ can be an inclusion, an equality, or a transitive asymmetrical relation), the concrete reasoning consists in concluding $A \to C$. This sort of inference is found in the reactions of HAL and OT. But this utilization of transitive relations does not give rise to an interpropositional operation such as an implication. It amounts only to combining classes or relations among each other on the basis of a certain order of class inclusion.

In contrast, propositional operations consist in combining various empirical associations on which multiplicative classes are based in all possible ways: implication, for example, would be defined as deriving from the combination $A_1X + A'_1X + A'_1X'$ the affirmation that $p \supset q$ (if $p =$ the affirmation of A_1 and $q =$ that of X), for if only $(p.q) \vee (\bar{p}.q) \vee (\bar{p}.\bar{q})$ occur and never $p.\bar{q}$ (corresponding to A_1X'), then q is always true when p is true. These are the new combinations which, as we shall see later, distinguish the thinking typical of the stage III subjects and which at the same time give rise to both deductive capacity for demonstrative reasoning and experimental capacity for the isolation of the relevant variables as each one may influence the end result.

§ Stage III (Substages III-A and III-B)

This level is characterized both by the incipient formal thinking revealed in the appearance of hypothetico-deductive reasoning and by an active attempt at verification. However, at first the subject is not able to handle the complete range of interpropositional operations; as a result, even though we may observe the genesis of implication, exclusion, etc., we do not yet find him able to organize a systematic proof conforming to the schema "all other things being equal" except in certain cases and even then not for all of the relevant factors.

PEY (12 ; 9) speculates that if the rod is to touch the water it must be "long and thin." After several trials, he concludes: *"The larger and thicker it is, the more it resists."*—"What did you observe?"—*"This one* [brass, square, 50 cm. long, 16 mm.² cross-section with 300 gram weight] *bends more than that one* [steel; otherwise the same conditions which he has selected to be equal]: *it's another metal. And this one* [brass, round] *more than that one"* [brass, square; same conditions for weight and length, but 10 and 16 mm.² cross-section].—"If you wanted to buy a rod which bends the most possible?"—*"I would choose it round, thin, long, and made of a soft metal."*

AULE (12 ; 10) wants to prove that a long rod bends more than a short one. He takes the two steel bars, one round and 22 cm. long, the other square and 50 cm. but, not noticing that they do not have the same cross-section form, he adjusts both of them to 22 cm. for length; *"This one* [round] *bends more because it is thin"* [they have the same width, but one is round, the other square].—"What have you proved?" —*"I don't think I've proved anything. Oh! Yes, that the round ones bend more than the square."*

DUR (11 ; 10): *"There are flat ones, wider ones, and thinner ones and longer ones. If they are both long and thin, they bend still more."*— "Could you show me that a thin rod bends more than a wide one?" —[He puts 100 grams on the round steel rod—50 cm. long and 16 mm.² cross-section—and 200 grams on the round steel rod—50 cm. long and 10 mm.² cross-section.] *"That one bends more"* [10 mm.² cross-section].—"I would like you to show me only that the thin one bends more than the wide. Is that way right?"—[He takes off the 100-gram weight and puts 200 grams on the 16 mm.² rod.] *"You see, this is the right way."*

KRA (14 ; 1): "Can you show me that a wide one bends less than the narrow?"—[He puts 200 grams on the round steel bar—50 cm. long and 10 mm.² cross-section—and 200 grams on the square brass rod—50 cm. long and 16 mm.² cross-section.] *"This one* [thin steel] *goes down more."*—"Why?"—*"It is round, more flexible, the steel is less heavy, it is round and narrower."*—"Fine, but I would like a rigorous proof that it's because it is narrower."—[He places 200 grams on the round steel rod—50 cm. and 16 mm.²—and 200 grams on the round steel rod 50 cm. and 10 mm.².] *"You see, this one bends more because it is less wide."*—"Bravo. Can you demonstrate the same thing with others?"—"Yes. [Steel, square—50 cm. and 16 mm.²—instead of round steel of 16 mm.²; thus the comparison is no longer exact.] *This one* [narrow and round] *bends more, it is less heavy."*—"And can you demonstrate the role of the form?"—He puts 200 grams on the rectangular brass rod, 50 cm. long and 16 mm.² cross-section.—"Why does this one [round, steel] bend more?"—*"Because it is round."*—"Is that the only reason?"—*"The brass is also heavier"* [he then spontaneously discards the steel rod and takes a square brass rod 50 cm. and 16 mm.²].

The subjects' set at stage III is essentially new in comparison to the set that characterizes concrete operations; it consists of not being satisfied with empirical events as directly given but in regarding them from the start as one aspect of a larger domain, the domain of the possible. In effect, stage II subjects are limited to recording the successive data in terms of all the relations and classes required by their diversity, but they neither separate out variables nor elaborate hypotheses or proofs. On the other hand, substage III-A subjects from the start conceive of reality as a product of various factors arranged in a set of possible combinations. This results in the appearance of two formerly insignificant behavior patterns: *the formulation of hypotheses,* which consists of the restructuring of these possible combinations as they might occur empirically, and *attempts at proof,* which consist of determining which of the possibilities in fact do occur.

No doubt, the initial reactions of each of the preceding subjects do not seem to differ from those observed at stage II. They consist of describing empirical givens by means of relations and classifications. But, whereas the stage II subject accepts everything pell mell, believing that in this way he has gotten to reality itself, stage III subjects use preliminary concrete descriptions only as material for setting up hypotheses and proofs; the result is a more active set.

This new behavior can be observed in the choice of rods to be compared—*i.e.,* in the tendency to compare them only from a standpoint bearing on a delimited question. Whereas the stage II subject compares any rod whatever to any other, limiting himself to a statement of the most obvious relations, the stage III subject understands that if he is to establish a given relationship, it is important to select certain pairs of rods rather than others. It is this choice, which is the most easily observed new reaction at stage III, that allows us to demonstrate the nature of the logical operations utilized.

The most important of these operations, or at least the one which nearly always orients the substage III-A subjects' analysis at the beginning of the experiment, is the formal operation of implication by which the subject assumes that a determinate factor produces the observed consequences in all cases. At stage II, a comparable causal relationship was established by simple correspondence—for example, the longer the rod, the more flexible—but this type of reasoning cannot be legitimately generalized to all cases. The operation of implication takes a similar statement of correspondence as its starting point (the conjunction $p.q$ translating AX). However, at substage III-A two new forms of behavior appear, resulting in three types of statement that distinguish the formal operation of implication from concrete correspondence. First, a more or less systematic effort is made to determine the consequences of eliminating or diminishing factor A, as compared to the simple search for association between factor A and its result X, which we found at the concrete level (although this effort is not completely systematic before substage III-B); thus subjects at the formal level can determine that in certain cases effect X is itself eliminated or diminished (the association $A'X'$ is found), whereas in certain others it is conserved because it can be produced by factors other than A (association $A'X$). The unified relation $AX + A'X + A'X'$ (or in propositions: $p.q$ v $\bar{p}.q$ v $\bar{p}.\bar{q}$) thus constitutes a system of interpretation which is broader than simple correspondence because it integrates three possibilities simultaneously (either AX or $A'X$ or $A'X'$) and because in this way it can bring into a single whole the results of several different groupings of classes and relations.

In this case we have an elementary example of the combinatorial system discussed above that handles "structured wholes": in the case of implication the three parts (or associations) AX, $A'X$, and $A'X'$ are integrated in the manner presented above, whereas in the case of disjunction or incompatibility they are linked in other ways. Moreover, the ability to handle the combinatorial system which appears at substage III-A is manifested not merely in the appearance of this or that operation but in their system as such—i.e., by all of the sixteen binary operations and by the possibility of linking a determinate number of them in such a way as to give rise to operations of an advanced sort.

Moreover, and this is the second behavior pattern new to substages III-A and III-B, the formation and the utilization of this total system are manifested in the development of proof and notably in the schema "all other things being equal." The latter assumes the utilization of a set comprising several distinct types of implications integrated with other operations. From substage

III-A on, we observe a search for demonstration which is oriented toward proof and the control of experimental conditions, but the difficulties which prevent its realization are equally evident. In fact, the subjects are not capable of more than partial proof. For example, in order to give proof of the influence of the type of metal, PEY compares two bars, one copper, the other steel, while holding all the other factors constant; but in order to prove the influence of width he compares rods of 10 and 16 mm.^{2} cross-section with unequal section forms (round and square) without realizing that he is then faced with two independent factors. Likewise DUR varies width and the weight simultaneously before correcting for the lack of equivalence between the two. We are certainly dealing here with a search for equivalence in the conditions of comparison, but we still find difficulties in achieving it. In order to have a better grasp of the nature of the operations required by the schema "all other things being equal" and of the dependence of these operations on the total combinatorial system referred to above, let us begin by comparing the reactions found at substage III-A with those of substage III-B; during this latter stage proof becomes rigorous for the experiment under consideration.

One good illustration will suffice:

DEI (16 ; 10): "Tell me first [after experimental trials] what factors are at work here."—"*Weight, material, the length of the rod, perhaps the form.*"—"Can you prove your hypotheses?"—[She compares the 200 gram and 300 gram weights on the same steel rod.] "*You see, the role of weight is demonstrated. For the material, I don't know.*"— "Take these steel ones and these copper ones."—"*I think I have to take two rods with the same form. Then to demonstrate the role of the metal I compare these two* [steel and brass, square, 50 cm. long and 16 mm.^{2} cross-section with 300 grams on each] *or these two here* [steel and brass, round, 50 and 22 cm. by 16 mm.^{2}]: *for length I shorten that one* [50 cm. brought down to 22]. *To demonstrate the role of the form, I can compare these two*" [round brass and square brass, 50 cm. and 16 mm.^{2} for each.]—"Can the same thing be proved with these two?" [brass, round and square, 50 cm. long and 16 and 7 mm.^{2} cross-section].—"*No, because that one* [7 mm.^{2}] *is much narrower.*"—"And the width?"—"*I can compare these two*" [round, brass, 50 cm. long with 16 and 7 mm.^{2} cross-section].

Our problem is to understand how the subject acquires such a systematic method—one whose apparent simplicity should not mislead us, for only at 14–15 years can subjects spontaneously organize and utilize it without error.

If we refer back to proposition (2) (on page 53), which gives the eight basic associations possible for two factors and their results X or X', we must first assume that the subject begins by establishing the facts, as at stage II, by means of concrete classificatory and correspondence operations. For example, for the factor B_2 (weight) and the factor B_3 (metal), he may obtain the following table of observations:

$A_2A_3X = 300$ gr., brass, inclination X (maximum);

$A_2A_3X' = 300$ gr., brass, inclination X' (because it is too short, etc.);

$A_2A'_3X = 300$ gr., steel, inclination X (sufficiently thin, etc.);

$A_2A'_3X' = 300$ gr., steel, inclination X';

$A'_2A_3X = 200$ gr., brass, inclination X (sufficiently long or thin, etc.);

$A'_2A'_3X' = 200$ gr., brass, inclination X';

$A'_2A'_3X = 200$ gr., steel, inclination X (sufficiently long, etc.);

$A'_2A'_3X' = 200$ gr., steel, inclination X'.

Such observations show the subject from the start that the factors A_2 and A_3 are not the only relevant ones, since the same combination A_2A_3 may give either X or X'. This table is actually extracted from the table of sixty-four associations corresponding to the subject's potential observations. But the innovation found at stage III is that, having organized a complex situation by means of concrete operations, the subject does not consider his sets of facts as a final ordering from which it would be sufficient to extract such and such relations and correspondences. Instead he views them as a starting point for new combinations such that, in associating each one of these eight base associations one-by-one, two-by-two, three-by-three, etc., he can extract a new set of operations corresponding to the "structured whole" of the initial tables. These are the new operations that make possible the separation of variables, owing to the utilization of a set of implications in combination with the simple conjunctions.

To state the new reasoning process in propositional terms, let us call p, q, r, s, and t the propositions which affirm the presence of factors A_1, A_2, A_3, A_4, and A_5 respectively and $\bar{p}$, $\bar{q}$, $\bar{r}$, $\bar{s}$, and $\bar{t}$ the propositions which deny their presence, and let us designate by x and by $\bar{x}$ the propositions which affirm the results X and X' respectively. The verification schema "all other things being equal" thus amounts to nothing more than varying one of the factors corresponding to p or q, etc., and leaving the others unchanged. For example, for A_1 corresponding to p, we have

$$(p.q.r.s.t.x) \text{ v } (\bar{p}.q.r.s.t.\bar{x}), \tag{3}$$

which amounts to saying that for two rods supporting 300 grams ($= q$), of brass ($= r$), thin ($= s$), and with round section forms ($= t$), it is sufficient to shorten the initial length of 50 mm. sufficiently (p transformed to give $\bar{p}$) if we are to modify the result (x transformed to give $\bar{x}$).

Thus we see that PEY compares two rods such that $(p.q.r.s.t.x)$ v $(p.q.\bar{r}.s.t.\bar{x})$ ($= r$): in order to demonstrate the role played by the factor metal in this case the equivalence $(r.x)$ v $(\bar{r}.\bar{x})$ acquires a demonstrative value because the other propositions $(p.q.s.t)$ remain unchanged—*i.e.*, that $(p.q.r.s.t.x)$ v $(p.q.\bar{r}.s.t.\bar{x})$ are chosen among the totality of possible propositions. Furthermore, this choice presupposes an understanding of the fact that, if propositions $(p.q.s.t)$ are not kept unchanged (thus if the facts that they express are not held constant), the effect x could result from a cause other than r: the equivalence (reciprocal implication) $(r.x)$ v $(\bar{r}.\bar{x})$ is thus actually derived from an implication $(r.x)$ v $(\bar{r}.x)$ v $(\bar{r}.\bar{x})$ and consequently is integrated with all the implications conceived of as possible between p, q, r, s, t, and x. In other words, to hold four out of five factors constant is equivalent to granting that each one could in turn give rise to the same combinations. That is why the process of verification based on the schema "all other things being equal" is so complex and actually involves the whole interpropositional combinatorial system.

The proof of this is that at substage III-A this type of demonstration is still only partially understood. The subject we have just cited, PEY, later reasons unsystematically when he tries to verify the role of the section form (round $= t$ or square $= \bar{t}$). He varies t and $\bar{t}$ and the width ($= s$) simultaneously—*i.e.* he sets up the proposition $(p.q.r.s.t.x)$ v $(p.q.r.\bar{s}.\bar{t}.\bar{x})$. In this case it is clear that nothing concerning the role of t alone can be deduced from $(s.t.x) \vee (\bar{s}.\bar{t}.\bar{x})$. This sort of error is found again and again throughout substage III-A. On the other hand, at substage III-B the proof is rigorous; for example, subject DEI applies the same schema (3) separately to all the factors which she distinguishes and does it without any error. In other words, the formal combinatorial system based on the "structured whole" (in contrast to the one-by-one multiplications which furnish it with its base associations) is under construction during substage III-A but is completed only at III-B.

Another acquisition that these same studies show to be specific to the formal level (substages III-A and B) is the capacity to determine qualitatively certain compensations between heterogeneous relations. We have already seen the operation of certain logical calculations for compensation which are based on the multiplication of concrete relations. Thus HAL (10 years) discovers that a rod made of the same metal as another but thicker rod may bend an equal amount providing it is lengthened. In this case, the compensation is explained by the following operations: if F designates the transition from thicker to thinner and L the transition from shorter to longer and F and L the inverse transitions, we have:

$$F \times L = F \times L. \qquad (4)$$
$$\rightarrow \quad \leftarrow \quad \leftarrow \quad \rightarrow$$

This compensation is easy to understand in terms of simple multiplication of inverse relations because these relations are homogeneous. Both thickness and length are spatial dimensions, and, since they work in opposite directions, it is easy to multiply them by each other in a compensatory manner to obtain the same product. We have already observed the same phenomenon in reference to the conservation of quantities: [5] a tall and narrow beaker may contain the same quantity of water as a low wide beaker because the increase in width may compensate for the loss of height. Although in both cases three dimensions (including two which are distinct) and an operation of logical multiplication are involved, in the case of quantities the relevant operation is more

[5] See Piaget and Szeminska, *The Child's Conception of Number* (Routledge & Kegan Paul, 1952), Chap. I.

additive than multiplicative. This is because of the subject's impression that it is possible to displace certain parts of the object taken from the width in the direction of the length and vice versa, thus leading him to an additive equalization of products.

Let us now examine the following cases of compensation:

1. For equal lengths, a round thin steel bar has the same flexibility as a round thicker brass bar;

2. For equal lengths, a round thin steel bar has the same flexibility as a flat brass bar with a larger cross-section surface;

3. For equal lengths, a round thick steel bar has the same flexibility as a square narrower steel bar.

These pairs of bars are shown to the child, who is asked to explain only why the rods bend equally for the same weights. Nevertheless, these three problems are correctly explained only at the formal level; problem 1 in substage III-A and problems 2 and 3 in III-B, with maximum difficulty for 3. Why this disparity among the three problems?

In problem 1 the fact that the first rod is thin compensates for the lesser flexibility of its steel composition. But since these two factors are dissimilar, the subject must first separate out the relevant variables. At the same time he must perceive them as acting concurrently if he is to multiply the concrete relationships between them. Here we see an analogy between this double requirement and the verification schema "all other things being equal." Actually, in both cases the subject must cancel the effect of one of the factors in order to determine the effect of the other. Since the two factors are always present simultaneously, in both cases he must limit himself to holding constant only the factor to be canceled out (mentally or experimentally). Thus, he actually cancels not the effect itself but rather possible variations in the effect.

However, at the concrete level relationships between the metallic composition of a rod and its flexibility or between thickness and flexibility are formulated in rough form simply by noting the data in varied situations without equalizing other factors. The result is that in situations where compensations are exact, such as in problem 1, stage II subjects cannot be certain that the difference in flexibility due to metallic composition is being compensated by thickness alone. On the contrary, ascertaining that the lengths and degree of curvature are the same, they are led to believe that the metallic factor (or eventually thickness) is less important than they had formerly believed. So it is only when the factors are both separated and integrated at the same time—*i.e.*, at the level where implication replaces simple concrete correspondence—that the subject is able to conceive of two factors as compensating each other exactly, even though he does not know how to determine the quantitative influence of each factor. When this equivalence is achieved and the subject has worked out the separation of variables, his thinking turns to the variations that are possible under pure, unmixed conditions, and it is not limited to actual and mixed variations. It is from this that formally deduced compensations derive ("if it were . . . that should be this case . . ."). They arise in cases where compensation by correspondences or concrete multiplications is inadequate.

The same holds for problems 2 and 3, but since in these cases form and thickness compensate each other while the forms themselves also differ, the thickness (section surface) is not given perceptually but must be formulated as a hypothetical possibility. Thus the greater difficulty of these latter problems, problem 3 in particular, is accounted for. Nearly everything must be deduced by the subject. In problem 3 the section surface is hard to discern.

As for the intellectual operations, there is (aside from implications) a sort of proportion mediating the subject's understanding of these compensations which is interesting because, since we have not given our subjects any metrical or numerical data, it is a pure qualitative or logical schema. The starting point is a double implication (which we write for statements p and q, which designate any two factors, it being understood that in the case of metal and thickness, r and s are used, or that in the case of thickness and section surface, s and t, etc.):

$$p.\bar{q} \supset x \quad \text{and} \quad \bar{p}.q \supset x \quad \text{or} \quad x \supset (p.\bar{q}) \vee (\bar{p}.q). \tag{5}$$

This double implication signifies that the presence of factor p, in combination with the absence (or diminution) of factor q gives the same result (designated by x) as the absence (or the diminution) of factor p and the presence of that designated by q.

In this case, the formulation is as follows: (1) conjunctions $(p.\bar{q})$ and $(\bar{p}.q)$, which individually express a relationship of re-

ciprocal exclusion between p and q (let $p \, \text{vv} \, q = p.\bar{q} \, \text{v} \, \bar{p}.q$), lead
to the same consequence, in the present case x: (2) thus, not only
are they reciprocal but factors p and q can be substituted for
each other without influencing the result. The notion of a certain
logical proportion by reciprocity (R) follows. It is general for
$(p.\bar{q}) \, \text{v} \, (\bar{p}.q)$. But here it serves as a schema for compensation
itself. For in this case reciprocity signifies an operation whose
value is equal but which is oriented in the opposite direction
(diminution or reinforcement):

$$\frac{\bar{p}}{q} = R\frac{p}{\bar{q}}. \tag{6}$$

This expression signifies (depending on whether it is read
diagonally, vertically, or horizontally): (a) that $\bar{p}.\bar{q} = R(p.q)$;
(b) that $\bar{p} \, \text{v} \, \bar{q} = R(p \, \text{v} \, q)$; (c) that $\bar{p}.q = R(p.\bar{q})$; and (d) that
$\bar{p}.p = R(q.\bar{q})$ since $o = Ro$.

We will come across many similar examples of logical proportionality either independent of all metrical data or prior to
numerical determination. For the moment it is enough to note
that the problem involves not only propositional reasoning but,
in addition, a formal structuring of the elements themselves. This
formal structuring is the subject matter of the second part of
this work.

4

The Oscillation of a Pendulum
and the Operations
of Exclusion[1]

WE HAVE JUST SEEN how the subject goes about separating out factors in order to determine their respective effects in a multifactor experimental setup. The present chapter takes up the reactions of the child and adolescent in an analogous situation [2] with the difference that only one of the possible factors actually plays a causal role; since the others have no effect they must be excluded after they have been isolated. Such is the case for the pendulum. The variables which, on seeing the apparatus, one might think to be relevant are: the length of the string, the weight of the object fastened to the string, the height of the dropping point (= amplitude of the oscillation), and the force of the push given by the subject. Since only the first of these factors is actually relevant, the problem is to isolate it from the other three and to

---

[1] With the collaboration of A. Morf, research assistant, Laboratory of Psychology and Institut des Sciences de lÉducation; F. Maire, former research assistant, Laboratory of Psychology; and C. Lévy, former student, Institut des Sciences de l'Éducation.

[2] The technique consists simply in presenting a pendulum in the form of an object suspended from a string; the subject is given the means to vary the length of the string, the weight of the suspended objects, the amplitude, etc. The problem is to find the factor that determines the frequency of the oscillations.

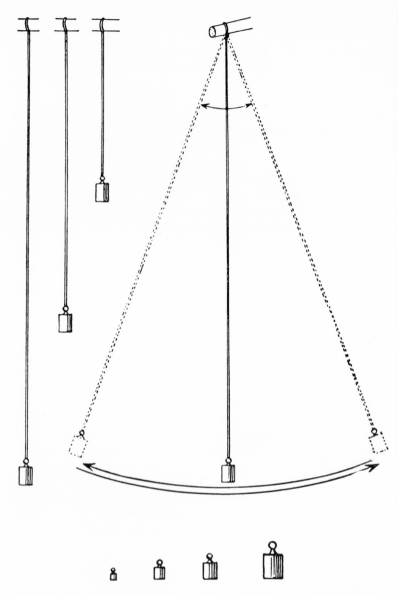

FIG. 3.   The pendulum problem utilizes a simple apparatus consisting of a string, which can be shortened or lengthened, and a set of varying weights. The other variables which at first might be considered relevant are the height of the release point and the force of the push given by the subject.

exclude them. Only in this way can the subject explain and vary the frequency of oscillations and solve the problem.

§ Stage I. Indifferentiation Between the Subject's Own Actions and the Motion of the Pendulum

The preoperational stage I is interesting because the subjects' physical actions still entirely dominate their mental operations and because the subjects more or less fail to distinguish between these actions and the motion observed in the apparatus itself. In fact, nearly all of the explanations in one way or another imply that the impetus imported by the subject is the real cause of the variations in the frequency of the oscillations:

HEN (6 ; 0) gives "some pushes" of varying force: *This time it goes fast . . . this time it's going to go faster."*—"That's true?"—"*Oh! Yes*" [no objective account of the experiment]. Next he tries a large weight with a short string: *It's going faster* [he pushes it]. *It's going even faster."*—"And to make it go very fast?"—"*You have to take off all the weights and let the string go all by itself* [he makes it work but by pushing]. *I'm putting them all back, it goes fast this time*" [new pushes]. As for the elevation: "*If you put it very high, it goes fast*" [he gives a strong push]. Then he returns to the weight explanation: "*If you put on a little weight, it might go faster."*—Finally we ask him if he really thinks that he has changed the rate.—"*No, you can't; yes, you can change the speed.*"

DUC (7 ; 3) is a little more advanced in that he finds several (nonsystematic) correspondences between the lengthening of the string and the increase in frequency. But he cannot prevent himself from pushing constantly and he counts the oscillations badly, always influenced by his expectations.

One can see, then, that because of the lack of serial ordering and exact correspondences the subject cannot either give an objective account of the experiment or even give consistent explanations which are not mutually contradictory. It is especially obvious that the child constantly interferes with the pendulum's motion without being able to dissociate the impetus which he gives it from the motion which is independent of his action.

§ *Stage II. The Appearance of Serial Ordering and Correspondences but Without Separation of Variables*

Stage II subjects are able to order the lengths, elevations, etc., serially and to judge the differences between observed frequencies objectively. Thus they achieve an exact formulation of empirical correspondences but do not manage to separate the variables, except insofar as the role of the impetus is concerned.

At substage II-A serial ordering of the weights is not yet accurate:

JAC (8 ; o) after several trials in which he has varied the length of the string: *"The less high it is* [= the shorter the string], *the faster it goes."* The suspended weight, on the other hand, gives rise to incoherent relationships: *"With the big ones* [= the heavy ones] *it falls better, it goes faster,* for example, *It's not that one* [500 grams], *it's this one* [100 grams] *that goes slower."* But after a new trial, he says in reference to the 100-gram weight: *"It goes faster."*—"What do you have to do for it to go faster?"—*"Put on two weights."*—"Or else?"— *"Don't put on any: it goes faster when it's lighter."* As for the dropping point: *"If you let go very low down, it goes very fast,"* and *"It goes faster if you let go high up,"* but in the second case JAC has also shortened the string.

Since the ordering serially of the other factors is accurate, the subject discovers the inverse relationship between the length of the string and the frequency of the oscillations at this and succeeding levels. However, since he does not know how to isolate variables, he concludes that the first variable is not the only relevant one in the problem. Moreover, if he attributes causal roles to the weight and the dropping point as well, it is because he varies several conditions simultaneously.

In spite of the marked progress seen at substage II-B, which is due to an accurate ordering of the effects of weight (in the raw data), the factors cannot always be separated:

BEA (10 ; 2) varies the length of the string [according to the units two, four, three, etc., taken in random order] but reaches the correct conclusion that there is an inverse correspondence: *"It goes slower when it's longer."* For the weight, he compares 100 grams with a length of two or five with 50 grams with a length of one and again concludes that there is an inverse correspondence between weight and frequency. Then he varies the height of the drop without changing the weight or the length [without intending to hold them constant, but by simplification of his own movements] and he concludes: *"The two heights go at the same speed."* Finally he varies the force of his push without modifying any other factor and again concludes: *"It's exactly the same."*

CRO (10 ; 2), likewise, cannot separate weight and length. However, in contrast to BEA, he does vary the dropping point. He begins with a long string and 100 grams, then shortens the string and takes 200 grams that he drops from a higher point: "Did you find out anything?" —*"The little one* [100 grams] *goes more slowly and the higher it is* [200 grams with a short string] *the faster it goes."* But afterwards he puts 50 grams on the same short string: *"The little weight goes even faster."* However, the subject neglects this last case: *"To go faster, you have to pull up the string* [diminish the length] *and the little one goes less fast because it is less heavy."*—Then: "Do you still wonder what you have to do to make it go faster?"—*"The little weight goes faster."*—"How can you prove it?"—*"You have to pull up the string"* [diminish the length].

PER (10 ; 7) is a remarkable case of a failure to separate variables: he varies simultaneously the weight and the impetus; then the weight, the impetus, and the length; then the impetus, the weight, and the elevation, etc., and first concludes: *"It's by changing the weight and the push, certainly not the string."*—"How do you know that the string has nothing to do with it?"—*"Because it's the same string."*—He has not varied its length in the last several trials; previously he had varied it simultaneously with the impetus, thus complicating the account of the experiment.—"But does the rate of speed change?"—*"That depends, sometimes it's the same. . . . Yes, not much. . . . It also depends on the height that you put it at* [the string]. *When you let go low down, there isn't much speed."* He then draws the conclusion that all four factors operate: *"It's in changing the weight, the push, etc. With the short string, it goes faster,"* but also *"by changing the weight, by giving a stronger push,"* and *"for height, you can put it higher or lower."*—"How can you prove that?"—*"You have to try to give it a push, to lower or raise the string, to change the height and the weight"* [He wants to vary all factors simultaneously].

MAT (10 ; 6) goes so far as to set up the simultaneous variation of factors as a principle.—"How do you know that it goes faster when there is more weight?"—"*When you put on a big weight, it goes faster.*" —"Did you find that out?"—"*Yes, by raising the string* [= by diminishing its length], *then you put on the big weight at the same time.*"

These cases are extremely instructive because they show the difference between concrete and formal operations. From the first standpoint, the subjects can handle all the forms of serial ordering and correspondence which make the variation of the four factors possible and assure the reporting of the result of these variations, but they know how to draw from these operations nothing more than inferences based on their transitiveness (from the model $A < C$ if $A < B$ and $B < C$). They remain inept at all formal reasoning. From the second standpoint, they commit the following two errors: (1) In varying several factors simultaneously so that $A_1 A_2 A_3 A_4$ are transformed to $A'_1 A'_2 A'_3 A'_4$ and in ascertaining the change from the result X to X', they think they have shown that each one of the factors in turn implies X'. Put into propositional language, the error amounts to concluding from $(p.q.r.s) \supset x$ that $(p \supset x).(q \supset x).(r \supset x).(s \supset x)$, without suspecting the existence of other possible combinations (see MAT for two factors: $p.q \supset x$ therefore $q \supset x$); (2) Reciprocally, subject PER, having varied all of the factors except one (the length of the string) and not being very sure whether or not the result has been modified, concludes that the single constant factor must be ineffective ("Anyway it's not the string . . . because it's the same string!"). In other words, from $p.q.r.s\,(x) \vee p.\bar{q}.\bar{r}.\bar{s}\,(x \vee \bar{x})$ he concludes $\overline{p \supset x}$.

Thus it is evident that the subjects still lack some logical instrument for interpreting the experimental data and that their failure to separate out the factors is not simply the result of mental laziness. Just as BAU (Chap. 3) varied two factors simultaneously in the comparison of the flexibility of rods so that the results would be "more different," so do MAT and the preceding subjects explicitly propose to modify all factors simultaneously so as to accomplish more impressive transformations. At this stage the subjects lack a formal combinatorial system. Since they are accustomed to operations of classification, serial ordering, and correspondences, they are limited to simple tables of variation and do not conceive of the multiplicity of combinations which can be drawn from them. Since they have no combinatorial system based on the "structured whole," they do not even begin to isolate the relevant variables.

§ Substage III-A. Possible but not Spontaneous Separation of Variables

At the lower formal level, substage III-A, the child is able to separate out the factors when he is given combinations in which one of the factors varies while the others remain constant. In this case he reasons correctly and no longer according to the kinds of inference of which we have just seen several examples. But he himself does not yet know how to produce such combinations in any systematic way—*i.e.,* formal operations are already present in a crude form, making certain inferences possible, but they are not yet sufficiently organized to function as an anticipatory schema.

JOT (12 ; 7) believes that *"you have to pull down* [lengthen] *the string."* He suspends 20 grams and varies the length: *"It goes more slowly when you lower* [lengthen] *the string and faster when it's high up."*—"That's all?"—*"Maybe the weight does something."* But to verify this, he takes 100 grams and lengthens and shortens the string, then 50 grams, lengthening and shortening the string again: *"Yes, it goes faster high up* [= when the string is short]; *it's the string."* In other words, he varies the string instead of the weight. Then he changes the weight while again varying the string in the same way. This process makes it possible to draw a conclusion, providing that the respective frequencies are remembered from one situation to another, but it complicates the matter uselessly. When the subject is asked to give proof of the influence of length, he is satisfied with a pure deduction: *"When the string is long, it takes more time to go from one end to the other. When it is short, it takes less time."*

ROS (12 ; 8) immediately discovers the role of length by lengthening and shortening the string with the same weight. Then he reduces the weight, but at the same time shortens the string [200 grams with the long string and 20 grams with the short string]. His conclusion is that "*the weight has an effect too.*" He proceeds in the same manner to control for the role of impetus and concludes that the "*impetus plays a part too.*" But he is doubtful about the weight and half sees the need to leave the other factor, that of length, invariant; he shortens the string, attaching 50 grams and 100 grams successively. The result does not change and his doubts grow: "*I have to do it over again to be sure it's right.*" Then he begins again, but once more he varies both weight and length. This time he doubts the role of the length and takes 20 grams while lengthening and shortening the string. "*When it's smaller* [= shorter], *the weight goes faster. It's because I didn't put on the same weight; that's why* [why nothing is proved]. *Now I'll put on the same weight.*" Nevertheless, he still believes that the weight has an influence.—Then we change the weights and lengths in front of him simultaneously: "Does that prove anything?"—"*No, because you have to put on the same weight.*"—"Why?"—"*Because the weight makes it go faster*" [!].

LOU (13 ; 4) also compares 20 grams on a short string to 50 grams on a long string and concludes that "*it goes faster with the little weight.*" Next, rather curiously, he performs the same experiment but reverses the weights [50 grams with a long string and 100 grams with a short one]. However, this time he concludes that "*when it's short it goes faster*" and "*I found out that the big weight goes faster*"; however, he does not conclude that the weight plays no role.—"Does the weight have something to do with it?"—"*Yes* [he takes a long string with 100 grams and a short one with 20 grams]. *Oh, I forgot to change the string* [he shortens it, but without holding the weight constant]. *Ah, no, it shouldn't be changed.*"—"Why?"—"*Because I was looking at* [the effect of] *the string.*"—"But what did you see?"—"*When the string is long, it goes more slowly.*" LOU has thus verified the role of the length in spite of himself but has understood neither the need for holding the nonanalyzed factors constant nor the necessity for varying those which are analyzed.

These transitional cases are of an obvious interest. They demonstrate, even better than the examples from substage II-B, the difficulty which arises in distinguishing factors and in the method "all other things being equal." In the first place, as among the substage II-B subjects, we find the tendency deliberately to vary two factors simultaneously, and even (as for LOU) the tendency not to vary the particular factor under consideration. But almost in spite of themselves and under the influence of nascent formal operations, these same subjects feel that in proceeding as they do they are not proving anything, so they manage either actually to transform the factor which they want to leave unchanged (as LOU) or to vary all factors by turns without knowing how to focus their analysis on the point being analyzed (as JOT). In such cases, the conclusion is accurate insofar as it relates to the factor of length, the only effective factor; but because the subjects lack combinations which would make exclusion possible, it is not accurate for weight or impetus, etc. In other words, the formal logic in the process of formation is for these subjects superior to their experimental capacity and has not yet adequately structured their method of proof; consequently, they are able to manipulate the easiest operations, those which state that which is and establish the true implications. But they fail in the case of the more difficult ones, those which exclude that which is not and deny the false implications.

§ Substage III-B. The Separation of Variables and the Exclusion of Inoperant Links

For the pendulum problem, as for flexibility (Chap. 3), substage III-B subjects are able to isolate all of the variables present by the method of varying a single factor while holding "all other things equal." But, since only one of the four factors actually plays a causal role in this particular problem, the other three must be excluded. This exclusion is a new phenomenon that contrasts sharply with substage III-A, where such an operation was still impossible, and with the flexibility experiment where it was unnecessary.

EME (15 ; 1), after having selected 100 grams with a long string and a medium length string, then 20 grams with a long and a short string, and finally 200 grams with a long and a short, concludes: *"It's the length of the string that makes it go faster or slower; the weight doesn't play any role."* She discounts likewise the height of the drop and the force of her push.

EGG (15 ; 9) at first believes that each of the four factors is influential. She studies different weights with the same string length [medium] and does not notice any appreciable change: *"That doesn't change the rhythm.*" Then she varies the length of the string with the same 200-gram weight and finds that *"when the string is small, the swing is faster."* Finally, she varies the dropping point and the impetus [successively] with the same medium length string and the same 200 gram weight, concluding for each one of these two factors: *"Nothing has changed."*

The simplicity of these answers is in contrast to the hesitation found at substage III-A, but this must not mislead us. The answers are the result of a complex elaboration whose operational mechanism must now be isolated.

Let p be the statement that there is a modification in the length of the string and $\bar{p}$ the absence of any such modification; q will be the statement of a modification of weight and $\bar{q}$ the absence of any such modification; likewise r and s state modifications in both the height of the drop and the impetus and $\bar{r}$ and $\bar{s}$ the invariance of these factors. Finally, x will be the proposition stating a modification of the result—*i.e.*, of the frequency of the oscillations—and $\bar{x}$ will state the absence of any change in frequency.

When EME varies the length of the string with equal weights (and successively for three different weights), she states the truth of the following combinations:

$$(p.q.x) \vee (p.\bar{q}.x) \vee (\bar{p}.q.\bar{x}) \vee (\bar{p}.\bar{q}.\bar{x}) . \tag{1}$$

This is to say that the modification of the length corresponds, with or without modification of weight, to a modification of the frequency and that the absence of the first transformation corresponds, with or without modification of weight, to the absence of the result x.

On the other hand, none of the four combinations $(p.q.\bar{x}) \vee (\bar{p}.q.x) \vee (p.\bar{q}.\bar{x}) \vee (\bar{p}.\bar{q}.x)$ is verified because when p is present $\bar{x}$ is never present and reciprocally when x is present $\bar{p}$ is never present.

But expression (1) can be broken down into two operations. First, when the subject says: "It's the length of the string which makes it go faster or slower," he expresses the reciprocal implication between p and x—*i.e.*, $p \gtrless x$. Secondly, between q and x there is no single linkage, since the four possible combinations $(q.x) \vee (q.\bar{x}) \vee (\bar{q}.x) \vee (\bar{q}.\bar{x})$ all occur. (This can be written in the form $(q * x)$, in which case we say there is a tautology or "complete affirmation.") This is what the subject expresses when he says: "The weight has no effect." As for the relationship between p and q, it can be written $p.(q \vee \bar{q})$ or, abbreviated, $p\ [q]$—i.e., there is affirmation of p with or without q; likewise, we have $\bar{p}.(q \vee \bar{q})$—i.e., negation of p with or without q. (The affirmation and negation brought together are the same as $p \ ^{\circ}\ q$).

Thus expression (1) can be written:

$$(p \supseteq x).(q \ ^{\circ}\ x) = p.(q \vee \bar{q}) \supseteq x \text{, or, abbreviated,}$$
$$p\ [q] \supseteq x. \tag{2}$$

We see in these formulae that the exclusion of weight as a cause of variation in the frequency of oscillations results simply from the subject's realization of $(q \ ^{\circ}\ x)$—i.e., from the fact that all of the combinations possible between q and x occur: to exclude weight means to exclude the choice of any particular linkage between q and x.

The reasoning process is the same for the exclusion of height of the drop and impetus. However, since the subject takes both the length and the weight into account when he analyzes the role of the height of the drop (r and $\bar{r}$), there are eight true combinations:

$$(p.q.r.x) \vee (p.q.\bar{r}.x) \vee (p.\bar{q}.r.x) \vee (p.\bar{q}.\bar{r}.x) \vee (\bar{p}.q.r.x) \vee$$
$$(\bar{p}.q.\bar{r}.\bar{x}) \vee (\bar{p}.\bar{q}.r.\bar{x}) \vee (\bar{p}.\bar{q}.\bar{r}.\bar{x}) = (p \subseteq x).(q \ ^{\circ}\ x).\ (r \ ^{\circ}\ x) \tag{3}$$
$$= p\ [q \vee r] \supseteq x,$$

where the expression $p\ [q \vee r]$ stands for $p.(q \vee r) \vee p.(\bar{q}.\bar{r})$.

Furthermore, when he studies the role of the impetus (s or $\bar{s}$) the subject also takes into account the length, the weight, and the height of the drop. In this case, he finds sixteen true combinations:

$$(p.q.r.s.x) \vee (p.q.r.\bar{s}.x) \vee (p.q.\bar{r}.c.x) \vee (p.q.\bar{r}.\bar{s}.x)$$
$$\vee (p.\bar{q}.r.s.x) \vee (p.\bar{q}.r.\bar{s}.x) \vee (p.\bar{q}.\bar{r}.s.x) \vee (p.\bar{q}.\bar{r}.\bar{s}.x)$$
$$\vee (\bar{p}.q.r.s.\bar{x}) \vee (\bar{p}.q.r.\bar{s}.\bar{x}) \vee (\bar{p}.q.\bar{r}.s.\bar{x}) \vee (\bar{p}.q.\bar{r}.\bar{s}.\bar{x}) \tag{4}$$
$$\vee (\bar{p}.\bar{q}.r.s.\bar{x}) \vee (\bar{p}.\bar{q}.r.\bar{s}.\bar{x}) \vee (\bar{p}.\bar{q}.\bar{r}.s.\bar{x}) \vee (\bar{p}.\bar{q}.\bar{r}.\bar{s}.\bar{x})$$
$$= (p \supseteq x).(q \ ^{\circ}\ x).(r \ ^{\circ}\ x).(s \ ^{\circ}\ x) = p\ [q \vee r \vee s] \supseteq x.$$

Thus we see that the exclusion of the three inoperant factors (which at first seemed so simple) as well as the reciprocal implications of the length and the result x actually presuppose a complicated combinatorial operation which the subject cannot master except by ordering seriately the factors which are to be varied one-by-one, each time holding the others constant. For example, in expression (4), the first two combinations $(p.q.r.s.x)$ v $(p.q.r.\bar{s}.x)$ are sufficient for the subject to deduce that frequency does not imply the operation of impetus $(s.x)$ and it is sufficient that he add the last two combinations to conclude $(s \circ x)$—i.e., to exclude completely the role of this factor. But it goes without saying that, in order to choose the conclusive combinations in this way, he must have at least an approximate idea of all of the rest. This fact explains why the isolation of variables by the method "all other things being equal" and the exclusion of inoperant factors appear at such a late date, being reserved for substage III-B.

The best proof that such a combinatorial system is needed is that the substage III-B subject is not satisfied with drawing exact conclusions from the demonstrative combinations that he conceives of in the course of the experiment with such apparent simplicity. He avoids as well all of the paralogisms that we have noted at substages II-B and III-A. But, in comparing the correct inferences found at substage III-B with the earlier false ones, we see that the choice is again dictated by the presence of one or two conclusive combinations. Once more they presuppose a degree of mastery of the system of all possible combinations.

For example, in the case of the hypothesized influence of weight (q), the subject may hesitate between operation (3), $p\ [q] \gtreqless x$ and the operations $(p \lor q) \gtreqless$ or $(p.q) \gtreqless x \ldots$ assumed at substage III-A and signifying that the change of frequency is due either to the length or the weight or to both at once $(p \lor q)$ or else that it is always due to both at once $(p.q)$. In such cases, we would have:

$$[(p \lor q) \gtreqless x] = (p.q.x) \lor (p.\bar{q}.x) \lor (\bar{p}.q.x) \lor (\bar{p}.\bar{q}.\bar{x}), \text{ and} \quad (5)$$
$$[(p.q) \gtreqless x] = (p.q.x) \lor (p.\bar{q}.\bar{x}) \lor (\bar{p}.q.\bar{x}) \lor (\bar{p}.\bar{q}.\bar{x}). \quad (6)$$

Here we see that expression (5) does not differ from expressions (1) and (2), themselves mutually equivalent, except for the presence of $(\bar{p}.q.x)$ and the absence of $(\bar{p}.q.\bar{x})$. And expression (6) does not differ except for the presence of $(p.\bar{q}.\bar{x})$ and the absence of $(p.\bar{q}.x)$. But the adolescent at substage III-B certainly knows how to exclude $(\bar{p}.q.x)$ and $(p.\bar{q}.\bar{x})$, since he verifies accurately the falsehood of $\bar{p}.x$ and $p.\bar{x}$ (= changes of frequency without modification of length or the reciprocal) even while admitting the truth of $q.x$ and of $\bar{q}.\bar{x}$ (= simultaneous variation of frequency and weight or invariance of both) when the length factor operates at the same time.

It should be clear that the fact that a mode of reasoning which was freely accepted up to substage III-B is then rejected again presupposes a certain choice among the possible combinations—*i.e.*, among those which are to be excluded as well as the true ones. To refer to a concrete case, the reader will recall that ROS (in III-A) varies weight and length simultaneously and concludes that the first is operant: from the combinations $(p.q.x) \vee (\bar{p}.\bar{q}.\bar{x})$ he extracts $q \supset x$ or $x \supset q$. But the distinctive feature of EME's experiment (in III-B) is that she is not satisfied with these two combinations and thus retains the truth of the four combinations contained in expression (1), notably, $(p.\bar{q}.x)$, which excludes $x \supset q$ (for $\bar{q}.x =$ variation of the frequency without modification of weight) and $(\bar{p}.q.\bar{x})$, which excludes $q \supset x$ (for $q.\bar{x} =$ variation of weight without result for the frequency). Of course a similar selection is found in connection with the height of the drops and impetus. Analyzing all the inferences accepted by a substage III-B subject and all those which he rejects, one must assume that he has knowledge of the combinations of expression (4). This knowledge itself presupposes a knowledge of the sixteen other rejected combinations—*i.e.*, a choice among thirty-two basic combinations.[3] Such choices imply, after all, a selection among a set of basic combinations. Once more we see that this selection implies the operation of the formal combinatorial system based on the "structured whole," whereas concrete operations amount simply to constructing correspondences from which these basic combinations are composed.

[3] In the case of flexibility (five factors and the result) there are even more—*i.e.*, sixty-four basic combinations. But to give proof of the influence of each factor it is sufficient to retain them separately by couples of combinations, whose model is furnished by operation (3) presented in Chap. 3, which can be taken in turn.

5

Falling Bodies
on an Inclined Plane
and the Disjunction Operations[1]

THE EXPERIMENTAL APPARATUS consists of a plane adjustable to various angles of incline. A ball can be rolled down the plane; it bounds when it hits a springboard at the base. The problem is to find the relationship between the height of the point from which the ball is released and the length of its bound. Naturally the subject will not be able to calculate the parabolic form of the curve the ball describes, but he will be able to discover that its length varies only as a function of the height of the release-point (learning to exclude the effects of the mass or weight of the ball). In part, the solution of the problem depends on the way in which the factors are presented.

§ Stage I. Global Intuition Without Operational Registering of the Experimental Data

Even before 7 years, the correspondence between the angle of incline and the length of the bound is perceived intuitively, but the height at which the ball is released is not separated from the angle of incline and weight is constantly assigned a role. However, this latter role is not always consistently formulated.

[1] With the collaboration of H. Aebli and L. Müller.

VER (5 years): *"That one goes to 2* [the second compartment from the lower extremity of the plane] *because it is too small. If it were big like that* [gesture], *it would go here"* [8].

STU (6 ; 5) discovers that a marble reaches the fourth compartment for a given slope, then the second *"because the gadget was lowered."*— "What are we going to do to make it go there?" [6].—*"Lower it more* [failure].—*No, you have to put it higher up. I want it to go here* [8]: *I have to put it way up* [approximate success]. *Yes; to go near you have to put it way down and to go far you have to put it higher up."* As for the mass, he believes that a small ball will not go as far.

PIT (6 ; 6): "Where will this ball go?"—*"Way down to the bottom: it's heavier."*—"Watch [we take a small ball which goes to the same place]. —*"It's because it's high up."*

MIC (6 ; 10): to make it go far, you have to *"raise up the trough."*— "And if you can't?"—*"You have to throw it hard* [it reaches the third compartment]. *It's because it isn't high up, it doesn't go fast."*

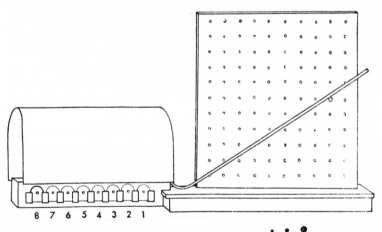

FIG. 4. The inclined plane can be raised or lowered by moving the peg on which it rests to different holes in the board. These also serve as an index for measuring height. Marbles of varying sizes are released at different heights on this plane, hit a springboard at the bottom, bound in parabolic curves, and come to rest in one of the compartments (numbered 1 to 8). These are the subject's index to the length of the bound.

VAL (7 ; 1): *"Because it rolls fast and it still has force."*—"This one?" —*"It will have enough force to get there* [4: failure]. *But it had force anyway; you have to go up a little more."*

WAG (6 ; 7): *"I'm going to put on that big one; I'm going to put it further down, otherwise it will go too far because it has more weight; when it is heavy, it goes too fast and it goes too far; it is heavy: that makes force."*—"And that one?"—*"The very tiny little one won't go so far because it won't have any force, it isn't heavy."*—Experiment: it does go far.—*"Because it was far! It should have fallen faster than the others because it's small. I am going to try a big one: maybe it will go all the way to the bottom* [experiment]. *Yes, because it is big it went far. I have to watch a middle-sized one* [it falls at the same point as the last two]. *Yes, it's because it's heavy: it falls faster* [new experiment: *idem*]. *It's because it's small, it's not heavy, so that's why! It didn't go very far"* [now he denies the fact]. Another ball: *"Because it is heavy, it falls faster because it has a lot of force. I am going to put on the big one: it wants to go far because it* [the slope] *is very steep."*

In each one of these cases we find some intuitive understanding obviously drawn from the child's experience (slides, sleds, small vehicles, etc.): the steeper the slope, the further and more quickly an object falls. But the height at which the ball is released is not separated from the angle of incline, and the weight (judged proportional to the size) is attributed a systematic role. But the specific role assigned to weight changes; in general a heavier marble is thought to roll further but if necessary this can also be the case for the smaller ones. In this respect inconsistent observations do not yet correct the subject, and when he is in a difficult spot he either contradicts himself or denies the facts (WAG uses the two processes alternately). This is the case because neither serial ordering nor correspondence operations, which can integrate separate statements coherently, are as yet organized. Restricting oneself to an untalkative subject like STU, one could gain the impression that an exact correspondence is formed between the angle of incline and the length of the bound. But when we consider a subject who says all that he thinks, or even a little more, we can see that this intuition does not go beyond the global level because it appears in a general form without differentiating operations.

§ Stage II. Attempts at Operational Correspondences and Usual Exclusion of Weight

Beginning at substage II-A, correct formulations of correspondences can be observed, but they are not yet systematic and of course they lack the formal procedures essential to the separation of variables. However, even at this point, depending on the way in which the balls are presented, the subject often manages to exclude the factor of weight insofar as it is incompatible with any serial correspondence:

GUI (7 ; 2). To make the ball roll further *"You have to put it higher up."*—"And to get [down] here?" [extremity].—*"Way up* [experiment]. *Ah! Yes. It's the last"* [compartment].—"And for this one?" [first compartment].—*"You have to lower it* [notch] *because it slides less."*— "And here?" [toward the middle].—*"Higher up, because it slides faster,"* etc.

LAU (8 ; 2) indicates same correspondences for the angles of incline. As for the sizes, LAU declares spontaneously: *"The balls will go in the holes* [at greater and greater distances] *in order of size"* [expected serial ordering].—"What do you mean by 'in order of size'?"—*"The smallest goes to the nearest and the biggest goes to the furthest hole; those in the middle go to the middle"* [he does the experiment].—"So?" —*"They go all over the place. Size has nothing to do with it; they were all about the same!"* At the end: *"According to where you put the slide* [inclined plane] *they go in the holes. You put it way up to make the marble go further: it depends on the height of the slide."*—"And the size of the marbles?"—*"The size doesn't do anything."*

SCHI (8 ; 8), likewise, *"You have to lower it, raise it,"* etc. At the end: "It depends on the size?"—*"Oh! No. They go in any old box, and then you raise it to make them go further,"* etc.

Here there is exact serial ordering of the slopes and lengths of the bounds, with approximate correspondence between the two ("the more . . . the more")—approximate because the subject does not think of the elevation and does not even consider the possibility of separating the distance covered on the downward path from the slope of the plane. But, since the deviations are not large, the correspondence works in a rough way.

But what is remarkable is the exclusion of weight, an exclusion which, though not commonplace, is easy to obtain, as is shown by the very clear cases of LAU and SCHI. But you will recall that in the pendulum problem weight was excluded only at substage III-B; the 12–14-year-old subjects (III-A) were not able to separate the relevant variables. On the other hand, the hypothesis that weight plays a role in the fall is very natural and is common even at the adult level among those who have forgotten their physics courses. Thus, the exclusion of this factor at substage II-A, when subjects are unable to make use of any formal propositional operation, poses a problem for us.

It seems to us that the explanation lies in the fact that in this particular case the factors of weight and slope dissociate themselves from each other without the subject's having to supply any operational activity. Actually, when LAU wants to verify his expectation that there is a correspondence between the size of the marbles and the length of their bounds, the idea does not occur to him to vary the slope at the same time because the slide is immobile unless it is intentionally moved. But in the case of the pendulum, where the problem is to estimate the frequency of oscillations and where the subject must adjust the weights to the strings, he will always be tempted to change the weight and the string at the same time as a way of obtaining clearer results ("more different," as LAU said). In addition, he has to use a systematic method to separate out the variables. The factors of slope and weight, however, are automatically dissociated. Consequently, in this problem it is easy for the child to see that balls of varied sizes may reach nearly the same place, in direct contradiction to his expectations.

The second reason for the ease with which weight is excluded has to do with the obvious lack of correspondence between weight and the length of the bound. In the pendulum problem, on the other hand, even after negative observations the subject could still ask himself whether or not the weight plays some role. The systematic experiments which result in a selection of crucial combinations among the total number of possibilities and which do not appear before substage III-B are needed for the exclusion of this factor.

In addition to serial ordering and more systematic correspond-

ences, substage II-B is distinguished by the beginning of dissociation of the height of the release-point from the slope. It is
interesting to note that it is at this same level in other experiments
as well (*cf.* the dumping apparatus in Chap. 13) that the height
factor is first differentiated and formulated in terms comparable
with the others. But this nascent differentiation does not go far
enough to allow the subject to exclude slope in favor of the height
alone; to do so would presuppose a systematic active verification
procedure designed to determine whether the two factors are
actually independent or not.

JEA (8 ; 10) orders the angles of incline systematically: "*Now 3 because I just tried 2,*" etc., then says, "*The more it goes down, the faster
it goes.*" Afterwards, he ascertains that with a gentler slope [4 instead
of 7]: "*If you put it further* [= higher], *it's as if you moved it a
notch.*" Thus, the attempt at systematic serial ordering forces him to
discover that the factor of height is distinct from the factor of angle of
incline.

MID (9 ; 9): "*It's combined; if you raise it* [he has successively raised
the slide to 3, 4, and 5], *it makes a bigger jump here. I'm going to
watch the bound.* [He takes a smaller ball and begins again: 3, 4, 5,
6, 7, and 8.] *It's the same for the big one and the little ones; it's the
height that does it* [determines the length of the bound]. *The lightness
has nothing to do with it.*" But he does not dissociate height and slope
further.

BLI (10 ; 2) varies the slope: "*If the slope is steeper, the ball goes further.*"–"And the sizes?"–"*All the balls will go in the same hole; that
can't change all of a sudden.*" He checks on a little one, then returns to
the slope, and, after an error in prediction, he says: "*I put it too far
backwards* [= too high], *so I have to put it further down* [experi-
ment]. *It's too low down* [new trial, still without varying the slope
again]; *you have to put it higher up because it has less force when it
slopes less*" [= he compensates for the small angle of incline by releasing the ball at a greater distance, thus at a greater height]. After several new trials: "*I know now. It always goes behind the same door
[= in the same compartment] for the same height.*" He tries to formulate a correspondence between the slope and the length of the
downward path so that he can reach the same hole each time [3]:
25 cm. for the incline 10, 30 cm. for 8, 35 cm. for 6, and 40 cm. for 4.

The experiment confirms his expectations and he concludes: *"The more you raise it, the more holes there are"* [= the longer the bound].

We see above that general correspondences of the type "the more . . . the more" no longer suffice at this stage; rather, the subjects become interested in organizing systematic correspondences, for example in following the ascending order 1, 2, 3, . . . for the angle of incline so that they are able to note the corresponding order of lengths for the bounds. (Moreover, neither of these series is numbered on the apparatus; the angles of incline are determined by a succession of holes in which the peg which fixes the slide is inserted and the compartments are distinguished by means of varied designs—house, pine tree, etc.)

Although the correspondence is accurately formulated at this stage, there are three reasons why it cannot be verified completely. In the first place, as was our intention, the holes determining the slope do not correspond exactly to the compartments. In the second place, there are possible chance fluctuations (due to jigglings, etc.). Thirdly, if he is not careful, the subject may vary the distances involuntarily; at the interior of the slide is a centimeter scale of such a sort that for a given slope one can still put the ball at either 25, 30, 35 cm., etc., thus varying the height of the release-point independently of raising or lowering the slide. Hence, another source of possible deviation from the initial correspondences.

Faced with these variations in the correspondence between the slope and the length of the bounds, the subject tries to determine which factors have influenced the result and in which ways. First, weight occurs to nearly all the subjects with very few exceptions (such as BLI). But this factor is discarded in the course of the experiment for the same reason as at substage II-A: absence of any observed correspondence (see MID).

The factor of height remains; at this substage the subjects generally discover its role as a result of the greater precision of their attempts at correspondence. For example, when JEA encounters irregularities in his correspondence, he notices the fact that "If you put it further away (thus higher), it's as if you moved it up a notch"—*i.e.*, that for a slope of 4 you can give the ball a higher starting point and obtain the same result as for a slope of 7 with a lower starting point. As for BLI, he goes so far as to determine a series of metrical equivalents according to the logical formula higher $\times$ less slope $=$ lower $\times$ greater slope, thus reaching the same compartment every time.

However, these subjects are far from the discovery that height, not slope and distance, is the only relevant factor, although height can be calculated from slope and distance combined (according to BLI's formula). The problem of the exclusion of slope in favor of height is quite different—at substages II-B, III-A, or III-B— from that of the exclusion of weight or amplitude in favor of length in the pendulum problem. For it is a question not of excluding one independent factor in favor of another, but rather of excluding a particular relationship in favor of another of which it is a part. Actually, at equal heights neither slope nor distance plays a role if it is varied; there is not, on the one hand, a factor *slope* and, on the other, a factor *distance*, or *height*; there is a logical multiplication, "slope $\times$ distance $=$ height," in which only the product (height) counts. The two multiplicands, in fact, never operate as separate factors. But this fact does not yet occur to the subject and it is understood only with difficulty at stage III. In other words, the subject at substage II-B thinks of slope and distance as if two independent factors were involved, one of which has a role that seemed obvious from the beginning, the other a role which he has just discovered. Moreover, he conceives of them as two factors that can compensate each other (*cf.* BLI). He has yet to see that height alone counts, and that in order to find a correspondence between the length of the bounds of the ball and the determinant causal factor, the height, it is sufficient to consider the latter without regard to slope or distance. It is true that the child sometimes seems to have understood ("It's the height that does it," MID), but we have here no more than inadequately differentiated statements.

§ *Stage III. Necessary Compensations Between Angle of Incline and Distance (III-A) Followed by Discovery of Height as the Sole Determining Factor (III-B)*

Substage III-A (12–14 years) hardly differs from II-B for this problem except in the method used. Subjects at substage II-B begin by finding systematic correspondences between slope and length of the bounds and discover the role of the distance only secondarily. But the preadolescents of substage III-A produce hypotheses more easily and from the start try to catalogue the factors. They do this in such a way that they are able to separate slope and distance as coexistent factors more quickly. But they do not discover (any more than the preceding subjects) the role of the height as the single sufficient factor because they fail to proceed according to the habitual method used at substage III-B— isolation of factors by one-at-a-time variation, "all other things being equal." The result is that, in reading the responses of substage III-A, one is especially struck by the widespread appearance of the idea of compensation between slope and distance, an idea which has already been found at substage II-B:

ROU (12 ; 1): *"The highest possible and it will get here"* [the furthest compartment]. But the slope continues to play a separate role: *"I thought that it would go with less force because it fell off steeply."* He then discovers the compensation: *"When it was higher* [angle of incline], *you had to put it one lower down* [distance], *and when it's lower* [angle of incline], *you have to put it one higher up"* [distance], and *"if you go up* [distance], *you have to take down the slide 5 or 10 degrees, and when you raise it you have to start lower down"* [distance]. He then takes a slope of 4 and [lowers] his starting point 5 cm. at a time to aim for progressively nearer compartments: *"When it stays fixed* [slope], *you have to lower* [the starting point] *in steps of 5 cm."* Conclusion: *"Each time the angle gets smaller by 5 degrees you have to go down 5 cm."*

STRO (12 ; 6): *"The more the slide is horizontal* [= less inclined], *the more you have to put the ball aside"* [= increase the distance]. Next he makes some complicated calculations: *"You can base it on the points* [slope] *and the intervals* [distance]; *you multiply each hole."* —"How?"—*"A little more, a little less"* [actually he does not get beyond the qualitative concept of compensation].

HER (13 ; 6) first tests the role of weight and concludes: *"That doesn't have too much to do with it; it's as if they were the same."*—"Sure?" —*"Quite sure."* Then, like the preceding subjects, he realizes that increasing the distance is equivalent to raising the slide.

As ROU, in particular, shows us when he analyzes the role of distance at equal angles of incline, as soon as these subjects proceed to a systematic study of the variables combined with the concept of compensation which is general at this level, they are led to the hypothesis that height is the single relevant factor. The hypothesis is actually proposed and verified at substage III-B:

SAL (13 ; 3) begins with the hypothesis that mass is the determining factor: *"The little one will certainly go faster."* But the facts do not confirm his expectations.—*"Does size have an effect?"*—*"No, I don't think so. The large one would naturally go further, but since the little one goes faster on the downgrade, they compensate each other."* He goes on to the variations in slope, then proposes *"to take the same slope with a higher starting point."* Next he varies the two simultaneously and discovers the compensation: *"Now I am going to vary the height [= slope] and the distance; they compensate each other!"* —*"And with extreme variations, would you get something?"*—*"Yes [trials]. That makes me think that it always has to take off from the same height—from the same horizontal point"* [= thus height independently of slope and distance!].—*"Are you sure or is it a hypothesis?"* —*"Whatever slope you take, a large or a small ball gets there [= to the same compartment] if it takes off from the same height."* The experiment that he devises as a control consists of taking the same height for slopes of 3 and 9: *"There you really have extremes!"*

HOW (16 ; 4) begins by discarding the weight hypothesis: *"I would have expected the difference in weight to have changed the distance"* [= the length of the bound]. Then he studies the role of slope, then distance: *"You have to make the ball start less high up,"* etc. Next he ascertains the possible compensation: *"If you raise [the slide] you have to start from lower down."* Finally he is asked to formulate the law: *"It depends on where you start the ball. The line is constant, but the angle moves."*—*"What line?"*—[He points out the guiding points which make it possible to determine common heights for different slopes.] *"The ball's starting point is constant."*—*"What do you mean?"* —*"The height."*

As usual, substage III-B subjects differ from intermediate substage III-A subjects in that they try to separate out the variables. In this task III-A subjects fail to dissociate them for two reasons aside from the usual ones. First, at equal slopes distance and height vary concurrently; thus they do not distinguish the two factors from each other and generally call "higher" or "lower" what they actually measure in distance covered on the inclined plane (*cf.* ʀᴏᴜ and sᴛʀᴏ). Thus, they believe that they have accounted for height when in fact they have not formulated a clear relationship. In the second place, in asserting that distance and slope compensate for each other, they actually limit themselves to a statement of covariance without looking for the invariant that results from it, since they partly confuse this invariant, height, with the distance itself. In contrast, substage III-B subjects try to separate out the variables by the usual method of varying each factor in turn while holding all of the other factors constant. In this case, where there is mutual compensation of slope and distance, they vary each relationship separately before varying them together (*cf.* sᴀʟ: "Take the same slope and start from higher up"; then, "Now I am going to vary height and distance"). Finally, this allows them to distinguish clearly among all three factors—slope, distance, and height—and not just between two of them as they have done up to this point. In addition, as the first two factors compensate each other, the subjects immediately look for the invariant that the compensatory mechanism presupposes; they are no longer satisfied with simple covariance.

But how do they come to decide that the constant is height and not either of the other two factors? Of course the discrimination is a result of their experimentation, but, as sᴀʟ shows, a preliminary deduction is involved. The starting point of this deduction is the compensation itself. The subject sees that if a given slope is conserved, distance and height increase or decrease simultaneously; if, on the other hand, the height is conserved, the slope increases while the height decreases or vice versa in such a way that the height, product of the compensation, is at the same time the invariant postulated to account for the occurrence of identical results even when the other two factors are modified. "Whatever the slope is," says sᴀʟ, you have to look for "the same height." Even more forcefully, ʜᴏᴡ states, "The line (height) is constant" even though "the angle moves." These seem to be the reasons for the discovery of the height factor; they also explain its late appearance.

In analyzing the reasoning of these adolescents, as usual we come first across a selection of the true combinations among the possible ones. Furthermore, since the subject does not make a trigonometric calculation but is restricted to observing the empirical covariations of the factors (both among themselves and with the experimental result), the combinations found will bear on the covariations as much as on the effect produced. (In fact, one may consider this as the innovation of the present experiment compared to those found in Chaps. 3 and 4.)

Let us call p_o the statement of the conservation of slope and $\bar{p}_o$ the statement of a variation in this factor; call q_o and $\bar{q}_o$ the same statements made in reference to distance; r_o and $\bar{r}_o$ the same in reference to height; finally, we can designate by x_o and $\bar{x}_o$ the statements affirming or denying invariance in the result obtained (length of the ball's bound).

In this case, the true combinations that the subject states (from the standpoint of invariance or variation of each factor in relation to the others) are the following:

$$(p_o.q_o.r_o) \lor (p_o.\bar{q}_o\bar{r}_o) \lor (\bar{p}_o.q_o.\bar{r}_o) \lor (\bar{p}_o.\bar{q}_o.r_o) \lor (\bar{p}_o.\bar{q}_o.\bar{r}_o). \quad (1)$$

Thus the excluded combinations are: $\bar{p}_o q_o r_o$ (when slope but not distance varies, the height must vary as well), $p_o \bar{q}_o r_o$ (reciprocally, if distance varies without slope, height also varies), and $p_o q_o \bar{r}_o$ (for if slope and distance do not change, height must also remain constant).

But from combination (1) a twofold consequence results which is correctly drawn by the subject when he is able to utilize the disjunction operation:

$$[\bar{r}_o \supset (\bar{p}_o \lor \bar{q}_o)] \lor [r_o \supset (p_o.q_o) \lor (\bar{p}_o.\bar{q}_o)]; \quad (2)$$

i.e., a modification of height (of the release-point) presupposes a modification of either slope or distance or both, whereas maintenance of the same height presupposes either variation of both slope and distance at the same time or conservation of both.

But it is clear that we also have:

$$[\bar{p}_o \supset (\bar{q}_o \lor \bar{r}_o)] \lor [p_o \supset (q_o.r_o) \lor (\bar{q}_o.\bar{r}_o)], \text{ and} \quad (3)$$

$$[\bar{q}_o \supset (\bar{p}_o \lor \bar{r}_o)] \lor [q_o \supset (p_o.r_o) \lor (\bar{p}_o.\bar{q}_o)]. \quad (3a)$$

But the subject assumes that height alone $(r_o.\bar{r}_o)$, not either of the two other possible factors, actually plays the causal role. The reason for this is that the three implications $r_o \supset (p_o.q_o) \lor (\bar{p}_o.\bar{q}_o)$; $p_o \supset (q_o.r_o) \lor (\bar{q}_o.\bar{r}_o)$ and $q_o \supset (p_o.r_o) \lor (\bar{p}_o.\bar{r}_o)$ contained in expressions (2) and (3a) are no longer isomorphic if the direction of the sign of the relevant changes is taken into account. Thus propositions $\bar{p}_0$, $\bar{q}_0$, and $\bar{r}_0$ can be broken down into two pairs of propositions that we shall call p, q, and r when they state respectively an increase in slope, in distance, and in height; $\bar{p}$, $\bar{q}$, and $\bar{r}$ when they state respectively decreases in the same factors. In this case we have:

$$p_0 \supset [(q.r) \vee (\bar{q}.\bar{r}) \vee (q_0.r_0)]$$
$$q_0 \supset [(p.r) \vee (\bar{p}.\bar{r}) \vee (p_0 r_0)] \qquad (4)$$
$$r_0 \supset [(p.\bar{q}) \vee (\bar{p}.q) \vee (p_0.q_0)];$$

i.e., the conservation of height r can be assured by the compensations $(p.\bar{q})$ or $(\bar{p}.q)$ as well as by the absence of change $(p_0.q_0)$, as is not true for either p_0 or q_0.

At this point, the subject hypothesizes (see SAL):

$$r \gtreqless x, \quad \text{or} \quad (\bar{r}_0 \gtreqless \bar{x}_0), \qquad (5)$$

which amounts to saying that either $\bar{r}_0.\bar{x}_0$ or $r_0.x_0$ must be true (height and result either vary together or are both conserved).

The experiment then gives the true combinations:

$$(p_0.q_0.x_0) \vee (p_0.\bar{q}_0.\bar{x}_0) \vee (\bar{p}_0.q_0.\bar{x}_0) \vee (\bar{p}_0.\bar{q}_0.x_0) \vee (\bar{p}_0.\bar{q}_0.\bar{x}_0). \qquad (6)$$

The following combinations are excluded: $\bar{p}_0.q_0.x_0$ (since change in slope without modification of distance transforms height and does not lead to the same result x_0); $p_0.\bar{q}_0.x_0$ (for reciprocally; $p_0.\bar{q}_0$ implies a change in height) and $p_0.q_0.\bar{x}_0$ (for conservation of slope and distance could not produce the change $\bar{x}_0$).

We see above how the true combinations (6) coincide with combinations (1); therefore the role of the height as the single necessary and sufficient factor is verified. It is worth noting that the above subjects are not satisfied with controlling the result of the variations in height $(\bar{p}_0.\bar{q}_0.\bar{x}_0)$ or $(\bar{p}_0.q_0.\bar{x}_0)$ or $(p_0.\bar{q}_0.\bar{x}_0)$, but also demonstrate the validity of $(\bar{p}_0.\bar{q}_0.x_0)$ as counterproof. Subject SAL even varies slopes from 3 to 9, concluding: "Here you really have extremes!"

6

The Role
of Invisible Magnetization
and the Sixteen Binary
Propositional Operations[1]

THE EXPERIMENTAL PROBLEMS set for the subjects in Chaps. 1 to 5 were designed to show a gradation in the sorts of difficulties overcome by the combinatorial method inherent in formal thinking and adolescent propositional logic. To conclude the first section, we should like to examine briefly another rather simple problem, one which has already been used in one of our previous studies; [2] it will serve to show how the stage III subjects utilize disjunctions and exclusions in integration with the entire set of binary operations. The problem is to determine why a metal bar attached to a nonmetallic rotating disk stops with the metal bar pointing to one pair of boxes instead of any other boxes placed around the disk; actually, the crucial pair contain several magnets concealed in wax. (Everything is placed on a board which is divided into sectors of different colors and equal surfaces.)

---

[1] With the collaboration of M. Denis-Prinzhorn, former research assistant, Laboratory of Psychology.

[2] J. Piaget and B. Inhelder, *La Genèse de l'idée de hasard chez l'enfant,* Chap. III. (Not transl.)

§ Stage I. Preoperational Disjunctions and Exclusions

We need not refer to the responses of the youngest subjects (sub-stage I-A), for they have been described in our previous study.

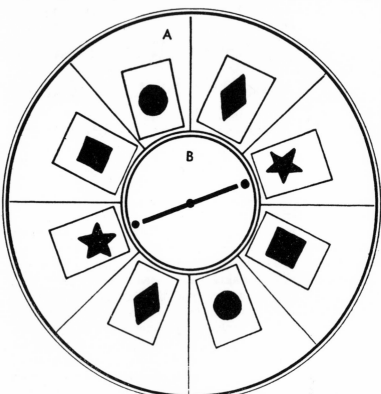

FIG. 5. One pair of boxes (the starred ones) contains concealed magnets, whereas the other pairs contain only wax. The large board (A) is divided into sectors of different colors and equal surfaces, with opposite sectors matching in color. A metal bar is attached to a non-metallic rotating disk (B); the disk always stops with the bar pointing to one pair of boxes. The boxes (which are matched pairs as to color and design) can be moved to different sectors, but they are always placed with one of a pair opposite the other. The boxes are unequal in weight, providing another variable.

But at substage I-B, a rough sketch of what at stage II we will call concrete disjunctions and exclusions (for example, the disk stops "here or there," "it's not that one," etc.) already appears in the form of intuitive representations:

voi (6 ; 5) thinks that the disk will stop on the blue *"because the blades* [on the disk, which act as brakes] *are blue,"* and, since it stops on the green [the magnet boxes have for the moment been placed on the green sectors of the board], he explains that *"green goes well with yellow"* [the boxes are yellow]. Next he predicts *"on the green, because it always stops on the green."* He opens the magnetized boxes: *"There is wax,"* but he does not find this fact helpful in explaining the phenomenon and adds, *"It doesn't come from that either"* [the designs decorating the boxes].—The magnets are put on the red sector. "Where will it stop?"—*"I don't know; here or there"* [red or green]. In the end he is limited to the explanation *"There is something in the boxes"* but without saying why some of them stop the disk while others fail to.

web (6 ; 9): *"Maybe there, because there is a star"* [decorating the magnet box]. Then: *"It will always stop on the red."*—"Why?"—*"I don't know. It's too heavy here"* [the blades serving as brakes].—"What could be done to see if it's really that?"—*"Take them off"* [this is done]. —"Where is it going to stop?"—*"On the blue."*—(Experiment: red.) "It was really that?" [the blades].—*"No."*—"Then how does it happen that it stops here?"—*"You push too hard* [force of the disk]. *I'm going to try gently* [it again stops on the red]. *Ah! I see. I'd say it's too heavy* [he opens the boxes and compares them]. *That one is the heaviest"* [the box with stars containing the magnets; it is not actually the heaviest]. —"And like that?" [the experimenter puts the magnets on the blue and the disk stops there].—*"Ah! I know, I'm happy I found out. Maybe the star is more useful, so it's heavier* [= strong], *because at night it lights the streets. Even between two houses you always see the star."*—"But is this one a real star?"—*"No, it's made of paper, but maybe it succeeded"*[!].

The elementary form of the operation which later becomes interpropositional disjunction is based on the observation that two classes are partially or entirely disjunctive. The subject need not possess concrete class operations before he realizes intuitively that the needle can stop on any color: "here or there," as voi says. His phrase expresses the beginnings of a development which leads both to the inclusion of partial classes in a total class and to the notion of *the possible*. The needle, he means to say, will stop on a color (B) which will be green (A) or red (A')—*i.e.*, an intuitive addition $A + A' = B$ with disjunction (either A or A'), although systematic operations are not yet present.

As for exclusion, it appears in its most elementary form as lack of correspondence. In this case as well, intuitive correspondences and noncorrespondences (based on perceptual configurations) may appear before operational correspondences (in which equivalences are conserved when the configuration changes).[3] But when voi rejects the explanation based on the content of the boxes (because not only the boxes in front of which the disk stops but all of them are filled with wax) or on the designs which distinguish the boxes ("it doesn't come from that either") he can do no more than note the lack of correspondence; he cannot organize his observations in a detailed way.

web's behavior is more advanced when he proposes to take off the disk brakes (blades) to see whether or not they have anything to do with the disk's stopping; this proposal is a preliminary type of verification. He discards other causes (force, etc.) in the same way because their removal does not eliminate the effect. We see here the beginnings of correspondence or the perception of non-correspondence with consequent reversals of behavior which forecast the transformation of this behavior into reversible operations. But the end of the interrogation shows that this nascent structuring is not carried very far yet. In the first place, when web tries to explain the stops as a result of weight, he does not compare all of the weights and is satisfied with two or three comparisons whose results are erroneous. Later, and more important, he goes so far as to attribute the stopping to the star design which he sees as "useful" (= efficacious) and "heavy" (= strong); he does this because the star has "succeeded," although it appears only symbolically as a paper representation.

§ *Substage II-A. The Beginnings of Concrete Disjunctions and Exclusions*

When concrete operations are organized by reversible coordination of behavior, the rough forms of disjunction and exclusion

[3] See *The Child's Conception of Number*, pp. 70ff.

which we have just noted begin to be systematically structured as a function of the nascent groupings of classes and relations:

KEL (7 ; 3) first says that the needle may stop *"here or there or there; you can't tell in advance."* Then he rapidly discovers that the needle always stops on the same color [violet]. Next the experimenter puts the box with the star design on the red sector; the needle stops there. *"It's because you changed that"* [blades serving as brakes for the disk]. The experimenter repeats the trial. *"No, it's the boxes! The stars were on the violet before, now on the red,"* etc. But he does not allow himself to say any more.

MAMB (8 ; 3). First: *"It depends on whether you turn it faster or slower."* He holds to this idea for a long time: *"Maybe you turned it too hard,"* etc. Finally, since the needle always stops on the star: *"It's because the [starred] boxes are heavier."*—"And these?" [the heavier boxes].—*"Maybe they are too heavy, so it doesn't work."*

BER (9 ; 9): *"One of them is light, the other heavy, and one a little less heavy."* He realizes that the starred boxes are the same weight as those marked with a circle, heavier than the squares, and less heavy than the diamonds. *"Yes, maybe it's the weight."* [4]

VOG (9 ; 10) has weighed all of the boxes: *"The square is lighter, then come the star and the circle."*—"So why does it stop in front of the star?"—*"Because the square is next to the star and it is lighter."*

KER (10 ; 0) also hypothesizes that it *"has to do with the weight in the boxes and in the disk,"* attempting to reconcile the ambiguities by using a notion of mean weight, defined in terms of the over-all distribution of the individual boxes.—"Do you want to see if you are right?"—He compares equal boxes: stars and circles. *"This one is pretty heavy; it's the same weight as the star."*—"But where does the needle stop?"—*"On the star."*—"Then it's the weight anyway?"—*"It must be the weight because they are in order* [= the weight is distributed in a certain manner which he describes in pointing out the boxes]. *That one* [diamond] *is heavier. The two round ones are the same weight; the two square ones too* [but lighter]. *So I'm sure it must be the weight."*—"Which are the heaviest?"—He indicates the diamonds.—"Then it stops there?"—*"No."*—"And the weight still has an effect?"—

[4] See *La Genèse de l'idée de hasard chez l'enfant*, p. 96 (Dan, Desp, and Tos), p. 97 (Ful, intermediate), and pp. 100–101 for similar cases.

"*Yes, of course*" [he is thinking of an equilibrium between all of the weights which results in the needle's stopping at the mean weights].

In our work on the child's notion of chance we have discovered a number of analogous cases in which subjects hypothesized that only the mean weights stop the disk—*e.g.*, "It's only the middle ones which stop it" (FUL 8 ; 4); "It's the weight of these boxes, because they are neither very heavy nor very light" (DUF 9 ; 4).

In these cases we see a curious mixture of concrete disjunctions and exclusions (or nonexclusions); it serves to show both the progress made over the preoperational level and the deficiencies of nonformal operations in comparison with the true exclusions of propositional logic.

We have already observed the child's behavior for disjunction in studying chance and lotteries.[5] At this level, when he draws elements from a collection *B*, he discovers that he may sometimes come across representatives of subclass *A* and at other times representatives of subclass *A'*, although at the preceding level he generally believed that he would come across one rather than the other. In the present problem we can see this elementary form of operational disjunction, based on the structuring of classes and relations, in part revealed by the way in which the notion of weight is employed. The subject believes that the choice of stopping point can be explained conclusively only in terms of weight; on the other hand, he discovers the diversity of weights in the experimental situation. Consequently, he assumes that the weight may act in one of three ways; the effect results from the heaviest or the lightest or from an intermediate value (not "too heavy," as MAMB says or "neither too heavy nor too light" as DUF says). We see that this type of disjunction is based on a simple approximated serial ordering. As KER expresses it: "It's in order" of three categories, ranging from the heaviest to the lightest.

This solution of the problem, based on the disjunction of relations, is a subtle one; still, it raises a delicate point with regard to exclusion. The fact that there is only correspondence between some of the weights and the disk's stopping points rather than between the weights as ordered serially and the degree of frequency or exactness of its stopping does not induce the subject

[5] *Ibid.*

to exclude the weight factor at this stage. However, at substage II-B and subsequently, weight is excluded for this reason. At substage II-A, on the other hand, absence of complete correspondence simply limits the effect of weight in the subjects' eyes to the effect of a hypothetical *optimum* weight which is assumed to be in the middle range. But why is this weak hypothesis maintained at substage II-A when it is rejected at II-B?

One very simple explanation can be offered. It is of particular interest from the standpoint of the psychology of exclusion; moreover, it relates to facts which we have studied extensively in another work.[6] At substage II-A conservation of weight is not yet organized and none of the concrete operations of serial ordering, equalization with transitiveness, correspondence, etc., which have already been acquired in a number of other areas, are as yet applied to it.[7] In contrast, at about 9–10 years (the beginning of substage II-B), all of the relevant operations are organized simultaneously and conservation is assured. Since at substage II-A weight is not always structured from the standpoint of concrete operations, it may still be conceived of as an active force giving rise to multiple and inconsistent effects. (We have observed exactly the same phenomena and the same inconsistencies in Chap. 2 for the floating bodies problem.) In other words, the child cannot make systematic exclusions because weight has not yet been given a place in the system of operations essential to the formulation of accurate correspondences.

In reading the responses of BER and especially of KER, we discover the surprising fact that they have recognized that the weights of the starred boxes (= those where the needle stops) and the circles (= where the needle does not stop) are equal, even though this discovery does not shake their belief in the effect of weight. However, if one remembers that they do not order the weights exactly and that the equalities are not transitive, the fact is less astonishing. A particular weight can be seen as possessing a potentiality for attracting the needle which cannot be replaced by any other equal weight.

[6] J. Piaget and B. Inhelder, *Le Développement des quantités chez l'enfant*, Chaps. II, VI, X and XI.

[7] We know, for example, that serial ordering of five distinct weights with equal volumes is accomplished only at the mental age of 10 as defined by the Binet-Simon tests.

voc's arguments are also interesting; after he has weighed all of the boxes, he maintains that the starred one stops the disk because of its medium weight but adds the provision that it is placed beside the square which is lighter. His combination of two weights has none of the distinctive characteristics of an operational composition; rather he describes an interplay of forces which cannot be translated into terms of conservation. The same holds for the notion of total action of all of the weights "in order" to which KER refers; moreover, KER is unable to make his conception more explicit.

But even if the weight explanation does not result in an accurate exclusion, the substage II-A subjects can still utilize concrete exclusions for various other factors which they first suppose to be causal—the disk brakes, force of pushes, etc. In such cases the incorrect hypotheses are more or less rapidly abandoned as soon as the lack of correspondence is seen.

§   Substage II-B. The Concrete Exclusion of Weight

At substage II-B concrete operations for handling weight have been structured (a delay of two to three years beyond the development of such operations for lengths and simple quantities); the result is that the subjects have quite a different attitude toward the present problem:

DUP (10 ; 9) begins with the hypothesis that weight is a causal factor: "*It depends on how the weight is placed. This box* [magnets] *is the heaviest* [he weighs all of them]. *Oh! It's the middle one! The heaviest is the diamond; that one* [square] *is empty and these two here* [the starred box containing the magnets and the box marked with a circle] *are the same* [he weighs them again]. *Yes, about the same.*" Then he spontaneously puts the circled boxes in the position of the starred boxes in order to verify his hypothesis, but again the disk stops in front of the star: "*You can't do anything about it, it's always the same thing! It's complicated.*" He finds no better explanation but does abandon weight.

SAN (10 ; 8) "Does the weight do anything?"—"*Oh, no. That one is heavier than the star* [than the box containing the magnets], *and it goes on the star!*"

PAU (11 ; 11): *"The round one and the star are the same weight, so it may fall on either one of the two* [he performs the experiment a second time]. *Yes, but there is something that does it because it always falls here"* [stars]. He weighs all of the other boxes and concludes by reasoning [in a form already hypothetico-deductive]: *"If it were the weight, it would fall on the heaviest and not the medium ones."*

The effects on these subjects of operations of serial ordering or serial correspondence as well as those of equalization with transitiveness, newly acquired for weight, are evident. From the fact that there is no term-by-term correspondence between the weight and the disk's stopping points, SAN and PAU conclude that the weight plays no role. Similarly, from the fact that two equal weights do not produce the same effect, DUP and PAU also conclude that this factor is ineffective. They both try one of two counterproofs. They either replace the magnet box with one of equal weight (though, of course, not containing a magnet), or they repeat the experiment to be certain that the disk stops only in front of one of the two. In sum, once operations are applied systematically to structuring a particular dimension such as that of weight, concrete operations are adequate to assure the possibility of excluding factors when there is neither correspondence between classes and relations nor transitiveness.

§ *Stage III. Propositional Disjunctions and Exclusions*

Although the exclusion of weight is already possible at substage II-B by the utilization of concrete operations, the formal operations of disjunction ($p \vee q = p.q \vee p.\bar{q} \vee \bar{p}.q$) and simple ($p.q \vee p.\bar{q}$) or reciprocal ($p.\bar{q} \vee \bar{p}.q$) exclusion present additional advantages. First, they allow some variety in the selection of disjunctions or exclusions; but more important, they locate these various possibilities in the total set of combinations. The combinatorial power of the structured whole then in itself determines implications and nonimplications or incompatibilities. To observe these advantages, we can turn to a case which will be referred to again in a later work by one of the authors from the standpoint of inductive strategy:

GOU (14 ; 11): *"Maybe it goes down and here it's heavier* [the weight might lower the plane, thus resulting in the needle's coming to rest at the lowest point] *or maybe there's a magnet"* [he puts a notebook under the board to level it and sees that the result is the same].— "What have you proved?"—*"There is a magnet* [he weighs the boxes]. *There are some that are heavier than others* [more or less heavy]. *I think it's more likely to be the content"* [in substance].—"What do you have to do to prove that it isn't the weight?"—He removes the diamond boxes which are the heaviest. *"Then I changed positions. If it stops at the same place again, the weight doesn't play any role. But I would rather remove the star boxes. We'll see whether it stops at the others which are heavier* [experiment]. *It's not the weight. It's not a rigorous proof, because it does not come to rest at the perpendicular* [to the diamond boxes]. *The weight could only have an effect if it made* [the plane] *tip. So I'll put two boxes, one on top of the other, and if it doesn't stop that means that the weight doesn't matter:* [negative experiment]. *You see."*—"And the color?"—*"No, you saw when the positions of the boxes were changed. The contents of the boxes have an effect, but it's especially when the boxes are close together; the boxes are only important when they are close* [he puts half of the boxes at a greater distance]. *It's either the distance or the content. To see whether it's the content I'm going to do this* [he moves the starred boxes away and brings the others closer]. *It falls exactly between the round ones which are near and the stars which are far off. Both things have an effect and it's the result of two forces* [experiment in which the star is moved away by successive steps]. *It's more likely to be distance* [new trial]. *It seems to be confirmed, but I'm not quite sure. Unless it's the cardinal points* [he takes off the stars]. *No, it's not that. The stars do have an effect. It must be the content. If it isn't a magnet, I don't see what it could be. You have to put iron on the other boxes. If the magnet is there* [disk], *it will come* [to] *these boxes. If it is in the boxes* [stars] *there is iron under the disk* [he removes the starred boxes]. *I'm sure that it's the boxes."*

We see here the great difference between substage II-B subjects, who are limited to serial correspondences or transitive equalities, and the stage III subject, who utilizes the formal combinatorial system and as a result does not experiment until he has made deductions from his preliminary hypotheses. Like the II-B subjects, GOU hypothesizes the relevance of weight, but he reasons from a set of possibilities as to ways in which it would be manifested if in fact weight had an effect (tilting the apparatus). This hypothetical reasoning not only gives him the idea of verifying whether the plane is horizontal but even the idea of placing two boxes together in order to increase the weight.

Moreover, cou uses propositional rather than concrete operations. Most important, they are based on the set of sixteen binary combinations in continuous transition from one to the next; their consistent integration is demonstrated with particular clarity. The following operations can be distinguished in his protocol:

(1) Disjunction $(p \vee q) = (p.\bar{q}) \vee (\bar{p}.q) \vee (p.q)$: "It's either the distance or the content (or both)";

(2) Its inverse, conjunctive negation $(\bar{p}.\bar{q})$: changing the position of the boxes verifies the hypothesis that neither weight nor color is the determining factor;

(3) Conjunction $(p.q)$: both content and distance are effective;

(4) Its inverse, incompatibility $(p.q) = (p.\bar{q}) \vee (\bar{p}.q) \vee (\bar{p}.\bar{q})$: the effect of the magnet is incompatible with moving the boxes from the center for the needle may stop without the boxes being moved and vice versa, or neither occurs.

(5) Implication $(p \supset q) = (p.q) \vee (\bar{p}.q) \vee (\bar{p}.\bar{q})$: if a magnet is attached to the disk, it will stop in front of the boxes containing iron;

(6) Its inverse $(p.\bar{q})$: when it does not stop, nonimplication is shown;

(7) Converse implication $(q \supset p) = (p.q) \vee (p.\bar{q}) \vee (\bar{p}.\bar{q})$: if there is a magnet in the box, it will stop the disk;

(8) Its inverse $(\bar{p}.q)$ operates in (1), (4), (10), etc.;

(9) Equivalence $(p = q) = (p.q) \vee (\bar{p}.\bar{q})$: to assert that weight has an effect is equivalent to asserting that the needle stops because of inclination of the plane;

(10) Its inverse, reciprocal exclusion $(p \vee\vee q) = (p.\bar{q}) \vee (\bar{p}.q)$: the fact that the plane is horizontal excludes the weight factor, for either the plane is horizontal and weight has no effect or weight has an effect and the plane is not horizontal;

(11) Independence of p in relation to q—i.e., $p [q] = (p.q) \vee (p.\bar{q})$: the stopping point may coincide either with a color or with its absence; thus color is excluded as a variable;

(12) Its inverse (which is also its reciprocal) $\bar{p} [q] = (\bar{p}.q) \vee (\bar{p}.\bar{q})$: failure to stop may also coincide with the color or its absence:

(13)–(14) Independence of q and $\bar{q}$ in relation to p—*i.e.*, $q\,[p]$ and $\bar{q}\,[\bar{p}]$: these operations are found in (15);

(15) Complete affirmation or tautology $(p*q) = (p.q)\,\mathrm{v}\,(p.\bar{q})\,\mathrm{v}\,(\bar{p}.q)\,\mathrm{v}\,(\bar{p}.\bar{q})$: all possible combinations, thus absence of particular links, for example between the box which contains the magnet and the colored sector on which it has been placed;

(16) Its inverse, complete negation or contradiction $(_0)$: to deny that weight has an effect and to reassert it would be a contradiction.

The above examples all come from the protocol of a single substage III-B subject; thus we are not exaggerating when we claim that it is possible for subjects at this level to work in turn with each of the sixteen binary combinations of propositional logic. Of course, at substage III-A the keyboard is not yet complete (for examples of this intermediate substage, see our previous study on the problem of magnets).[8] But when formal equilibrium has been attained the combinatorial system which characterizes the "structured whole" pays off in full, and the subject is no longer satisfied with reasoning based on simple concrete correspondences. For example, when GOU has observed the noncorrespondence of the stopping points with weights, he does not feel that his proof is adequate ("rigorous") because he realizes that if weight acted to produce an inclination of the plane, it could be combined with other factors.

In sum, even in a problem as simple as the present one (chosen to conclude the first section because of its very simplicity), the transition from concrete to formal operations is distinguished by the appearance of a complete combinatorial system whose various types of disjunction and exclusion are continuously linked to implications. Lacking in even the most advanced children at substages II-A and II-B, this is what gives the hypothetico-deductive new look to the responses of stage III subjects; it manifests itself even in the small details of experimentation.

[8] *La Genèse de l'idée de hasard*, pp. 101–106.

Part II

THE OPERATIONAL SCHEMATA

OF FORMAL LOGIC

THE TRANSFORMATIONS of thought that characterize the first stages of adolescence during stage III (notably at substage III-B) can in no way be reduced to the formation of propositional operations of the sort we found in connection with the sixteen binary operations (after having analyzed more complicated examples involving ternary operations, etc.) On the contrary, the analyzable facts of the growth of experimental reasoning are interesting because they show us that a number of new operations and concepts emerge in close linkage with the establishment of propositional logic; they require intellectual capacities greater than those of the concrete level and derive from the operational transformations entailed by the total structures ("groups" and "lattices") inherent in propositional logic rather than the propositional operations themselves.

Thus far we have seen that propositional logic is always bound up with a combinatorial system based on the "structured whole" as opposed to the simple class inclusions that make up the "groupings" of classes and relations of concrete logic. But this "structured whole" and the combinatorial system it presupposes form more complex structures which, in contrast with these elementary groupings, fuse the two great modes of reversibility into a single whole—*i.e.*, *inversion* (or negation) characteristic of "groups" and *reciprocity* (or symmetry) characteristic of "lattices." Thus the operations or new notions which we have just mentioned and which we are going to study in this second section have this common characteristic of deriving from specific properties of these total structures as such—*i.e.*, their general transformations—and no

105

longer only from the particular operations to which they give rise.

Thus we will designate by the term "operational schemata" those operations and new notions which are relative to total transformations of a system as opposed to the particular operations analyzed in the first section. In the first instance they will be the combinatorial operations themselves, no longer conceived of in their purely propositional aspect but in their general form. Next, they will include the notions relative to inversion and reciprocity which appear in all the problems that relate to the physical notion of equilibrium or of action and reaction. In addition, they will consist of certain notions of conservation, whose discovery requires the use of formal thought. They also include the notions of proportions, whose mathematical form derives from a more general qualitative logical form. And last we have the notions of correlation, in certain respects close to the notions of proportion.

In sum, we are dealing with a set of schemata whose dual nature stems from the fact that, whereas their structuring presupposes formal reasoning, they also derive from the most general characteristics of the structures from which this same formal thought arises.

7

Combinations
of Colored and Colorless
Chemical Bodies[1]

We have constantly seen that the formation of propositional logic, which itself marks the appearance of formal thought, depends on the establishment of a combinatorial system. The *structured whole* depends on this combinatorial system which is manifested in the subjects' potential ability to link a set of base associations or correspondences with each other in all possible ways so as to draw from them the relationships of implication, disjunction, exclusion, etc. Faced with an induction problem in which subjects at concrete stage II would be limited to classifications, serial ordering, equalizations, and correspondences, the substage III-B adolescents combine all of the known factors among themselves in terms of all of the possible links. But the problems given the subjects up to this point have involved factors which can be disassociated and combined at will or simply made to correspond without going beyond the level of observation or of "raw" experiment. One may wonder what would happen if we posed a problem that involved combinations directly—*i.e.*, one that involved elements or factors whose combination is indispensable if variable results are to be obtained. Will subjects at

[1] With the collaboration of M. Noelting, research assistant, Laboratory of Psychology, and doctor in chemistry.

substage II-B or even II-A discover a combinatorial system to meet the requirements of the experiment, one which would demonstrate the independence of this combinatorial system in relation to propositional logic? Must one await the formal stage for the establishment of this experimental combinatorial system, and will the stage II children accomplish nothing more than scattered empirical combinations without a total system such as we have seen elsewhere (in studying the formation of the mathematical operations of combinations, permutations, and arrangements)?[2]

The best technique with regard to this matter is to ask subjects to combine chemical substances among themselves. In experiment

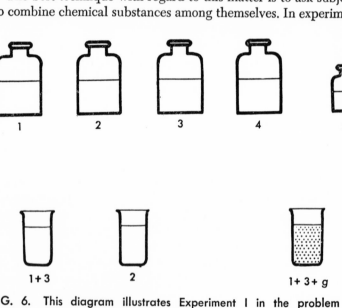

FIG. 6. This diagram illustrates Experiment I in the problem of colored and colorless chemicals. Four similar flasks contain colorless, odorless liquids: (1) diluted sulphuric acid; (2) water; (3) oxygenated water; (4) thiosulphate. The smaller flask, labeled g, contains potassium iodide. Two glasses are presented to the subject; one contains 1 + 3, the other contains 2. While the subject watches, the experimenter adds several drops of g to each of these glasses. The liquid in the glass containing 1 + 3 turns yellow. The subject is then asked to reproduce the color, using all or any of the five flasks as he wishes.

[2] *La Genèse de l'idée de hasard chez l'enfant,* Chaps. VII to IX.

I, the child is given four similar flasks containing colorless, odorless liquids which are perceptually identical. We number them: (1) diluted sulphuric acid; (2) water; (3) oxygenated water; (4) thiosulphate; we add a bottle (with a dropper) which we will call g; it contains potassium iodide. It is known that oxygenated water oxidizes potassium iodide in an acid medium. Thus mixture (1 + 3 + g) will yield a yellow color. The water (2) is neutral, so that adding it will not change the color, whereas the thiosulphate (4) will bleach the mixture (1 + 3 + g). The experimenter presents to the subject two glasses, one containing 1 + 3, the other containing 2. In front of the subject, he pours several drops of g in each of the two glasses and notes the different reactions. Then the subject is asked simply to reproduce the color yellow, using flasks 1, 2, 3, 4, and g as he wishes.

A second experiment (II) made use of combinations which were not between substances alone but between some substances and an indicator. Take Ac = a burette containing sulphuric acid $N/4$; B = a burette containing caustic soda $N/4$; E = three glasses of pure water, and Ind = a little phenolphthalein in three other glasses of water. The combinations in this case are:

$$(Ind \times B) = \text{pink}$$
$$(Ind \times Ac) = \text{colorless}$$
$$(E \times B) = \text{colorless}$$
$$(E \times Ac) = \text{colorless}$$
$$(Ind \times B \times Ac) = \text{colorless}$$
$$(Ind \times B \times E) = \text{pink}$$
$$(E \times Ind) = \text{colorless}$$
$$(B \times Ac) = \text{colorless}$$

In practice the ternary combinations are rare and serve only as a counterproof for the older children; they are not needed to produce color. As for $B \times Ac$, this combination is excluded in practice because two burettes are involved.

The result obtained by means of these two experiments demonstrates that a systematic combinatorial system appears only at substage III-A. At substage II-A the subject is limited to multiplying all of the factors 1 to 4 by g. At substage II-B a preliminary attempt at combination by trial-and-error is observed, but it is unsystematic.

§ Stage I. Empirical Associations and Precausal Explanations

At the preoperational level subjects are limited to randomly associating two elements at a time and noting the result in explaining it by simple phenomenalism or by other forms of prelogical causality:

NOD (5 ; 5): [*Ind* × *B*] *"Syrup!"*—[*Ind* × *Ac*] *"Water."*—"Can you make some more syrup?"—*"Yes. You have to do this* [he shakes the water, then reproduces *Ind* × *B*]. *It's syrup again."*—"Can you [change it back] to water?"—*"Yes* [*Ind* × *Ac*]. *It's water again."*—"Why?"—*"The syrup has gone away"* [he points to a bottle of methylorange one meter away].

MAM (5 ; 9): [*Ind* × *B*] *"It's like wine.* [*Ind* × *Ac*] *It's like water."*—"Is there any color?"—*"It went down to the bottom, it went away like that* [gesture, then *Ind* × *B*]. *Some red* [*Ind* × *Ac*]. *The red runs away in the glass. The color disappeared at the bottom. You don't see it any more. It melted."*

EG (6 ; 6): [*Ind* × *B*] *"It turned pink. Maybe there is paint in the glass.* [*Ind* × *Ac*] *Maybe the piece melted. Maybe the paint flattened out completely in the glass.* [*Ind* × *B*] *Rose. Maybe it's because the water changes. Maybe it changes at the surface of the water* [*Ac* × *Ind* × *E*]. *Maybe it's because when you have taken some white water in the tube, there is a bar that stops it and it's the pink water that runs out."*

AR (6 ; 9): [*Ind* × *B*] *"This time you put some red water inside.* [*Ac* × *Ind*] *It can't get red because the red has gone away in the water over there* [the first]. [*B* × *Ind*] *It's formed, it's getting colored. It can come back better in that water over there than in this.* [*E*] *It can't ever come back there* [*E*] *because it's the same color but it's not the same water."*

Preoperational thinking of this type contains neither proof nor even hypothesis. The apparent hypotheses of EG—"maybe it's because . . ."—are nothing more than fictions, for they make no reference to verification and she simply replaces or fills in the real world with imagery. Since they are not placed in a precise context of actions, these representations remain precausal; the color is a sort of active element that emanates from the water (it's the water that "changes") but may "go away," "go down to the bottom," "flatten out" to the point where it becomes invisible, or fly away to a beaker more than one meter away. The color can also "come back" but only to certain beakers of "water" and not to others. Appropriately, the subject may even shake the uncooperative beakers (subject NOD, for example).

§ Substage II-A. Multiplication of Factors by "g"

At the time concrete operations appear it is interesting to note the extent to which subjects spontaneously and systematically associate the element g with all of the others (in the case of experiment I) but without any other combination. If the subject is directly encouraged to combine several factors simultaneously, a few tentative empirical procedures are elicited but they are not followed up:

REN (7 ; 1) tries 4 × g, then 2 × g, 1 × g, and 3 × g: "I think I did everything. . . . I tried them all."—"What else could you have done?" —"I don't know." We give him the glasses again: he repeats 1 × g, etc.—"You took each bottle separately. What else could you have done?"—"Take two bottles at the same time" [he tries 1 × 4 × g, then 2 × 3 × g, thus failing to cross over between the two sets (of bottles), for example 1 × 2, 1 × 3, 2 × 4, and 3 × 4].—When we suggest that he add others, he puts 1 × g in the glass already containing 2 × 3 which results in the appearance of the color: "Try to make the color again."—"Do I put in two or three? [he tries with 2 × 4 × g, then adds 3, then tries it with 1 × 4 × 2 × g]. No, I don't remember any more," etc.

GAY (7 ; 6) also limits himself to 4 × g, 1 × g, 3 × g, and 2 × g, and discovers nothing else. "Could you try with two bottles together?"— [Silence.]—"Try."—[4 × 1 × g] "It doesn't work."—"Try something else."—[3 × 1 × g] "There it is!"—"And that one [2], do you think that it will be as yellow?"—[No trial.]—"What do you think makes the color, the three together or only two?" "Here" [3].—"And that one?" [1].—"There isn't any color."—"And that one?" [g].—"Yes, it's there inside."—"Then what good are 1 and 3?"—"There isn't any color."

IM (7 ; 6) also begins with 4 × g, 2 × g, 3 × g, and 1 × g, but since nothing happens he adds more drops to 3 × g, then to the entire series.

After this [which is new] he mixes the four together, but in the sequence $3 \times 2 \times 1 \times 4$, adding the drops each time: *"It didn't come. It's gone away again"* [the color appeared after $3 \times 2 \times 1 \times g$, but he did not stop at this point and the color disappeared with 4].— "What made it go away?"—*"Because I put in too much water"* [= liquid from the four bottles.)—"I'll take away that bottle [4]. Begin again." Again he makes the whole mixture, without understanding the suggestion about exclusion. *"It didn't come back because I put in too much*

cur (8 ; 11) also proceeds one by one with g: *"Nothing happens. You can't do it unless you put everything in the same glass."* He mixes the four without success, then hypothesizes not that he has put in too much but that he should have chosen another order: *"Nothing happens. I should have started with that one"* [2]. He does this, but since he does not control the permutation operations any better than the combination operations, he follows the sequence 2, 3, 4, 1, g, then he adopts any sequence whatsoever: *"The color doesn't come because I did it in reverse."* Finally [always with the intention of blending the four] he follows the sequence $1 \times 2 \times 3 \times g$: *"It's becoming yellow!"* But immediately he adds 4 and has to begin all over again. "Put in as few as possible."—*"The fewest possible, that's two."*

The reactions at this stage are of interest because, although these subjects are in possession of logical multiplication operations of one-by-one correspondence, the idea of constructing combinations two-by-two or three-by-three, etc., does not occur to them.

From the standpoint of combination operations, the only spontaneous reactions of the subject are either to associate each one of the bottles 1 to 4 in turn to the dropper g or to take all four at the same time. In both cases combinations are involved, but only the elementary and limited combinations that operate in multiplicative "groupings" of classes and relations (*i.e.*, associations or correspondences between one term and all the others).

Even when he sees he has failed, the subject does not use two-by-two combinations without prompting by the experimenter. On the other hand, his two hypotheses are either purely quantitative ("too much water" or not enough, the result being a new distribution of drops) or have to do with serial ordering (cur). But this appeal to order is also prompted by a grouping structure, since the serial ordering which is acquired from 7 years on rests on sequences. But here again the subject's reaction is to introduce a single change in order or to invert this sequence; he fails to try all of the possible sequences that combinatorial permutation operations allow. In sum, no true combinatorial operation has appeared as yet, but only correspondences and serial ordering— *i.e.*, first-degree combinations based on fixed class inclusions.

Another interesting point is that, with the color already formed, the subject is well aware that liquid 4 is "a kind of water which takes away the color." But when he is in the process of bringing together the four elements $1 \times 2 \times 3 \times 4 \times g$ and the color, which has appeared after he has combined the first three elements with g, disappears under the influence of 4, he no longer has the idea of a possible exclusion between 4 and the color; he simply declares that the color has disappeared for various reasons.

Because it helps to point up the opposition between noncombinatorial and combinatorial structures, we should note that at this level the child does not think of attributing color to the combination of several elements as such. Rather, he thinks in terms of such and such an element taken by itself, whether or not it combines with others. For example, GAY thinks that the color is in 3; then he withdraws it from 3 in order to assign it to g, as if it could be linked to only one liquid at a time.

The experiments in which indicators are used differ from those involving solutions in that in the former the two-by-two combinations are sufficient for a complete classification of the color-producing cases (base $B \times Ind$) and of the three nonproductive cases. Can we then say that the concrete operation of logical multiplication can alone furnish the solution of the problem? Interestingly enough, it is adequate for the extraction of the law (favorable and unfavorable cases) but not sufficient for its explanation. This is so for exactly what we have just seen in reference to the color attributed to a single liquid.

At substage II-A the elements of this second apparatus give rise to the following associations. In certain cases, the glasses (E and Ind conceived of as identical) are associated successively with the base B and the acid Ac. But from 7 years on complete tables of four cells, $(B \times Ind) + (B \times E) + (Ac \times Ind) + (Ac \times E)$, are obtained just as often, thus making possible discovery of the law:

SEHNE (7 ; 5) associates [*Ind* × *B*] pink, [*Ind* × *Ac*] colorless, [*E* × *B*] colorless. *"There isn't any more pink";* again [*E* × *B*]: *"The pink doesn't come any more."* [*Ind* × *B*] *"Pink again."*—"Can you take away the red?"—*"You put in a little white"* [he associates *Ind* × *Ac* and *E* × *Ac*].—"They're all the same, these four glasses?" [*E* and *Ind*]— "No. . . . Yes [he tries *Ind* × *B* and *E* × *B*]. No."—"Why?"—*"It comes only in two glasses. These two are pink and these two white"* [correct].—"And the burettes?"—*"No, here pink* [*B*] *and here white"* [*Ac*].—"Can you take the pink away from this glass?" [*Ind* × *B*].— *"Yes"* [he pours in some acid].

However, these multiplicative operations differ from complete combinatorial reactions in two ways. First, it does not occur to the subject to combine the two glasses between themselves (*E* × *Ind*) even after having established their differences, nor to combine the two burettes: thus he is restricted to the four base combinations (glass *E* × burette *B*) + (glass *Ind* × burette *B*) + (glass *E* × burette *Ac*) + (glass *Ind* × burette *Ac*). Secondly, even in discovering the law (*Ind* × *B* = pink), the subject does not conclude that the color is due to the combination; rather, he thinks that the base and the indicator contain it, thus making use of the notion of the potential in the sense of a "disposition to produce pink" and not yet in the sense of a possible combination, with a resultant which is distinct from the effects linked to each of the elements of the combination.

§ Substage II-B. Multiplicative Operations with the Empirical Introduction of n-by-n Combinations

The substage II-B reactions are analogous to the preceding ones but with a visible progress, namely, the appearance of *n*-by-*n* combinations. However, the subject does not as yet discover any system; only tentative empirical efforts are involved:

KIS (9 ; 6) begins with [3 × g] + [1 × g] + [2 × g] + [4 × g], after which he spontaneously mixes the contents of the four glasses in another glass; but there are no further results. *"O.K., we start over again."* This time he mixes 4 × g first, then 1 × g: *"No result."* Then he adds 2 × g, looks and finally puts in 3 × g. *"Another try* [1 × g, then 2 × g, then 3 × g]. *Ah!* [yellow appeared, but he adds 4 × g].

Oh! So that! So that's [4] *what takes away the color. 3 gives the best color.*"—"Can you make the color with fewer bottles?"—"*No.*"—"Try" [he undertakes several 2 by 2 combinations, but at random].

ALB (10 ; 4) begins with 1 × 2 × 3 × 4 × g, then changes the order: 3 × 1 × 4 × 2 × g. *"It's different, because the first time I went in order and this time I didn't.* [He puts 2 × 4 × 1 × 3 × g.] *Gives nothing*" [he tries several more permutations at random, then abandons the effort].—"Do you have to take all of them?"—"*No, you can take 2 or 3 if you want* [he tries unsystematically and succeeds by chance]. *It changes!*"

TUR (11 ; 6) begins with 1 × g, etc. *"That doesn't work. You have to mix all four* [he does this]. *That doesn't work either* [he changes the order several times without success, then tries two-by-two combinations: 1 × 4 × g, 2 × 3 × g, 3 × 4 × g, then 2 × 1 × g]. *I wonder if there isn't water in all of them!*" Then he spontaneously moves on to three-by-three combinations [× g], but without order: 3 × 4 × 1 × g, then 2 × 3 × 4 × g, then 1 × 4 × 2 × g, then 3 × 1 × 2 × g. *"That's it."*—"What do you have to do for the color?"—"*Put in 2.*"— "All three are necessary?"—*One at a time* [always with g] *it doesn't work. It seems to me that with two it doesn't work; a liquid is missing.*" —"Are you sure that you have tried everything with two?"—"*Not sure* [he tries in addition 2 × 1 × g, already attempted, then 3 × 1 × g]. *It works! It's 1 and 3!*"—"Tell me what effect the bottles have."— "*1 is a colorant, 2 prevents the color; no it doesn't prevent it because it worked. 3 takes away the effect of 2, and 4 doesn't do anything.*"

We see that, as at substage II-A, these subjects begin by multiplying each element by g or by taking them all at once, but finally they spontaneously use two-by-two or three-by-three combinations (each time with g). This is the true innovation of this substage, since at substage II-A this type of combination had to be called forth by the experimenter. But the fact that these combinations are not systematic defines the upper limit of this substage: TUR, who is the most advanced of the cases cited, does not even attain the six possible two-by-two (× g) combinations.

As for the cause of the color, it is still sought in particular elements rather than in their combination; TUR locates the color in 1 only and misinterprets the roles of 2, 3, and 4. Others discover the negative effect of 4 but by direct (and fortuitous) formulation and without having a specific method of proof.

In connection with the indicators (experiment II), the only notable innovation is the appearance of combinations between two glasses ($E \times Ind$); this shows that there is no longer only a double-entry multiplicative table but a search for all of the possible combinations. But the explanation remains the same:

MER (9 ; 3) tries $E \times B$; $Ac \times Ind$; $Ind \times B$; $E \times Ac$: *"There [B], it makes it get red and in that glass [Ac] it stays white."*—"And in the glasses, is it the same?"—*"The water isn't the same."*—"Which one gives the color?"—*"In both of them, in the glass [Ind] and in there"* [B].—"Are they the same?" [B and *Ind*].—*"Yes."*—"Can you show me?" —[He combines $Ind \times B$ (red), then $Ind \times E$ (colorless)] *"Oh! No, it's not the same and that [B] isn't the same as that" [Ac].*—"In the four, is it different?"—*"Yes.*—"Tell me what there is in that one" [B]. —*"It makes it get red."*—"And there?" [Ac].—*"That bleaches the water."*—"And that?" [E].—*"There isn't any pellet"* [with red dye].— "And there?" [Ind].—*"There is a pellet."*

Notice here the new combination $E \times Ind$, devised to see whether B and Ind are similar (a proof which, moreover, is not complete). But the explanation remains the same as at substage II-A: the color is thought to be virtually contained in B and in Ind (potential) or that there is a "color pellet" hidden (invisible content).

§ Substage III-A. Formation of Systematic n-*by*-n Combinations

The two innovations which appear at the formal level are the systematic method in the use of *n*-by-*n* combinations, and an understanding of the fact that the color is due to the combination as such:

SAR (12 ; 3): "Make me some more yellow."—*"Do you take the liquid from the yellow glass with all four?"*—"I won't tell you."—[He tries first with $4 \times 2 \times g$, then $2 \times g \times 4 \times g$] *"Not yet.* [He tries to smell the odor of the liquids, then tries $4 \times 1 \times g$] *No yellow yet. Quite a big mystery!* [He tries the four, then each one independently with g; then he spontaneously proceeds to various two-by-two combinations but has the feeling that he forgot some of them.] *I'd better write it down to remind myself:* 1×4 *is done;* 4×3 *is done; and* 2×3.

Several more that I haven't done [he finds all six, then adds the drops and finds the yellow for 1 × 3 × g]. *Ah! it's turning yellow. You need 1, 3, and the drops."*—"Where is the yellow?"— . . . —"In there?" [g]—*"No, they go together."*—"And 2?"—*"I don't think it has any effect, it's water."*—"And 4?"—*"It doesn't do anything either, it's water too. But I want to try again; you can't ever be too sure* [he tries 2 × 4 × g]. *Give me a glass of water* [he takes it from the faucet and mixes 3 × 1 × water × g—*i.e.,* the combination which gave him the color, plus water from the faucet, knowing that 1 × 2 × 3 × 4 × g produce nothing]. *No, it isn't water. Maybe it's a substance that keeps it from coloring* [he puts together 1 × 3 × 2 × g, then 1 × 3 × 4 × g] *Ah! There it is! That one* [4] *keeps it from coloring."*—"And that?" [2].—*"It's water."*

CHA (13 ; 0): *"You have to try with all the bottles. I'll begin with the one at the end* [from 1 to 4 with g]. *It doesn't work any more. Maybe you have to mix them* [he tries 1 × 2 × g, then 1 × 3 × g]. *It turned yellow. But are there other solutions? I'll try* [1 × 4 × g; 2 × 3 × g; 2 × 4 × g; 3 × 4 × g; with the two preceding combinations this gives the six two-by-two combinations systematically]. *It doesn't work. It only works with"* [1 × 3 × g].—"Yes, and what about 2 and 4?"— *"2 and 4 don't make any color together. They are negative. Perhaps you could add 4 in 1 × 3 × g to see if it would cancel out the color* [he does this]. *Liquid 4 cancels it all. You'd have to see if 2 has the same influence* [he tries it]. *No, so 2 and 4 are not alike, for 4 acts on 1 × 3 and 2 does not."*—"What is there in 2 and 4?"—*"In 4 certainly water. No, the opposite, in 2 certainly water since it doesn't act on the liquids; that makes things clearer."*—"And if I were to tell you that 4 is water?"—*"If this liquid 4 is water, when you put it with 1 × 3 it wouldn't completely prevent the yellow from forming. It isn't water; it's something harmful."*

We see the complete difference in attitude between these subjects and those at substage II-B, in spite of the fact that the latter attempt some *n*-by-*n* combinations. The new attitude found at substage III-A can be noticed both in the combinatorial methods adopted and in the reasoning itself.

From the point of view of method, two achievements are worthy of note. The first is the establishment of a systematic *n*-by-*n* combinatorial system complete for the numbers involved in this experiment. For example SAR, who is afraid of forgetting certain associations, makes out a written list, and CHA works out the six two-by-two combinations without hesitation. We again encounter (though in a form which is all the more significant since it is more spontaneous) what we have seen in another work in studying the operations of combination with instructions which themselves suggest the operation.[3] The second achievement is just as important from the point of view of the utilization of these combinations (for it is obviously the needs linked to this use or, in other words, functional considerations which determine the completion of the corresponding structure): once the combination $1 \times 3 \times g$ which brings about the color is found, the subject, not satisfied with a single solution to the problem, does not stop there but looks for others. Thus his main interest is not success by the intermediary of a particular combination but an understanding of the role which this combination plays among the total number of possible combinations.

This leads us to the advances made in reasoning. The way subjects use combinatorial operations demonstrates that they are not concerned with particular mathematical operations at this point (moreover, the required operations have not yet been taken up in class by these subjects); but certainly we are dealing with a general logical structure, analogous to that of the multiplicative groupings utilized at substage II-A and tending to round out the structure after substage II-B.

At the same time as they combine the factors involved in the experiment among themselves (the liquids presented in the four flasks), stage III subjects form their judgments according to a combinatorial system having the same form, that of the sixteen binary propositions (combinations one-by-one, two-by-two, three-by-three, four, or zero of the four base possibilities $p.q \vee p.\bar{q} \vee \bar{p}.q \vee \bar{p}.\bar{q}$). In other words, when these subjects combine factors in the experiment, by the same token they generate a combinatorial system which corresponds to the observed facts. This is how they determine the links of conjunction, implication, exclusion, etc., by means of which they interpret the experimentally established combinations. Moreover, this fact explains the progress—correlated with that of the combinatorial operations themselves—which is noted in their deductive reasoning and in the formulation of verbal statements.

[3] *Ibid.*, Chap. VII.

This reasoning bears on elements 2 and 4 in particular. Element 2 is judged neutral because it is sometimes present, sometimes absent, in a colored combination as well as in others. If p designates the presence of color and q the presence of element 2, then the subject sees that one can have:

$$(p.q) \text{ v } (p.\bar{q}) \text{ v } (\bar{p}.q) \text{ v } (\bar{p}.\bar{q}) = (p * q), \qquad (1)$$

thus excluding the possibility of any positive or negative effect for 2: "It's water," conclude SAR and CHA. On the other hand, between liquid 4 and the color there is reciprocal exclusion or incompatibility, as CHA says clearly:

$$(p.\bar{q}) \text{ v } (\bar{p}.q) = (p \text{ vv } q), \text{ or} \qquad (2)$$
$$(p.\bar{q}) \text{ v } (\bar{p}.q) \text{ v } (\bar{p}.\bar{q}) = p/q \qquad (3)$$

(where q now designates liquid 4).

But, from the fact that he has formulated the association $p.q$ (in combinations 1×4; 3×4; etc.), at first SAR believes that 4 is neutral, just as is 2, so he replaces 2 with 4 in a combination $(1 \times 3 \times 2 \times g)$ and perceives that $1 \times 3 \times 4 \times g$ fades, whence the associations $(p.\bar{q}) \text{ v } (\bar{p}.q)$ which characterize reciprocal exclusion.

Secondly, this formal mode of reasoning—*i.e.*, founded on the combinations of factors and consequently on combinations of the statements themselves—naturally leads the subject to a new conception of the cause of the color. This cause is no longer sought in one or another of the elements but in their being brought together —or, more precisely, in the very fact of their combination. For example, SAR refuses to locate the color in g because "they go together" (= it's the whole [mixture] $1 \times 3 \times g$ as such which is the cause); CHA refers to elements which make "or don't make any color together"; and another subject, SIE (12 ; 6), declares: "This one (3), joined to 1 and to g, gives the color: 3 all alone does nothing and 1 alone does nothing either." From this, if p, q and r = the statements concerning the effects of 1, 3, and g—and if x = the statement that the color appears:

$$x \supset (p.q.r) \text{ and no longer } x \supset r. \qquad (4)$$

As for the reactions of the subjects at this level to experiment II (in which indicators are used), they add nothing new to the preceding. Nevertheless, it is interesting to note that even after having carried out no more than the four base combinations corresponding to a double-entry multiplicative table, the substage III-B subject already concludes that the color is a result of the combination as such, according to the schema which we have just described:

VIR (13 ; 4) associates *Ind × B*, *E × Ac*, *Ind × Ac*, and *E × B*: "What do you think about it?"—*"Simply that there is chemical water in two glasses and ordinary water in the other two . . . with one column* [burette] *it turned red and with the other nothing happened."* —"So where does the color come from?"—*"It's only the contact of the two waters . . . when they touch each other the color appears."* Then he passes on to combinations *Ind × E* and even *Ac × B* and to threefold combinations to study successive reactions.

It is evident that even before passing on to the combinations beyond his initial double-entry table schema, VIR already had a combinatorial interpretation of the color.

§ Substage III-B. Equilibration of the System

In experiment I the difference between substages III-A and III-B is only one of degree, actually it is not at all necessary in this case to apply the method "all other things being equal," since the factors are already presented in a dissociated state. Thus, the only innovations of substage III-B are that the combinations, and more particularly the proofs, appear in a more systematic fashion—*i.e.,* this level appears as a point of equilibrium in relation to the preceding level which is a phase of organization:

ENG (14 ; 6) begins with $2 × g$; $1 × g$; $3 × g$; and $4 × g$: *"No, it doesn't turn yellow. So you have to mix them."* He goes on to the six two-by-two combinations and at last hits $1 × 3 × g$: *"This time I think it works."*—"Why?"—*"It's 1 and 3 and some water."*—"You think it's water?"—*"Yes, no difference in odor. I think that it's water."*— "Can you show me?"—He replaces *g* with some water: $1 × 3 ×$ water. *"No, it's not water. It's a chemical product: it combines with 1 and 3 and then it turns into a yellow liquid* [he goes on to three-by-three combinations beginning with the replacement of g by 2 and by 4—*i.e.,* $1 × 3 × 2$ and $1 × 3 × 4$]. *No, these two products aren't the same*

as the drops: they can't produce color with 1 and 3 [then he tries
1 × 3 × g × 2]. *It stays the same with 2. I can try right away with 4*
[1 × 3 × g × 4]. *It turns white again: 4 is the opposite of g because
4 makes the color go away while g makes it appear."*—"Do you think
that there is water in [any of the] bottles?"—*"I'll try* [he systematically
replaces 1 and 3 by water, trying 1 × g × water and 3 × g × water,
having already tried 1 × 3 × water]. *No, that means 3 isn't water
and 1 isn't water."* He notices that the glass 1 × 3 × g × 2 has stayed
clearer than 1 × 3 × g. *"I think 2 must be water. Perhaps 4 also?* [He
tries 1 × 3 × g × 4 again] *So it's not water: I had forgotten that it
turned white; 4 is a product that makes the white return."*

Thus the results are the same as in III-A (save that the neutral
character of 2 had not been established systematically at the
earlier level). But they are discovered by a more direct method
because, from the start, the experiment is organized with an eye to
proof. This method may be described as a generalization of substitution and addition. For example, having established the fact
that the color is due to 1 × 3 × g, the subject replaces g by 2
then by 4 to see if they play equivalent roles; then he immediately
goes back to 1 × 3 × g and adds 2 and 4 alternately to the mixture in order to determine the effects of these additions. But it
should be understood clearly that substitution as well as addition
is already operating in the stage III-A combinatorial system.
When the subject constructs the combinations 1 × 2, 1 × 3, and
1 × 4, the very construction of these associations implies the substitution of 3 and then of 4 for 2; and when he makes the transition from two-by-two to three-by-three combinations, he adds the
alternative elements 3 and 4 to a given couple (for example 1 × 2)
—*i.e.*, 1 × 2 × 3 and 1 × 2 × 4. Moreover, as we have seen, substage III-A subjects already use these substitutions and additions
to prove certain effects. Thus, the only innovation appearing at
substage III-B is the greater speed with which the subject understands the use he may make of these substitutions and additions
in the determination of the respective effects of the elements during the actual construction of these combinations. Thus, first and
foremost, progress is to be sought in the organization of the proof
and in the integration of methods of discovery and methods of
proof. From the start, the combinatorial system becomes an instrument of conclusive deduction.

On a more general level, the lesson to be drawn from this experiment is that it points up the close correlation that exists between the mode of organization or the over-all structure of the combinatorial operations on the one hand and those of the formal or interpropositional operations on the other. At the same time that the subject combines the elements or factors given in the experimental context, he also combines the propositional statements which express the results of these combinations of facts and in this way mentally organizes the system of binary operations consisting in conjunctions, disjunctions, exclusions, etc. But this coincidence is not so surprising when we realize that the two phenomena are essentially identical. In other words, the system of propositional operations is in fact a combinatorial system, just as from the subject's point of view the only purpose of the combinatorial operations applied to the experimental data is to make it possible for him to establish such logical connections. Nevertheless, we had to show empirically that such an intimate relationship between the combinatorial operations and the propositional operations does exist, and in order to do this we have had to examine the reactions of the child and the adolescent to an experimental situation that did not impose either kind of operation by any sort of instructions but in which they would have to be discovered and organized in a completely natural and spontaneous way.

8

The Conservation of Motion
in a Horizontal Plane[1]

THE FIRST formal operational schema we described had to be the schema of combination operations, since the lattice structure which characterizes the system of propositional operations implies a combinatorial system. On the other hand the second operational schema, which we are now going to study, derives from the group structure and the reversibility by inversion which is its distinctive feature. As we will elaborate at greater length in the following discussion, the system of formal operations constitutes both a lattice and a group and thus unites transformations by reciprocity and transformations by inversion into a single cluster.

The experimental problem involves a ball[2] launched by a spring device and rolling on a horizontal plane. If no external obstacle interferes, it will maintain a uniform rectilinear motion (principle of inertia). Actually, a number of factors prevent the free operation of inertia—friction, which slows the ball down as a function of weight, air resistance, which slows it down as a function of volume, the irregularities of the plane, etc. As a result, two interesting problems arise which must be resolved by formal thought: (1) the problem of what is ideally or theoretically pos-

[1] With the collaboration of A. M. Weil and J. Bal, former student, Institut des Sciences de l'Éducation.
[2] The material consists of a set of balls of various weights and volumes.

sible—*i.e.*, not realizable in fact. In other words, how does the subject come to understand the conservation of motion by inertia given that it is never observable? From the physical and mathematical viewpoint, conservation of motion is a group invariant; but we would like to know whether an understanding of conservation also presupposes mediation of the reversibility by inversion that characterizes the groups of transformations from the purely logical and qualitative viewpoints of our subjects. We will try to show that this is the case. (2) The problem of the relative possible—*i.e.*, of the possibilities which are realizable in fact—modification of the movement by retarding factors and interferences among these factors explaining the irregularities and fluctuations of the course of a particular ball.

The subject's task is to predict the stopping points while varying the size and weight of the balls and to explain the observed movement. Our interest in the problem lies in the fact that, if concrete operations of serial ordering and correspondence formation allow the establishment of some relationships between the

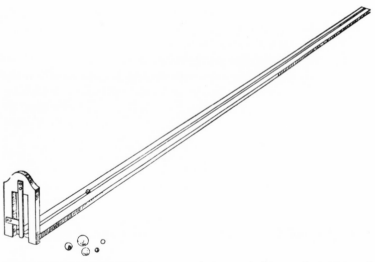

FIG. 7.   Conservation of motion in a horizontal plane is demonstrated with a spring device which launches balls of varying sizes. These roll on a horizontal plane, and the subjects are asked to predict their stopping points.

properties of the balls and the stopping points, the idea of conservation of movement by inertia escapes the realm of the "concrete," for such conservation cannot actually be achieved under
ordinary experimental conditions.

§ *Stage I. Absence of the Operations Necessary for
an Objective Account of the Experiment and the
Use of Contradictory Explanations*

The very young subjects react to this experiment as they react to
the problem of floating bodies (Chap. 2)—*i.e.*, with a group of
precausal predictions and explanations possessing certain regularities but mutually contradictory: the light balls will go further
because they are easier to set in motion and the large ones because
they are stronger; or there is no motion without force (the force
residing in the moving body or the force of the mover) and the
motion stops of itself by extinction of the force imparted by the
initial push, by fatigue, or by a tendency to rest.

RA (5 ; 4) tries to prolong or to stop the motion of the ball by framing
it with his hands, which are placed parallel to it without touching.
Sometimes the small and sometimes the large balls are supposed to go
the furthest, the first because they are light and the second because
they are heavy, but when a heavy one does not go far, it is *"because
it's too heavy."*

BREI (6 ; 4): "Will they all go the same distance?"—*"No, there are
some that will go further."*—"Which ones?"—*"That one"* [small wooden
ball].—"Why?"—*"Because it's smaller."*—"Are there others which will go
further?"—*"That one* [also a small wooden ball], *because it's smaller,
and that one"* [large, copper].—"Why that one?"—*"Because it's bigger,
and that one* [large, aluminum] *because it's big."* We ask the child
to show where these four balls will stop and he answers: *"There* [7–8
units for the small wooden ball], *because it's smaller. That one* [large,
aluminum] *there"* [13–14]. The small aluminum ball is also placed at
13–14 as is the small copper one; the large wooden ball at 5–6 *"because
it's bigger, and that one there* [large, aluminum, at 19–20] *because it's
big. This one here* [small wooden, at 24] *because it's smaller."* It is
evident that the small ones are expected to go near or far [from 7–8
to 24] because they are small and the large ones near or far [from 5–6

to 19–20] because they are large. Next we ask for explanations, which we find similar but with a certain note of finality about them: *"It didn't get very far because it didn't have a flag."*

MEY (6 ; 8). The little wooden ball *"won't go very far because it's small."*—"And that one?" [large wooden ball]—*"It can't go very far because it's big."* Then: *"The two big ones will go less far because they're big. . . . The three little ones won't go as far as the big ones."*

The contradictions among the predictions bear witness to the absence of any law in the child's mind. His explanations do not achieve a greater coherence but relate all types of motion to a sort of animated force.

§ Substage II-A. Attempts to Eliminate Contradictions and Corrections after the Experiment

Although the conservation of motion may not always be seen (the motion is regarded as being due to a *force* in the Aristotelian sense, and the cessation of movement is spontaneous) and although the predictions are based on variable factors (false or correct), henceforth there is a certain internal consistency in the assertions as well as in the utilization of experimental results:

PIR (7 ; 6): *"Some of them will go further than others."*—"Why?"—*"This one will go further because it is big and that one less far because it is small* [the first one is put in motion]. *It's less far than I thought."*—"Why?"—*"Because it's heavy."*

NIC (8 ; 0): *"The big one will go further because the little ones have more weight."* And *"that one won't go as far because it's big, heavy, and made of iron."*

HAL (8 ; 3): *"The big ones won't go as far because the little ones are lighter."* When a ball comes to rest close to the starting point: *"It's because it is heavier than I thought"* and, comparing a small copper ball to a large aluminum one, *"They go to the same place because they have the same weight."*

But the difficulty with an explanation in terms of force, such as used at this level, is still that of reconciling the force with which the object is launched with the force of the moving body and under conditions when the latter is heavy and when it is light.

HOR (8 ; 6): *"This one* [large, aluminum] *will go further because it is heavy"* [force itself is tied to the weight]. She rolls the copper ball. *"It doesn't go as far because it is small."*—"And the other?"—*"I didn't push it hard enough."* Next the large wooden ball: *"It will go all the way to the bottom because it's light."*

In spite of the effort to eliminate them, a residue of contradictions is left from the fact that the heavy balls have a greater force when in motion but are less easily set in motion, whereas the light ones have less force but are more easily launched.

§ *Substage II-B. The Beginning of the Reversal of the Problem in the Direction of the Causes of Slowing Down*

The explanations used at this level are not different from the preceding ones, in spite of the increasing but fruitless effort to unify the factors. However, since the child is increasingly sensitive to chance variation in the results, he exhibits a tendency to reverse the problem and to explain the causes of the slowing down rather than the cause of motion. He is not aware of this tendency. More particularly, little by little weight ceases to be perceived as a cause of motion and comes to be thought of as the (indirect) cause of the balls' coming to rest. Moreover, to the extent that subjects understand that the variability of the stopping points is due to the factors of volume, weight, and force of launching, they are more likely to think that weight and volume have a braking effect and even less likely to maintain that light weight and small size are causes of the prolongation of motion. These two kinds of assertions seem to be equivalent; the following will show that this is not true in the least:

JAD (10 years), referring to a zone of dispersion of about 20 cm., says of one ball, *"It is too heavy to go any further"* [than the extreme point] but at the same time *"it is too light"* to come to rest before the zone.

This kind of assertion shows clearly that the subject tends to invert the problem of motion. But he is that much less likely to suspect that his explanations remain the same as at substage II-A. In particular, he thinks of the air as promoting the motion by current backlash (ἀντιπερίστασιζ) rather than as an obstacle.

§ Substage III-A. Explicit Reversal of the Problem of Motion During the Experiment

The great difference between this level and the preceding ones is that from this point on the objective of the explanation is reversed; the problem is no longer to understand why the ball advances but what blocks its movement at a given moment.

As we have just seen, this reversal begins in substage II-B, but unconsciously. In contrast, although at first the III-B subjects are preoccupied in their predictions with motion, the experiment immediately leads them to focus their attention on the causes of the balls' slowing down or stopping. Thus, for these subjects the cessation of motion is no longer a positive state, the repose or aim of movement; instead, it becomes a negative state which must be explained by the intervention of new factors working in opposition to the positive state of motion.

MAL (12 ; 3): "For a ball to go far?"—"*You have to pull the trigger* [spring] *hard*" [experiment]. "So, why didn't it go further?"—"*Yes, but it's a bad stretch* [plane insufficiently smooth]; *it won't go so far.*"

CHAP (13 ; 3) predicts that the large ones will go further because they are heavier. After the experiment, he reverses his explanation.—"Why do the light ones go further?"—"*It depends on whether there is wind.*" —"What?"—"*It's the wind* [= air] *that stops them from going on. When there isn't any wind, the light ones go far because nothing stops them.*"—"And the heavy ones?"—"*I don't know.*"

MET (13 ; 3): "*The air keeps it back and it doesn't go as far.*"

Thus, starting with substage III-A, subjects touch on two causes of the cessation of motion: friction (terrain) and air resistance.

Without doubt the progress involved in reversing the explanation is due to the need to unify nascent formal thought. Since neither weight nor volume are causes of motion and (in contrast to explanations based on this conception) the ball goes further in proportion as it is both small and light, it follows that there is no simple cause for the continuation of motion. But it is more difficult to acknowledge multiple causes for motion itself (which may be considered the prototype of any simple phenomenon) than for the factors relating to the cessation of motion. However, even here the subject begins by looking for a unified explanation. He does so in spite of having seen the spread in results and chance fluctuations, which themselves were one of the reasons for his reversal of the question. That is why he does not succeed at first in this new line of attack. Time after time he fails to determine all the relevant variables simultaneously. Thus, CHAP discovers the factor of air resistance but fails to think of the friction for the heavy balls. MAL does the opposite, etc.

§ *Substage III-B. Conservation of Motion*

Finally, substage III-B leads to the fundamental explanation which results from the reversal of the positions taken at substage III-A: the conservation of motion by inertia. It should be said that all of the subjects do not solve the problem. Naturally, culture plays its diffuse role here. (Society had to wait for Galileo and Descartes with the "intellectual mutation," as A. Koyré called it, which resulted from their discovery.) But for certain subjects the rediscovery of the principle of inertia seems quite spontaneous, whereas for others there is, at least, a personal reconstruction of what they had learned:

DEV (14 ; 6) from the first experiment [large wooden ball]: *"It stopped because the air resists."*—"And this one?" [a small wooden ball, prediction].—*"It's about the same, but the ball is smaller: there is less resistance from the air and it will go further."*—"Is it the same for all of the balls?"—*"No, the bigger they are, the stronger the air resistance."* —"And for the small, heavy one?"—*"A heavier ball takes off less easily, but goes further because it has force in itself"* [weight = force!].— [Experiment] "So?"—*"That comes from the surface and the friction. The resistance varies with the substance the balls are made of: the wood is rougher, it scrapes more; the metal balls are smooth and will scrape less."*—[Experiment: small aluminum and large wood.]—*"Air resistance is proportional to size and weight* [!].*"*—"And if you compare this large aluminum ball with the small brass one?"—*"Oh! No, they take off with the same force. Only air resistance and friction come into play. . . . This ball* [brass] *is heavier and there will be more friction."* Conclusion: *"And if there were no air resistance, the ball would continue to roll."*

RAS (14 ; 4) *"Theoretically it should go to the end, but it's completely illogical"* [he means by illogical that which is contrary to the facts of direct experience]. Comparing a small and a large ball, he says again: *"The friction is less for the little one. Air resistance also plays a role. Theoretically, you would have to move it in a vacuum."*

DESB (14 ; 9): *"If you send them off with a push of the same strength, it* [resting point] *depends on weight, friction, and volume."* Next, he doubts that volume plays any role, but in comparing a small and a large ball, he says: *"The small one will go better because it has less friction, less air resistance."*—"That's all?"—*"If it's truly horizontal."* [3]

The protocols show that the reasoning which leads to the conservation of motion is extremely simple and is furnished in the most explicit form by DEV. The first stage consists of establishing the causes of the balls' slowing down or stopping. If we let p be the statement concerning slowing down or stopping, and let q, r, s, t, etc., be statements of friction or air resistance, irregularities of the track, of an eventual lack of (perfect) horizontality, etc., then:

$$p \supset (q \vee r \vee s \vee t \ldots). \tag{1}$$

Inversely, at the second stage the subject asks himself what should be the result of the negation of all of these factors, this negation implying a corresponding negation of statement p, that of slowing down. This is equivalent to the assertion of the continuation of motion:

$$\bar{q}.\bar{r}.\bar{s}.\bar{t} \ldots \supset \bar{p}. \tag{2}$$

It is interesting to compare this form of conservation, which is specific to formal thinking, with numerous concrete forms of conservation (wholes, lengths, weights, etc.; conservation of volume

---

[3] See other cases of this stage or other protocols from the same subjects in the third section of Chap. 15.

and surface area imply formal thought only because of the proportions). In both cases, conservation is achieved because of the role played by reversible operations (reversible by inversion or negation). When a modification arises as the result of the experimental actions, they allow a correction to be made for it by a transformation in the opposite direction (and thus a return to the null transformation). But in the case of concrete thinking this inverse transformation, even if it occurs only mentally, is of the same order as experimental modifications which alter the system and could in fact be carried out by the subject. For example, the transformation of a stretched-out section of modeling clay can be annulled by pushing it into a more compact mass, for what the object has gained in length it has lost in thickness. Thus it is possible to restore it by actions involving inverse modifications.

In contrast, in the case of the conservation of motion, operational reversibility occurs at the mental level only and does not correspond to any transformation which can be realized in full by the subject even in a laboratory situation. Even if one could eliminate all the causes of slowing down (though it is in fact impossible), one would still have to make use of an infinite amount of space and time to verify the principle of inertia completely. Nevertheless, the substage III-B subject manages to discard mentally the causes of stopping by thinking in terms of what is theoretically possible (but which cannot occur in fact) or, in other words, in terms of purely hypothetico-deductive implications.

Having done this, once more a reversible operation—(1) and (2)—suffices; here it is the counterposition (equivalence of $p \supset q$ and $\bar{q} \supset \bar{p}$), but in this case it rests on the double negation of $(p \vee q \vee r \vee \ldots)$ resulting in $(\bar{p}.\bar{q}.\bar{r}. \ldots)$ (thus of p or q or $r \ldots$ resulting in neither p nor q nor $r \ldots$) and of p resulting in $\bar{p}$.

One may, if one wishes, say that this reversibility comes back to the famous principle *tollitur causa, tollit effectus,* but on the one hand in order to eliminate the causes in the particular case the subject must think in terms of what is theoretically possible; on the other hand, since these causes cannot be eliminated in fact, the operation amounts to inverting an implication to give its converse by changing signs. Thus, the subject is proceeding on the basis of pure implications and no longer on the basis of trans-formations which can actually be effected.

We now see both the similarity and the differences between the several forms of conservation: all are based on a group prin-ciple (which is qualitative or logical before becoming quantita-tive or metrical), but conservation may be achieved either by concrete operations of classes and relations [4] (or at an even earlier stage by the integration of parts into a whole object) or, as at the formal stage, by the use of interpropositional operations alone.

---

[4] In this case the group aspect corresponds to the reversibility of the "grouping"—*i.e.*, to the nontautological transformations (identical with those of Boolean algebra).

9

Communicating Vessels[1]

IN THE PROBLEM of the conservation of motion, we encountered
the simplest form of the operational schemata relating to group
structure, for the construction of this notion by the adolescent
rests directly on formal reversibility by inversion. In the equilibrium problems, of which the problem of communicating vessels gives us a first example, we come to a more complex variety
of schema resting on group structure. In every equilibrium the
two possible forms of reversibility operate simultaneously: *inversion,* which corresponds to the additions or eliminations effected
in the parts of the system which come into equilibrium, and *reciprocity,* which corresponds to the symmetries or compensations
between these parts (thus to actions which are both equivalent as
regards their respective products and oriented in opposite directions). But, inversions and reciprocities also form a group between
themselves.

In order to illustrate our point and, more particularly, in order
to understand more clearly in what way the operational schema
corresponding to the notion of equilibrium is at the center of the
mechanisms of formal thought, we have to remind ourselves that
beyond the operations themselves in the strict sense of the term

[1] With the collaboration of F. Pitsou, former research assistant, Institut
des Sciences de l'Éducation, and A. M. Weil.

(or "operators")–*i.e.*, the operations of propositional logic, such as disjunction $(p \vee q)$, implication $(p \supset q)$, etc.–there are more general transformations which transform particular operators into others. Thus an operator such as $p \vee q$ can be transformed by inversion or negation into $\bar{p}.\bar{q}$, a transformation that we may designate by N, so that $N(p \vee q) = \bar{p}.\bar{q}$. But $(p \vee q)$ can also be transformed by reciprocity R, so that $R(p \vee q) = \bar{p} \vee \bar{q} = p/q$. Again $(p \vee q)$ can be transformed, by correlativity C (*i.e.*, by permuting the $\vee$ and the .), so that $C(p \vee q) = p.q$. Finally, the operator $(p \vee q)$ may be transformed into itself by identical transformation I, so that $I(p \vee q) = (p \vee q)$. Thus, one can see that I, N, R, and C form a commutative group of four transformations among themselves, for the correlative C is the inverse N of the reciprocal R, so that $C = NR$ (and $C = RN$ as well). Likewise, we have $R = CN$ (or NC) and $N = CR$ (or RC). Finally, we have $I = RCN$ (or CRN, etc.).

This group is of psychological importance because it actually corresponds to certain fundamental structures of thought at the formal level, for inversion N expresses negation, reciprocity R expresses symmetry (equivalent transformations oriented in opposite directions), and correlativity is symmetric with negation. This explains why the notion of equilibrium, which at a very early age gives rise to certain rough intuitions (balance, etc.), is not really understood before the formal level, when the subject can both distinguish and coordinate inversions, reciprocities, and correlativities (*inversions:* for example, increase or diminish a force in one of the parts of the system; *reciprocities:* compensate for a force by an equivalent force, thus assuring symmetry between the parts; *correlativities:* reciprocity in negation).

Although they may be relatively simple in certain concrete cases, these transformations actually require thinking and statements of a very abstract sort in most problems involving action and reaction, for here the difficulty is to grasp that X is at the same time equal to Y and acting in the opposite direction from it. In such cases, the instruments necessary for thinking go beyond propositional logic to include its fundamental group I N R C. This is what we shall see in the following chapter in reference to the problem of the equilibrium between the pressure of a piston and the resistance of liquids; but at this time take note of the same question as it relates to the preliminary problem of the equilibrium of communicating vessels.

In the case of communicating vessels, *reciprocity* serves to express the compensatory actions between separate vessels; transformations by *inversion* express the rise and fall of the water level. (Changes in water level are brought about not by adding or taking away water but by raising and lowering the receptacles.) In apparatus A, the subject raises or lowers the vessels by hand by adding or taking away the stands on which they rest. In apparatus B, he raises or lowers the two vessels with levers, and in apparatus C, he can only move one of the vessels, the other being stationary.

Since the receptacles have neither the same shape nor the same volume, in some cases one has to exclude these two factors to find the law. But air pressure can be disregarded, for it is equivalent for the two columns of liquid.[2] On this last point acquired knowledge may intervene, but we still want to know how well the adolescent can understand and make use of this knowledge, so the problem of formal operations remains decisive here and the influence of school is no bar to our analysis.

§   *Stage I. Lack of Differentiation Between the Actions of the Subject and the Transformations of the Object and Absence of Reciprocity*

The stage from 4–5 to 7–8 years is highly interesting from the point of view of the development of operations. No operation is yet possible at this stage, for the child fails to dissociate his subjective action from objective transformations and there is no reversibility between successive actions. The result is that the subject succeeds neither in predicting nor in understanding the

---

[2] The liquid used was only water and there was no difference in density between the contents of the two vessels. In systems of communicating vessels, the pressures are proportional to the weights of the liquids (pressure is the quotient of force divided by surface area). The fundamental principle involved is the following: the difference between two pressures $p_1 - p_2$ exerted at two points by liquid of density d in equilibrium is equal to the weight zd of a cylinder of liquid having as a base a unit of surface area and for height the vertical distance between the two points: $p_1 - p_2 = zd$ (where zd represents the pressure force measured in grams).

symmetry of the objective effects in the relations between the two containers:

GUY (5 ; 6). *Apparatus* B: He pulls the lever at the two sides alternately and concludes: *"If I pull there* [I] *the water goes away, and then it comes up there* [II]. *If you pull there* [II] *the water goes here"* [I]. *Apparatus* A: He lowers the vessel to the left and holds the one on the right in an inclined position: *"There is more water here* [right]. *Look, there is more water there* [left]: *I lowered it here and the water went there."*—"Why does it go down?"—*"I don't know; it's because it doesn't want to go back up."* *Apparatus* C: *"I get it: the water has to go there"* [when he pulls at the other side].

We see that the child has a perfect understanding of the fact that, when he pulls on one side, thus raising the receptacle, the water passes to the other side; but he does not understand that a difference in height is involved. For apparatus A, designed so as to make the differences in water level more visible, he thinks that it is enough to tilt a beaker to make the water flow into the other. And when we insist that the water level goes down, he is limited to saying that it does not want to go back up. Furthermore, it should be noted that this reaction is not specific to the problem of communicating vessels. For example, during this same stage the child does not know that the water in rivers always flows downward.[3] But in this particular case this lack of understanding is reinforced by the subject's failure to dissociate his own actions (pulling, tilting, etc.) from the objective process. The effect is that reciprocity is understood only when it occurs between the act of pulling and its results and not when it occurs between the rise and fall of the liquid which tends towards an equalization of levels.

§   *Substage II-A. The Translation of Actions into Objective Operations and the Discovery of the Elevation Relationships*

The role of elevation is discovered. The higher one beaker is in relation to the other, the more the water level rises in the latter as it falls in the former. This observation is certainly based on the

[3] See Piaget, *The Child's Conception of Physical Causality,* pp. 104–114.

subject's actions as he raises or lowers the beakers with levers or directly, but these actions are translated into operations which bear on the results obtained and describe those results in terms of objective relations dissociated from the subject's own activity:

NEL (7 ; 11): *"It always goes up more in that one when I raise this one and it goes down when I go down."*

TAC (8 ; 10). *Apparatus* B: We hide the right beaker and we ask the subject to bring the water level up to the third marker. The subject succeeds but only in a rough way and says, *"I looked about here"* [she looks at the height of the other beaker]. *Apparatus* A: *"Before you had to pull; now you have to put on the stands and that raises the beakers."* Then she draws the same conclusions as NEL.

But as far as the explanation itself goes, it does not at all deal with the equilibrium between weight and pressure; instead, subjects assume that water merely descends in the higher beaker to enter the lower one simply because that one is lower. As for explaining how the water level rises in this latter beaker (since it enters at the lower end of the receptacle), at this stage subjects refer to the impetus, rate of speed, air, etc.:

MIC (7 ; 10): "Did you understand how the water moves?"—*"You have to lower* [the beaker in which the water is to rise], *then the water comes . . . it flowed."*

TEA (8 ; 1): *"The water went through the pipe and into the other one."* —"What did you see?"—*"To see the speed. The water doesn't diminish as fast* [in the small one] *as in the big tube, because the tube is thinner and has less air than the big one. I'm surprised that the big tube, which has more air, works, and that one which has less works too: maybe it's because it's longer. If there is a big block of air, the water would move less quickly."*

Thus the air aids or blocks depending on whether it pushes or already occupies the place.

The progress made at this level over stage I is quite clear. The subject now describes the rise and fall of the water level and no longer only his own actions of pulling the lever or moving the beaker. On the other hand, it is hard to see the difference between these reactions and those of substage II-B unless we refer to the spatial structures which are available to the subjects. As we have demonstrated elsewhere,[4] at substage II-A the child is not yet able to represent the horizontality of the water level in a tilted receptacle because he does not try to base his observations on reference points outside this receptacle and limits himself to the interior relations. It is only toward 9–10 years, at the stage when coordinate systems are structured, that the horizontal and the vertical acquire a precise representative meaning. But it so happens that in the present experiments the subjects of substage II-A do not yet perceive the equality of level attained by the water in the two receptacles; they discover simply that the water goes down in one as it rises in the other until it stops moving in both. Furthermore, they know that the water level drops in the beaker in the higher position and rises in the lower one, but these relations of elevation are applied only to the receptacles themselves and do not always imply that the two water levels will finally be equal— *i.e.,* that the line that unites the two levels will be horizontal. That is why, when one of the beakers is hidden and the child is asked to attain a certain elevation (see TAC: the third stage for apparatus B), he can only succeed in a rough way and focus on the height of the other beaker and not its water level.

All in all, then, subjects have just about begun to get a glimpse of the notion of system equilibrium. And what notion they have boils down in essence to raising and lowering the beakers with a view toward raising or lowering the water level. Doubtless, a preliminary inversion (raising and lowering the beakers) and reciprocity (the water goes down in one vessel as it rises in the other) are present. But lacking is the condition of equivalence which alone would allow the child to coordinate these transformations— the final equality of the two water levels.

§ Substage II-B. Final Equality of Water Levels but Without an Explanation

As we know, subjects become able to handle concrete operations at substage II-B. This substage also marks a kind of upper limit in the structuring of the equilibrium schema, that is, insofar

[4] Piaget and Inhelder, *The Child's Conception of Space,* Chap. XIII.

as the subject does not bring in any formal operations. Owing to the construction of systems of spatial reference (natural coordinate axes) the child discovers the law of the equality of water levels in the two receptacles for conditions of system equilibrium (the water line in both beakers falls on a single horizontal line). But in this way subjects can only enunciate the law without discovering its causes, for a statement of the law depends on class and relational operations alone, and these are sufficient to determine the relevant correspondences, but an *explanation* of the law requires the intervention of the four groups of interpropositional transformations cited at the beginning of this chapter.

XI (8 ; 9): *"When I pull here, the other* [beaker] *fills up; when I pull there, the other one fills up too."*—"The water is at No. 2; put it at No. 3."—He succeeds.—"How did you do it?"—*"I saw that when the water goes up here it goes down over there; so I did the opposite . . .* [etc.]." At one point, he makes a mistake [the point to be reached is hidden each time]. He then takes the ruler and measures [the distance] from the table to the number indicated. Then he refers to the same elevation on the visible beaker at the other side in order to determine the water level. Another time, he places the ruler horizontally to assure the equality of the levels.

MIC (9 ; 11) succeeds in determining the water level correctly when one beaker is hidden: "How did you know?"—*"Because I calculated the height here and I looked there for the same thing."*

SOC (10 ; 9): *"The water is at the same level. When I raise it here, the water goes up there, but there is always the same capacity, even if it goes up."*—"What do you mean by capacity?"—*"There is always the same amount of water* [he knows well that the volumes differ]: *the water stays at the same height on both sides."*

DOM (11 ; 4): *"The level is exactly the same. The water rises quickly in this tube and falls less quickly in the bottle. That comes from the volume of the beakers, but in contrast the water will always stay at the same level."*—Apparatus A: "And the level?"—*"It will change. No, rather it will always be the same* [on both sides], *but the* [absolute] *height will change."*

GAS (10 ; 6) measures the elevations and verifies the horizontal level. He is presented with a long tube communicating with a very large crystallizer: he predicts that the levels will still be identical.

Thus, one can see that these subjects discover both the equality of water levels and the means of verifying this equality once a reference system based on the coordinates of immediate physical space (vertical and horizontal) is established. Verification is effected either by checking whether the line uniting the two surfaces is horizontal (xi and gas) or by measuring their respective heights (xi, mic, and gas). This transition from the qualitative to the metrical shows us well enough how preoccupied the subject is with coordinate axes and with substituting the concept of equilibrium based on the equality of water levels for the concept of equilibrium found at substage II-A—*i.e.*, based on rising and falling. In this case transformations by *inversion* amount to the raising or lowering of the level in one of the beakers, whereas, henceforth, *reciprocity* includes the whole set of transformations which relate the level of liquid in one receptacle to the level of the liquid in the other.

What is the nature of the mechanism of these transformations? The concrete operations available to the subject at this stage do not allow him to answer this question; by their temporal and spatial serial ordering and correspondences, they allow him to determine the conditions of equilibrium, but by no means do they allow him to grasp the play of forces involved. In the course of proceeding from a statement of the law to its explanation, some subjects, like soc, invoke an equality of "capacities" or "amounts," but since it is evident that the volumes differ, in the final analysis this quantity amounts to nothing more than the equality of the elevations themselves—*i.e.*, the equality of water levels. A ten-year-old subject specified that the water always goes "as low as possible." This demonstrated that henceforth the tendency of the water to fall is accounted for in terms of its weight (as we had known as the result of other experiments); but weight itself is no longer often called upon as an explanation of the equilibrium, since the volumes involved are clearly different. In sum, from this point on the equilibrium is well described, thanks to the concrete operations which make it possible for spatial and temporal inversions and reciprocities to be established. But by no means is it explained, for the child fails to make use of inversions and reciprocities bearing on the actions and reactions themselves.

§ Substage III-A. Preliminary Explanation and Formal Structuring

At this first formal level we can observe, in contrast, an important reworking of the operations and the explanation. The conservation of volume is finally acquired and the volume is finally distinguished from the quantity of matter and the weight.[5] This leads to the paradoxical fact that substage III-A subjects seem to find it impossible to accept a situation that did not bother substage II-B subjects at all—*i.e.*, the equality of levels when the volumes (as well as the shapes of the receptacles) differ. Far from generalizing the law to all cases, as at substage II-B (where, by the way, the generalization is limited, since it bears on the levels alone), substage III-A subjects start by restricting the scope of the law to those cases in which shapes and volumes are equal. They expect that the equality of levels will no longer hold for unequal forms and volumes. When the experiment contradicts their expectations, they limit their conclusions to the cases actually observed and refuse to make any generalization that would admit of what seemed to them to be an exception. We have here a neat example of mutual interference between the operations constitutive of the law and explanatory or causal operations. More precisely, in becoming explanatory the stage III operations lead the subject to limit the generalization based on concrete or legitimate operations (legitimate because here they bear on the levels alone and not yet on the equilibrium of actions and reactions).

What goes into these new operations? At this stage equilibrium in communicating vessels is no longer conceived of as the simple flow of water from a higher level to a lower one until equality of levels is achieved, but as a system of actions and reactions whose inversions and reciprocities are stated in mechanical and not merely in spatiotemporal terms. That is why the subjects require equality of weight based on equal volumes before they are willing to talk about equality of levels, and that is why they deny that two vessels of unequal capacity can verify the law. They have failed

[5] See Piaget and Inhelder, *Le Développement des quantités chez l'enfant*, Chaps. III and VIII–IX.

to understand the compensation resulting from the relationship between the weight of the vertical column of water and the surface area of the base of this column.

AND (12 ; 9) establishes the equality of levels in all positions.—"And if instead of this one we used a bottle with a conic shape?"—"*It wouldn't work because the shape is conic.*"—"But you told me that the levels were always the same?"—"*They're the same providing the diameter is the same at all heights of the beaker. Here* [with conic and cylindrical forms] *the elevations . . . you couldn't manage to compensate* [one with the other]. *It's the shape of the beaker that plays a role.*" He does the experiment and is astonished to discover the same level.— "How do you explain it?"—"*Probably that the shape of the vessel doesn't matter.*"—"Why 'probably'?"—"*Because the facts are there!*" [*cf.* the opposition between facts and theory so characteristic of formal thinking].—"Why is it the same level?"—"*Because it* [the conic beaker] *widens toward the top*" [the opening is at the bottom].—"But if it were?" [turned upside down].—"*It wouldn't be the same level* [!]."— "What happens when the water is in the pipe?"—"*It isn't the same level because the tube is thinner than the beaker.*"

BON (12 ; 8) affirms that the levels are the same. With *apparatus* b, he takes exact vertical measurements when one of the beakers is hidden: "*Whenever you lift one container, the water rises in the other, so the water should rise or fall in both.*"—"Does it always happen like that?"—"*Yes, always. . . . No, not in all cases, not when the beakers are not of the same width.*"—"But in this apparatus [C] are the beakers of the same width?"—"*No, but the length of the pipe and the width of the beaker can contain the same quantity of water.*"—"And here [long tubular beaker to the left and large crystallizer on the right (*cf.* the case of GAS at substage II-A)]?"—"*The water here* (crystallizer) *will only go up to here*" [much lower level than the other].—"Why?" —"*Because the beaker is larger.*"

EAN (13 years). *Apparatus* A: He raises both beakers "*to see if by raising the two together I get the same level in both as when they were down below.*" Next: "*To add some water in one, you have to take out the same amount in the other. When I put the beakers in different positions* [in relation to each other], *I can always see* [that the level is] *the same for both.*"—"And if you put a narrow bottle in that one's place?"—"*No, if I have a large one and a small one because the volume is larger . . . the level is always higher.*"

PIE (14 ; 3), identical beakers: "*If one bottle is* [placed] *higher than the other, the water goes into the other*" because "*for the quantity, it's the same thing.*" But with unequal beakers, "*When there is the same quantity in the two glasses, the level in the tube will be higher, because it is thinner.*" When one of the beakers is hidden, he answers: "*I can't figure it out, because the diameter of the two glasses isn't the same.*"

We see the difference between these subjects and those of sub-stage II-B even though they often use the same words. Henceforth "compensation" (AND) is a matter of "quantities" understood in the sense of sources of equal forces (because of the equality of weight and volumes) as if it were a balance scale. Thus unequal levels should correspond to unequal quantities. When they perceive facts to the contrary, the subjects resign themselves to them, as for example AND ("because the facts are there"), but refuse to generalize to other cases. In sum, they do not know the details of the explanation. But if we consider only what they do know, we find that they reason in a coherent manner and furnish a very revealing example of the logical subordination of a general law to the concrete case and of the assimilation of that case to the formal transformations of inversion and reciprocity projected into the real world. In fact, one might say that these subjects are of interest to us precisely on account of their ignorance of the exact explanation: though they have received no academic instruction about communicating vessels, they still sketch out an interpretation based on compensation (as AND says)—*i.e.*, on the fact that each of the two quantities of liquid exerts a pressure on the other, the two pressures being, by this very fact, oriented in opposite directions.

Certainly we have here a differentiation and a coordination of the transformations of inversion (raising or lowering the levels) and reciprocity (actions and reactions of one of the quantities of water on the other). The only limitation of this explanation is that the subject does not yet know how to generalize it to the case of unequal quantities; still, the principle is accurate. Before trying to give a precise statement of this reasoning, let us examine the reactions of substage III-B, which we have not yet considered. Unlike the earlier reactions, these are influenced by academic knowledge (which, moreover, has been assimilated only to the extent that it fits into the schema whose development we have just noted).

§ Substage III-B. Formal Generalization of Acquired Knowledge

Finally, at substage III-B, the spontaneous schema of explanation outlined during substage III-A is filled in with information gained through education; thus the contradiction between the equality of water levels and the eventual inequalities of the amounts of liquid is eliminated. But one can easily see that this contribution from without does not modify the structure of the reasoning:

PIC (13 ; 6): *"These levels are always equal because the forces compensate each other";* according to PIC these forces are air pressure and the weight of the water.

MIN (14 years): *"If you have two beakers of the same sizes or different sizes, the water will come up to the same level in both, because the larger the beaker, the more air presses on a large surface; and vice versa, the smaller the beaker, the more the water will act on a small surface, so an equilibrium is reached."—"Always?"—"No. When you have two chambers, if there were more air pressure in the chamber where you put the bottle on the left and less pressure in the chamber where you put the bottle on the right, the level at the left would be lower."*

In other words, having a more or less clear understanding of the fact that the pressure of the liquid is relative to the surface area of the vertical column at its base, the subject explains the phenomenon of communicating vessels in a fashion analogous to that used at substage III-A, but generalizes to the case of unequal quantities. Thus the essential point in the explanation is that even in the case of unequal volumes the pressures compensate each other in function of the height of the columns "so equilibrium is reached," as MIN says, this time referring to beakers having different capacities.

§ The Notion of Equilibrium and the Group of Four Interpropositional Transformations I N R C

In order better to understand the nature of the formal structuring which culminates in the operational schema of equilibrium, it seems worth while to compare what can be called the concrete reciprocities of stage II with the formal reciprocities of stage III.

The preliminary form of reciprocity appears for the first time at substage II-A with the discovery (inaccessible to stage I subjects) that the higher one of the beakers (I) is in relation to the other (II), the more the water rises in the second beaker. The discovery of this relationship entails the following operations:

Serial ordering of heights:

$$A_1 < B_1 < C_1 < \ldots \tag{1}$$

taking the other beaker as a point of reference;

Serial ordering of the levels of elevation as they increase in the lower beaker (with interior references to this beaker):

$$A_2 < B_2 < C_2 < \ldots ; \tag{2}$$

An (ordered) correspondence between the two sets of serial orderings:

$$A_1 < B_1 < C_1 < \ldots , \\ \updownarrow \quad \updownarrow \quad \updownarrow \\ A_2 < B_2 < C_2 < \ldots ; \tag{3}$$

In the case of reversal of the situation, the elevations occupied by beaker I may be ordered serially in descending order:

$$\ldots C_1 > B_1 > A_1 > , \tag{4}$$

as may be done for the levels in II:

$$\ldots C_2 > B_2 > A_2 \quad . \tag{4a}$$

The correspondence is in this case established in reverse order.

Thus concrete reciprocity consists of a symmetry between the two correspondences:

$$(A_1 \leftrightarrow A_2) \rightleftarrows (A_2 \leftrightarrow A_1). \tag{5}$$

At substage II-B, a system of external reference is added to these relationships, allowing the introduction of the notions of the horizontal and of the equality of levels in terms of rate of flow. We have treated the operations needed to construct this spatial system elsewhere.[6] Here we may limit ourselves to noting that the

[6] Piaget and Inhelder, *The Child's Conception of Space*, Chap. XIII.

serial orderings and correspondences (1)–(3) are replaced by a correspondence between the actual water levels in the two beakers. If we call $+ A$, $+ B$, $+ C$, etc., the increasing elevations included between the horizontal (the line of final equality) and the levels in the higher beaker, and $- A$, $- B$, $- C$, the corresponding increasing heights included between the line of equality and the levels in the lower beaker, the substage II-B child establishes the correspondence:

$$\ldots (+ C) > (+ B) > (+ A),$$
$$\ldots (- C) > (- B) > (- A). \tag{6}$$

Thus the reciprocity depends on the equality of the differences $+ X$ and $- X$ and their continuous compensation, which occurs until the difference is zero (when the line which eventually unites the two levels is horizontal). So reciprocity boils down to a spatial symmetry (but without an adequate causal explanation). As for the inversion operations, they consist of increasing or diminishing the differences $\pm A$, $\pm B$, $\pm C$, etc. This operation may be effected by addition or by elimination of quantities of liquid in one of the two vessels; this is accomplished easily by raising or lowering this beaker. Hence (if A' is the difference between the increasing heights A and B):

$$A + A' = B, \text{ etc., and } B - A' = A. \tag{7}$$

But no total operational system as yet allows the subject to fuse reciprocities by correspondence and increases or decreases of differences into a single whole. That is why the subject is limited to describing the equilibrium and cannot manage to understand it as a single causal system. When the notion of compensatory actions and reactions appears at stage III, two innovations come into play: the spatial reciprocity of levels becomes a reciprocity of pressures; this constitutes a single operational system with inversion operations.

This coordination of inversions and reciprocities can be formulated in the following manner: Let us call p and q the statements concerning the effects of any two pressures exerted at separate points on the liquid contained in beaker B. Let us call $\bar{p}$ and $\bar{q}$ the statements that these effects are canceled out, either by inversion of their causes (thus of diminutions in beaker A until the initial elevations are eliminated) or by compensation under the influence of pressures operating in opposite directions.

Thus there are four possibilities:

$$I(p \text{ v } q), \tag{8}$$

direct transformation or effects of the pressures exerted by liquid A on liquid B;

$$N(p \text{ v } q) = \overline{p \text{ v } q} = \bar{p}.\bar{q}, \tag{8a}$$

inverse transformation or elimination of the effects p and q;

$$R(\overline{p \text{ v } q}) = \bar{p} \text{ v } \bar{q}, \tag{8b}$$

reciprocal transformation or effects of the pressures exerted by liquid B on liquid A;

$$C(p \text{ v } q) = \overline{\bar{p} \text{ v } \bar{q}} = p.q, \tag{8c}$$

correlative transformation—*i.e.*, inversion ($=$ negation) of the reciprocal, thus canceling out the opposite (negative) effects $\bar{p}$ v $\bar{q}$, which is equivalent to the simultaneous assertion of p and q— *i.e.*, $p.q$.

Such is the reasoning schema which the stage III subject uses. He understands that the point of equilibrium is reached when values x and y, corresponding respectively to the pressures represented by p v q and by $\bar{p}$ v $\bar{q}$, are equal. As long as one has $x > y$ or $x < y$, the liquids are actually still in motion in tubes A and B. On the other hand, any movement ceases when the liquids reach the same level (represented by r) because:

$$r \supset [x(p \text{ v } q) = y(\bar{p} \text{ v } \bar{q})]. \tag{9}$$

Although substage II-B subjects perceive the horizontal level common to tubes A and B, they are unable to explain it. Stage III subjects interpret it as due to an equality between pressures, stated by the double reversibility of transformations I N R C.

10

Equilibrium
in the Hydraulic Press[1]

THE INCREASINGLY more advanced explanation which the subjects give of the phenomenon of communicating vessels has just shown us the importance of the formal transformations of inversion and reciprocity and of the I N R C group that they form among themselves, according to the possible combinations, for the establishment of the operational equilibrium schema. But the drawback of the experiment with communicating vessels is that the pressure intrinsic to the liquid is completely overlooked by our subjects. A detailed account of the explanation is not found until it is given in terms of acquired knowledge. In the apparatus dealt with in this chapter, two communicating vessels again appear, but one of them is provided with a piston which may be loaded with varying weights; thus, the pressure exerted on the liquid is directly proportional to the weights. (It is to be noted that the piston is propelled not by an external force but by its own weight.)

Now, there is a reaction of the liquid corresponding to the action of this weight (the displacement of the liquid under pressure is inversely proportional to its resistance), but here, too, the resistance reaction can be made tangible by varying the density

---

[1] With the collaboration of A. M. Weil; A. Tissot, former research assistant, Institut des Sciences de l'Éducation; and M. Wikström, _élève diplômée_, Institut des Sciences de l'Éducation.

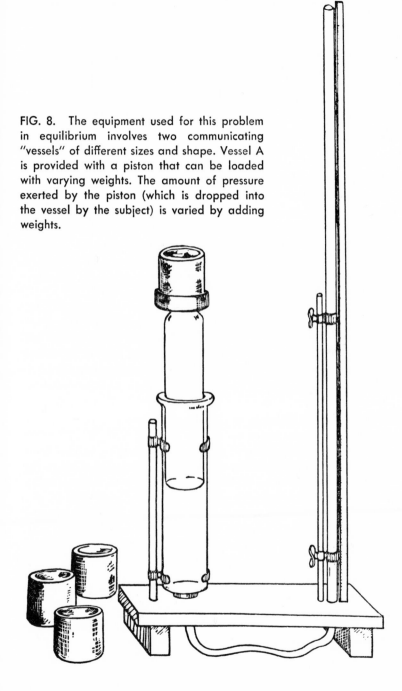

FIG. 8. The equipment used for this problem in equilibrium involves two communicating "vessels" of different sizes and shape. Vessel A is provided with a piston that can be loaded with varying weights. The amount of pressure exerted by the piston (which is dropped into the vessel by the subject) is varied by adding weights.

of the liquid—*i.e.*, by using water, alcohol, or glycerine in turn. To the question of equilibrium is added that of the actual transmission of forces; the problem is to understand that the force exerted by the piston is transmitted in a uniform manner through the entire liquid and that the equilibrium between action and reaction relates not only to the surface of the liquid (or the lowest point in the apparatus) but to the entire system.

Therefore, in this particular case the problem of the relationship between concrete and formal thought is to understand how the subject makes the transition from simple observed correspondences between the weights and the displacements of the liquid to an explanation expressing the complete transmission of force as a function of weight and inversely as a function of density. Thus it is not only the general operational schema for equilibrium which reappears here but, more particularly, the equality of action and reaction. The subject matter is especially promising for the study of the role played by the I N R C group in thinking.

§   *Stage I. Lack of Understanding of the Role of Weight (Substage I-A) Followed by Global Understanding Without Either Serial Ordering or Operational Correspondences*

At substage I-A, the subject does not even make an unequivocal prediction that the water will rise in the thin tube (B) as a result of the weight of the piston because, if the "heavy one" has force ("leaned on" or "pushed" by assimilation to his own action), the "light one" is likely to "rise" (by assimilation to raising itself). Moreover, no conception of conservation of quantities is present at this level. Thus the water in the tubes will not necessarily be conserved and the tubes may be filled or emptied without adequate reason.

KOT (5 ; 6): "*The water will go up* [he points out tube B]; *it has to fill up to go up.*"—"If I use this piston?"—"*It's going to fill up.*"—"And if I put on this box?" [500 grams]—"*It could fill up more* [this is done]. *It's because it* [the box] *presses.*"—"If I take off the box?"—"*It will go down.*"—"Where?"—"*Like before.*" He puts on the 2-kilogram box. "*A little higher.*" He replaces the box by the 1,500-gram box. "*A little lower. It's the heavy ones that go the lowest* [in B!]. *The little ones can't lift anything high.*[2] *The big one and the fat ones can bring it up close* [small displacement]. *The heavy ones can't, but the little ones go high up"* [contradiction].—"Take the little ones."—[He tests their weight.] *"This one must be the smallest* [1,500 grams; he puts it on]. *It went up a little anyway! I think the lightest is this one* [500 grams] *and I think that this one is the heaviest* [2 kilograms; the two extremes are evaluated accurately, but the rest of the serial ordering is wrong: 1,000 > 1,500 grams, etc.]. *Oh! This one* [2 kilograms] *is the smallest* [he puts them all on in turn]. *This one goes the lowest"* [500 grams]. —"Why?"—"*It's the biggest* [he puts on 2 kilograms]. *Oh! But that one that goes up highest: you see that it's the smallest."*—"Why the smallest?"—"*It's the lightest, and also the smallest."*—"It's the lightest?"—"*I feel it when I carry it* [he lifts them in turn without comparing them two by two]. *Yes."*—"Why does the water go up most with that one?" —"*Because it presses, it makes the water go up."*—"If you put on a small one?"—"*That makes the water go up very high."*—"And a light one?"—"*It's like the fish in the water: they can raise themselves* [he puts 1,500 and 1,500 grams together]. *It goes up high."*—"Why?"— "*Because that* [he points out one of the two 1,500-gram boxes] *is the heaviest"* [!].

JOG (6 years): *"It's the water that went up again!"*—"Why?"—"*Because the tube* [piston] *was put on top; that made the water go up."*—"And if you put this box on top, will it change?"—"*It will go up more"* [he points out a level that is too low].—We do the experiment and point out the predicted level. "Why not there?"—"*It was too heavy"* [thus the weight is seen as the cause of the rise].—"Next?"—He uses 1,500 grams and 1,000 grams and sees that the first results in a greater rise in the water level.—"Why?"—"*Because it's less heavy than the yellow"* [this is not true; moreover he now sees the lack of, rather than presence of, weight as causing the rise!].

These cases were worth citing because they show the initial incoherence of the reasoning which later reaches the level of formal logic. In the first place, we see that these subjects can neither serially order the weights correctly nor allow for the equality of size of the boxes which contain them (although they are all exactly alike). Secondly—and in connection with this preoperational perspective—the heavy weight may cause the water to rise more or

---

[2] The boxes (from 500 to 2,000 grams) all have the same volume.

less and the light one may in turn fail to press on the water or, on the other hand, cause it to rise. It may even happen that two superposed weights cause the water to rise quite high because one of the two is erroneously judged lighter than the other; thus, nonadditivity is an additional source of contradictions. In sum, the child cannot possibly formulate any law yet, for he lacks operational coherence.

At substage I-B, the child understands in general terms that the heavier the weight, the more the water rises in tube B, but he is not yet using serial ordering, correspondence, or exact predictions, for completed operations are still lacking:

JAC (6 years), after several trials, predicts that the 1,500-gram box will produce the same result as the 2-kilogram box *"because it's just as heavy."* The water does not rise as high. Next the experiment is done over again with the 2-kilogram box. He predicts once again that the level will be the same *"because they are the same."* The water rises higher.—"Why?"—*"Because it's heavier."*—"And that one [500 grams] will go where?"—*"I don't know."*—"To the same place?"—*"No, because it is less heavy."*—"And that one [1,500 grams]; can you guess where it will make the water go?"—*"No"* [however he weighs it].—"Why does the water come up here?" [B].—*"Because there is something that touches the water, the pipe* [piston] *presses on it. It presses more or less and it always makes it go up."*

MON (6 years): *"Because there is more weight and that makes it go higher up. The piston sinks in more* [in A] *and there will be more water there"* [in B].

DEL (7 years) gives the same explanations but also tries to recognize the weights by the sizes. But the 500-gram box seems to him larger than the 1,500-gram box, whence spring several contradictions: "heavy" goes with "big"; the "heavy one" makes the water rise more, but the "less big," too.

Thus there is progress in comparison to substage I-A in the sense that as a general rule the weight becomes the cause of the rise of the water. However, since the weights are badly ordered and the equalizations badly established, a certain amount of indeterminacy often renders the predictions inaccurate.

§ Substage II-A. Exact Serial Ordering of the Weights and Correct Correspondence with the Water Levels

With the appearance of concrete operations, the weights and the water levels are made to correspond; the weights being found to be either equal to or different from 500 grams, their serial ordering does not cause any more difficulty (which is not the case before substage II-B when the differences are less perceptible). But progress stops at this point. For, as regards the density of the liquid, up until 9 years (on the average) the child thinks that the heavier the liquid, the higher it will rise, because its weight is added to that of the piston which makes the water rise. Thus, at this first operational level there is no notion of liquid resistance—*i.e.*, of a reaction oriented in the opposite direction from the pressure action—thus no notion of reciprocity in the realm of forces:

SOL (7 ; 11): *"If the piston sinks in further, the water goes up more."* The 2-kilogram box will cause the water to rise more than the one of 1,500 grams *"because it's loaded."*—"What does that depend on?"— *"On the weight that's inside."*—"If it is heavier?"—*"It goes up higher."* —"And lighter?"—*"Lower. If you had a very big weight, it would go* [gesture], *it would gush out from up above."*

COR (7 ; 11) same reactions with water, after which we substitute alcohol: "Why doesn't this box go up as high now?"—*"The box got lighter."* —"Why?"—*"It's not the same liquid."*—"What difference does that make?"—*"Maybe the liquid goes up higher. It's because the liquid is heavier, it has more weight."*—"This [alcohol] is heavier?"—*"Yes, it's heavier* [since] *it goes up higher; that makes the weight and then it makes it go up."*

PAL (8 years). Experiment with glycerine: *"The water is heavier, so everything will go a little higher because it's heavy."*—"Why doesn't it go as high with the glycerine?"—*"Because the glycerine is lighter."* —"But weigh it yourself" [two equal volumes].—*"I was wrong, it's the opposite."*—"So why does it rise when it's heavier?"—*"Because it doesn't have enough force to rise higher."* Thus there is a contradiction; weight involves force, but the glycerine does not come up to the level.

These facts are interesting from two points of view: the child's physics and his logical operations. From the first point of view, he does not perceive action and reaction, but thinks that the force of the pressure exerted by the piston and the boxes releases in the liquid a force due to their own weight and oriented *in the same direction*. We have here a new example of the influence, which is general at this age, of the two motors schema.[3] The external motor sets off the action of the internal motor, and the two work together for the execution of the same movement. (This Aristotelian schema is frequently used in the child's explanations, even for the movement of projectiles.) These same subjects do have a notion of action and reaction in the case of balance, but it is intuitive (for it is simply linked with visible displacements) and lacks operational generality. In the present situation, they have none, although they may compare the increasing weights of alcohol, water, and glycerine at equal volumes and see the inverse correspondence between the weights and the rise of the liquid in tube B (with the same boxes placed on the piston of tube A).

The reason for this is that the principle of action and reaction cannot be understood in terms of concrete operations alone. Class groupings presuppose inversion and relational groupings presuppose reciprocity, but neither taken by itself provides any mechanism for the integration of the two in a single operational system such as the I N R C group. But the principle of action and reaction is based on this group: it entails the intervention in the subjects' reasoning of logical transformations which include both reciprocal and inverse operations and coordinations between them. Inverse operations alone may assure the coordination of operations in the same direction (*e.g.*, understanding that taking off weight diminishes the pressure), but reciprocal transformations are required for equating operations which are oriented in opposite directions (*e.g.*, for understanding that a greater liquid density *compensates* rather than *adds force* to the weight on the piston so that the more dense the liquid, the *less* the rise of the water in tube B). The integrated group by definition implies formal operations or a "structured whole" as opposed to elementary groupings of classes and relations.

Thus, because he lacks formal operations, the subject at this

[3] See J. Piaget, *The Child's Conception of Physical Causality*, Chaps. I–V

level does not come to understand the relations between action and reaction, just as in the case of communicating vessels he gets as far as a spatial reciprocity of rises and falls without discovering the reciprocal relationship of operative forces themselves. It is clear that, at this substage (II-A), equilibrium between liquid and a piston of variable weight pressing on it is not conceived of as an equilibrium process but as a one-way process; the piston exerts its force on the liquid in the departure tube (A), and the liquid acts not in return but (even in tube B) in the direction of the piston itself and adds its own force (because of its weight) to that of the piston. But the transmission of force does not raise any problem because, strictly speaking, it is not a transmission. It still consists of releasing or activating the force of the liquid with the force of piston and weight placed on it. Thus the serial ordering of weights and the correspondence between them and the attained levels are still a long way from being an expression of the law or even *a fortiori* an understanding of it.

§   *Substage II-B. Intuitive Formulation of the Notion of Resistance*

The stage from 9 to 11 years is a transition stage, during which the subject begins to get a glimpse of the fact that the liquid resists as a function of its density. Some children say even at this point that in tube B the water rises less than the alcohol because the alcohol is lighter. But that does not yet mean that they have mastered the problem from the point of view of action and reaction. In particular, they wait for the piston to fall to the bottom of large tube A, as if it fell to the bottom of any receptacle whatever and as if counter pressures were irrelevancies. Thus, the column of water in narrow tube B is always conceived of not as exerting a reaction in a direction opposite to that of the piston pressure and coming into equilibrium with it but only as resisting the rise somewhat as a function of the liquid's weight.

HID (10 ; 3) predicts that *"it's going to rise here* [B], *and here* [A] *it will go down."*—Experiment.—*"Oh! I thought it would go up higher."* —"Why?"—*"Because the piston didn't go all the way* [as he expected] *and the water didn't go all the way up."*—"And with that box?" [1 kilogram].—*"Way up there"* [higher].—"And if we take it off?"—*"It will go down again."*—"And with that box?" [2 kilograms].—*"It will go higher because it's heavier."* The water is replaced by alcohol; he expects the same levels: *"It was higher."*—"Why?"—*"Because it* [the alcohol] *is lighter."*—"And with that box?"—*"There"* [a little lower than with water].

FRA (10 ; 10): *"The tube* [piston] *is going to fall and the water will overflow because when you put something heavy in a container full of water, then there is more volume and that makes it overflow."*

HAF (11 years), contradicting HID, attributes to the lightness of the water the fact that it comes to rest lower than the alcohol: *"It's not the same liquid now."*—"Why?"—*"Maybe it weighs less"* [the water; we began with alcohol].—"Why?"—*"But it doesn't go as high. . . . It's the liquid that weighs less. It's surely the liquid since the boxes are the same. The first time* [alcohol] *the liquid was heavier, since it went higher. When the liquid is heavy, it has more weight, more pressure: it goes down faster here"* [in A].—"And here [in B] the liquid doesn't press?"—*"No, since it's this one* [in A] *that goes there"* [in B].

This reasoning is extremely clear and less contradictory than it seems, for sometimes the subject thinks of the liquid which drops in tube A, sometimes of the liquid that rises in tube B. When considering tube A, he expects the heaviest liquid to fall the most easily, with the aid of the piston, and consequently the level to be highest in B (HAF). On the other hand, in considering tube B, he thinks that the heaviest liquid rises with the most difficulty and consequently that the level will be lower (HID). In both cases, it is less a question of resistance offered by the liquid in the sense of an equal and opposite reaction in relation to the piston than a variable resistance to the rise with a variable facility for the descent; whence the idea that the piston "sinks in" to the bottom of tube A without any resistance, as HID and FRA expect.

Thus, at the last of the stages prior to formal operations, the subject can make accurate predictions of the effect of the weights on the piston and sometimes even of that of the densities of the liquids used, but he does not as yet formulate (from his predictions) any total explanation in terms of an equilibrium principle. The reason is that he lacks the operational instruments which would permit coordination of inversions and reciprocities (the

I N R C group). But there is no doubt of the fact that weight is still conceived of in absolute terms (as a "pressure" says HAF) and is not adequately related to volume although its conservation and rudimentary quantification are present. Even at this point, density gives rise to the intuition of "filled" or full but not as yet to the operational relations which will constitute the notion of compression-density found at stage III.

§ *Stage III. Reciprocity of Action and Reaction*

The best index of the appearance of the notion of reaction, or of resistance oriented in the opposite direction from that of the pressure of the weight, is the subject's attitude toward the drop of the piston in tube A. Whereas up to this point the piston was seen as sinking into a liquid without resistance and even as tending to descend with it as a function of its own weight, from this point on the descending piston is seen as meeting resistance proportional to the density of the liquid. Density, in turn, comes to be conceived of as a relationship between weight and volume—*i.e.*, as the result of a more or less great compression of matter into an equal volume. But to conceive of a resisting force, capable of equilibrating the force of the pressure according to a set of varied compensations until it stops the piston in its descent, the subject must introduce a reciprocity between the density (conceived of as a compression capable of resistance) and the pressure of the weights. In other words, the subject joins a reciprocal transformation in a single I N R C group with the inverse transformations (of adding or subtracting weights). This integration is made initially at substage III-A:

TRI (11 ; 2): *"It's the weight of the boxes. It's that* [piston] *that pushes the water."*—"And if you change the liquid, will that have an effect?" —*"Yes. Some liquids are heavier than others."*—"If you use alcohol?"— *"I think that alcohol is heavier* [simple factual error]. *So it will rise less because it's harder to move"* [resistance!]. He does the experiment without comment on our part. *"No, the alcohol is lighter."*—"Why do you think that now?"—*"Because the weight of the box makes it rise more."*—"What does that?"—*"Because the weight of the box makes it go up higher. The weight* [of the box] *can push better."*

DUM (11 ; 2). After the water, we perform the experiment with alcohol: *"It will go up less because the weight of the liquid is heavier."*— "The alcohol is lighter than the water. So?"—*"It will go up higher."* —"Why?"—*"The lighter the liquid, the more the piston will be less . . ."* [less resisted].—"Less what?"—*"Since the alcohol is lighter than the water, the piston will descend more."*—"You said 'less' "—*"No, the piston descends more. The liquid is lighter, so the piston sinks in more."*—"Why?"—*"Because the liquid is lighter than the water."*—"And it follows that the piston sinks in more?"—*"Because the piston isn't held back as much by the weight of the alcohol."*

YA (11 ; 6). Same reactions.—"Why doesn't the piston go all the way to the bottom?"—*"Because the piston no longer has enough force to bear down. It is held back because the liquid is heavier than the piston."*

RIV (13 ; 0): The water goes up to there *"because it has to come back to the same weight in both tubes."*—"But why doesn't the water rise any higher?"—*"Because the piston can't come down any more."*— "Why?"—*"Because the water holds it back."* Having pressed down on the piston with his hand, he sees that it returns to the initial position and says, *"If you press on the piston, the water has more force."*

These responses show clearly that from this point on the subject is aware of the existence of action and reaction. The weight of the liquid or, more exactly, its density (for henceforth the subjects speak only of weight relative to volume) is no longer a factor promoting the pressure of the piston but, to the contrary, is an obstacle to this pressure and thus a factor whose action is oriented in the opposite direction; thus it is a reaction. The liquid, when it is lighter than the water, actually rises more in B, because the piston, reinforced by the weights, "can push better" (TRI) or is "less held back" (DUM). With respect to tube A, on the other hand, the piston "can't go down" below a certain point "because the water holds it back" (RIV and YA).

Thus, the discovery characteristic of substage III-A is that the system involved is an equilibrium of opposed forces and no longer a one-way process. But before trying to formulate the reasoning involved, let us examine the further problems which the subjects have yet to solve—the way in which the force of the piston is transmitted and the place in which the action and reaction attain equality and come into equilibrium.

The first question is solved from substage III-A on, since the pressure of the piston is no longer considered a triggering device or an excitation of the force intrinsic to the liquid (due to its weight). Rather, it is viewed as an action exerted through the entire liquid from the beginning (descent of the liquid in tube A) up to the arrival (rise and coming to rest in tube B) and bringing on a reaction in the opposite direction resulting in a final state of equilibrium.

But at what point does equilibrium come about? Can it be localized at a particular point, or is the total quantity of the liquid involved by degrees? This particular question cannot be clarified prior to substage III-B. The reactions of this latter substage are roughly analogous to those of the preceding. As regards this question, several substage III-B subjects still imagine that there is a particular place where the opposing forces meet each other. This would be at the bottom of the rubber tube connecting the two glass tubes—*i.e.*, the lowest point of curvature of the system. In contrast, other subjects come to understand that, from the point of contact between the piston and the liquid up to the level reached by the liquid in tube B, there is action and reaction. On the one hand, the pressure of the piston makes itself felt throughout the liquid. On the other hand, since the reaction is a function of the density (which is conceived of as a compression), it surges at every point of the volume occupied by the liquid in such a way that throughout there are both action and resistance, the latter tending to repulse the action exerted on the liquid; the action and reaction are thus equal at every point.

The following are examples of each of these response types:

BOI (14 ; 6): "If you put on that box?"—"*It will rise higher because of the pressure*" [experiment].—"Why not higher?"—"*The pressures at the bottom of the tube* [of the rubber tube] *are equal.*"—"How do you know?"—"*Because the apparatus* [piston] *doesn't fall and doesn't rise, and, reciprocally, because the water neither rises nor falls.*"—"And when I put on a box?"—"*It's heavier. The result is a higher column of water* [in B] *and also a larger weight* [in A], *and it comes into equilibrium at the bottom of the tube.*"—"Why?"—"*Because the pressure is the same at the bottom of the tube.*"—"But that doesn't explain the combination of the weights?"—"*Yes, it does, because the water is dislodged by the weight* [in A]: *It comes into equilibrium at a certain moment be-*

cause the weight of the water [of the column of water in B] *increases when it rises."*—"What makes the water rise?"—*"The weight of the boxes; it makes a greater pressure at the bottom of the tube and that dislodges the water."*

IAC (14 ; 6): *"The pressure is the same* [from both sides] *at the bottom of the tube* [*cf.* BOI's conception]. *No, the water comes into equilibrium if it communicates by a tube and the pressure is transmitted in full."*—"Does the elevation play a role?"—*"No, the water will transmit the pressure the same way if both columns are high or low."* The resistance is also conceived as the same throughout.

Thus we see that the substage III-B responses (in which the influence of acquired knowledge may occasionally be perceived) add little to those of substage III-A, which are more spontaneous.

§ Conclusion. Stage III Reasoning and the Formal Operational Schema for Equilibrium

In order to analyze the formal operations needed to understand the notions of equilibrium or equality between action and reaction, one must first remember that causality is a system of operations applied to transformations in the real world in such a way that each one of these transformations can be assimilated to an operation of the subject while at the same time is conceived as accomplished by the objects themselves. Thus, we must first establish the transformations which our subjects attribute to the system under consideration and then look for the operations or logical transformations to which these real modifications are assimilated. But the principal transformations involved in the system are the four following:

I. The action exerted by the pressure of the weight of the piston and by the weights added to its own weight;

II. The suppression or diminution of this action by eliminating the weights added or the weight of the piston itself;

III. The reaction due to the resistance of the liquid, which itself constitutes a pressure, one which, however, is oriented in the opposite direction and which is dependent on the height and density of the liquid;

IV. The elimination or diminution of this resistance by eliminating part of the liquid or by substituting a liquid of less density.

But we notice that although transformations (I) and (II) can be assimilated to direct and inverse operations, transformations (III) and (IV), in contrast, are symmetrical with the first two; they too consist of operations which are direct and inverse in relation to each other, but which act in the opposite direction from the first two. Transformation (III) thus constitutes a reciprocal of transformation (I) and transformation (IV) a reciprocal of transformation (II). Whereas (IV) is the inverse of (III), in contrast (III) is not the inverse of (I), since it does not cancel it but simply neutralizes the effect by compensation. In other words, the composition of (I) and of (III) results not in the cancellation of the pressures but in their equivalence, and thus in an equilibrium.

Consequently, here again we find a mechanism isomorphic to the group of four transformations I N R C. Therefore we should not be astonished if we find this same I N R C structure in the very reasoning of the child in a form analogous to that which we have already seen at work in the experiment of the communicating vessels, although here we find it in slightly different form since in this case the reasoning bears directly on pressures and resistances:

I. The first operation consists of stating the intervention of a pressure in tube A under the influence of one weight or the other: let this be $(p \vee q)$;

II. The inverse operation consists of stating the cancellation of this action: let this be $(\bar{p}.\bar{q})$;

III. According to the stage III subjects, each pressure p or q has a corresponding resistance which we may designate by p' or q'—*i.e.*, $(p' \vee q')$—and which is expressed in the column of liquid B by the weight of the portion going beyond the level of the liquid in column A;

IV. The inverse of III will consist in stating the cancellation of p' and q'—*i.e.*, $(\bar{p}'.\bar{q}')$.

The discovery unique to stage III is that transformations (II) and (III) are reciprocal—*i.e.*, that there is compensation between them. But to hold that compensation occurs is tantamount to regarding the intervention of a resistance in B (expressed by p') as equivalent to the elimination of a pressure in A. The adolescent realizes that, without the resistances in B, the pressures exerted in A would cause the liquid to rise much higher and that to each pressure p corresponds an equal resistance p'. Thus $(p' \vee q')$ can be written in the form $(\bar{p} \vee \bar{q})$. Whence:

$$I(p \vee q)$$
$$N(\bar{p}.\bar{q})$$
$$R(\bar{p} \vee \bar{q}) \tag{1}$$
$$C(\overline{\bar{p} \vee \bar{q}}) \,(= p.q).$$

Thus one can see that transformation (III) is clearly the reciprocal R of transformation (I) and that (IV) is the correlative C— *i.e.*, the NR or the RN of transformation (I). Thus, the equivalence between pressures and resistances is expressed in two ways (positive R and negative C). Since the reciprocal $(\bar{p} \vee \bar{q})$ of the operation $(p \vee q)$ is the symmetrical operation in which the same combinations $(p.q \vee p.\bar{q} \vee \bar{p}.q)$ and $(\bar{p}.\bar{q} \vee \bar{p}.q \vee p.\bar{q})$ are found, though with a change of signs, it expresses the fact that equivalent forces oriented in opposite directions are involved, and thus the operation is distinguished from the inverse N. In other words, the inverse cancels the direct operation while the reciprocal does not cancel it; instead, it compensates it by a symmetrical operation of the same value but with a change of signs.

This distinction between the reciprocal and the inverse causes all the difficulty in the problem of action and reaction, and the subject cannot make the distinction before the formal stage. Until then, although the child clearly understands the inverse operation (*i.e.*, can cause the water in B to drop again by cutting down the weight in A) he does not understand the reciprocal operation. He conceives of it as a simple prolongation of the direct operation and not as a symmetrical operation oriented in the opposite direction and compensating for the direct operation (= the liquid does not resist the pressure but acts in the same direction). This is so because, if the subject is to understand the four transformations (I-IV) when he has not intuitively distinguished between their respective actions (as in the case of the balance), he must possess an operational mechanism made up only of formal operations. Thus, the late appearance of this discovery.

As for fixing the equilibrium at a particular point (stated in the proposition r, subject RIV describes its conditions when he says that the water in B stops at a fixed level "because it has to come back to the same weight in both tubes." In other words:

$$N \supset [x(p \vee q) = y(\bar{p} \vee \bar{q})] \qquad (2)$$

where x and y are the values assigned to $(p \vee q)$ and to $(\bar{p} \vee \bar{q})$.

Unlike those in communicating vessels, the equilibrium points do not form a horizontal line from one tube to the other but vary according to weights $p \vee q$ and to the resistances, which themselves depend on the density of the liquids. With that exception, the explanation of the phenomenon presupposes the same formal schema. Thus, it is not by chance that the mechanisms of action and reaction are discovered at the same substages, III-A and III-B. We have seen this to be true with respect both to a piston acting on a liquid and to communicating vessels. And we shall soon see that it holds for the case of the balance scale. In these three cases, an understanding of the physical processes presupposes an operational schema putting into effect simultaneously the inverse and reciprocal transformations which remain separated at the level of concrete groupings and which are linked into a whole only through the I N R C group. But in this, as in all the other cases examined up to this point, the difference between concrete and formal thinking relates to the construction of the "structured whole." It is the double reversibility characteristic of this whole (which at the same time allows for the lattice and group structures) which constitutes the I N R C group, whereas the two forms of reversibility remain separated in the elementary concrete groupings—inversion is found only in class groupings and reciprocity in relational groupings.

11

Equilibrium in the Balance[1]

IN A PROBLEM using a simple balance-type weighing instrument, a seesaw balance, we again find the operational schema of equilibrium between action and reaction. But the experiment was set up in a way that would force the question of proportionality. When two unequal weights W and W' are balanced at unequal distances from an axis L and L', the amounts of work WH and $W'H'$ needed to move them to heights H and H' corresponding to these distances are equal. Thus, we have the double (inverse) proportion:

$$W/W' = L'/L = H'/H$$

The result is that finding the law presupposes the construction of the proportion $W/W' = L'/L$ and spelling out its explanation implies an understanding of the proportion $W/W' = H'/H$. It seemed to us that it would be interesting to study how this proportionality schema develops as it is linked with the equilibrium schema. As a result of previous research we know that in all realms (space, speed, chance, etc.) the notion of proportions does not appear until formal substage III-A. Now we are going to find out why this is so.

[1] With the collaboration of F. Matthieu, former research assistant, Institut des Sciences de l'Éducation, and J. Nicolas. In reference to the same subject, see the previous study of Mme. Refia Ugurel-Semin, *Istanbul University Yayinlari* (1940), 110, pp. 77–211.

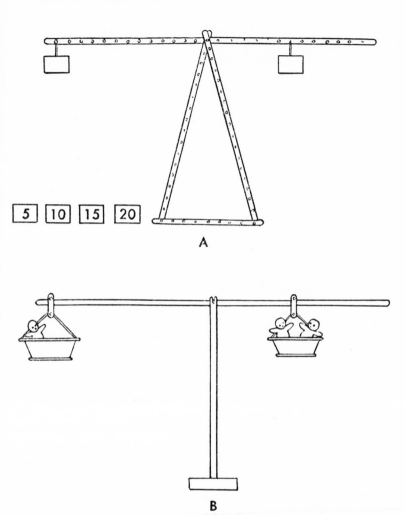

FIG. 9. The balance scale is here shown in two forms: (A) a conventional balance with varying weights which can be hung at different points along the crossbar; (B) a balance equipped with baskets which can be moved along the crossbar to different points and in which dolls are used as weights.

§   *Stage I. Failure to Distinguish Between the Subject's Action and the External Process (I-A) Followed by Integration of Intuitions in the Direction of the Compensation of Weights (I-B)*

From about 3 to 5 years, subjects give responses which, given our interests, are instructive. As we have said before, in general causality is an assimilation of external processes either to the subject's own actions or to his operations, but with the delegation of one or the other to reality itself. In the case of an apparatus such as a balance scale, the notion of an equilibrium between one's body weight and other weights is constructed very early, but the notion is undifferentiated and extends beyond weight itself to include the muscular force of an upward or even a downward push. (Moreover, the weight is thought to be linked with the actions of pushing up or pressing down.) The balance is first assimilated to this sort of undifferentiated action and not to a system of compensation operations between weights nor *a fortiori* to weight × length. In fact, no form of concrete operations exists at this level for there are only representational regulations—*i.e.*, instruments of global compensation without systematic reversibility. The result of this situation is that the substage I-A subjects cannot guarantee equilibrium simply by distributing weights but intrude in the working of the apparatus with their *own* actions, which they fail to distinguish from the actions of the *objects* that they are trying to control.

MIC (4 ; 6), presented with two equal weights at distances of 14 and 9: "Why is one way down and the other up high?"—He continually raises and lowers the arms of the apparatus, believing that they will maintain the forces and positions he delegates to them.—"Can you make it straight [horizontal gesture] so it will stay there all by itself?"—Neither "yes" or "no."—"How was it before?"—"*Like that*" [horizontal]—"You can't do it with the weights?"—He shakes his head and tries to maintain the horizontal position with two unequal weights, raising and lowering the arms several times.—"Can you do it without your hand?" —[We have him weigh the weights with his hands, then he works at a new set of trials. We suggest that he add weight to one side or the other, etc.] Conclusion: "*You can't!*" [attain the horizontal position].

MAR (4 ; 8) suspends two weights on one side without putting anything on the other, with the aim of reaching horizontal equilibrium!

One can see that, in constantly interfering with the apparatus in order to correct the position of the balance arm, the subject expects the apparatus to conserve the results of his manipulations. Thus, the instrument and his own actions are not distinguished. But, although barring the notion that the balance constitutes an independent equilibrium, the lack of differentiation does not preclude his making predictions about some more or less constant effects. It is true that for our purposes the most striking aspect of these predictions is their negative aspect. For example, at this level the child does not yet think that equilibrium implies the equality of weights (even at equal distances); thus MAR puts two weights on one side and none on the other in order to attain the horizontal. The heavy side moves upward and the light one downward, as well as the reverse. The relationship between weights is not formulated; the epithet "too heavy" may be applied to a single weight suspended to one arm without its counterpart, "too light," being used. There is no conservation of weight. The subject tries constantly to repeat with new weights what he has just accomplished by chance with others, without paying any attention to differences in weight. However, through improved regulation these subjects come to see that weight has a relative influence. Generally they suspend at least one weight at each side for purposes of symmetry. Often they add new weights to the others to improve the equilibrium, but they add them not to the side where weight is lacking (which would tend to equalize the weights) but to the side where the weight already is largest with the idea that several weights will improve the situation.

But the adding characteristic of this level is not yet operational. Although it does constitute the beginning of the additive operation, the operation is not achieved because of lack of equalization between parts $(A + A')$ and the whole B (compensating $A + A'$ on the other arm). Most important, it is not an operation, because reversibility is lacking; at this point elements are not removed with the deliberate aim of equalizing the weights. When the subject removes a weight, it is only in order to try a new and different course of action after earlier attempts have failed.

Generally, the subject is not concerned with the question of the distances from the axis and does not look for any equality or coordination between the distances and the weight. Nevertheless, operations may start to take form here in that the subject may establish a preliminary form of symmetry. However, once again this is not an operation in the strict sense. First, coordination with the weight is lacking; secondly, this symmetry is related generally to the two extremities of the arms and does not include equalizations for the intermediate distances.

In contrast, from about 5 to 7–8 years (substage I-B) one can see increasing integration of these intuitive representations moving toward reversible operations.

MAL (5 ; 8) notes that the arms are not horizontal: *"You have to put another* [weight] *on the other side. I know what has to be done; put still another one there because there isn't any weight here* [she adds it]. *These here must be lighter than those over there. You have to take two that have the same weight."* Next: *"You could take one off"* [because it is too heavy on one side]. MAL does not spontaneously discover the influence of distance, but when a weight is moved in front of her she says: *"You brought that one up closer, that makes more weight. If it were at the end, it wouldn't work and there it makes more weight."*

GAS (5 ; 9): *"You could put one at the other side: the same* [he takes a weight of the same color but having a very different weight]. *That doesn't work: maybe there is a little too much weight there."*

Thus, from this point on the child understands that weight is needed on both sides to achieve a balance and even that the weights should be approximately equal. But he does not yet know how to proceed toward this equalization in a systematic way. Similarly, henceforth he succeeds in adding and subtracting, but without accurate equalizations. His actions are successive corrections, (thus regulations) and are not yet strictly reversible.

We see how these two sorts of regulations—by equalizations and by addition or subtraction—furnish the starting point for future transformations by reciprocity (symmetries) and by inversion, relative to the weight. As for the distances, there is progress in the tendency toward symmetry (the weights are no longer put at equal distances only at the extremities but also in the region close to the axis). Sometimes the subject discovers the role played by changes in distance (*cf.* MAL). But there are as yet no systematic correspondences of the type further = heavier.

§ Substage II-A. Concrete Operations Performed on Weight and Distance but Without Systematic Coordination Between Them

From this point on, weights are equalized and added exactly, while distances are added and made symmetrical. But coordination between weights and distances as yet goes no further than intuitive regulations. The subject discovers by trial-and-error that equilibrium between a smaller weight at a greater distance and a greater weight at a smaller distance is possible, but he does not yet draw out general correspondences: [2]

MAS (7 ; 7) begins with E 3 and D 3, then replaces them with G 3 and F 3 (thus equal distance and an attempt to find equal weights), adds two other weights, takes off some, then all, and finally weighs two equal weights [E] in his hands, counts equal numbers of holes [14] and places E 14 at each side. Afterwards he looks for other forms of equilibrium; he adds the weights, moves them, takes off some, and finally has GED on one side and P 3 on the other: "*That's it* [empirical compensation of weight and distances]. *It's just like when there weren't any* [when the arms were horizontal without weight]; *it's the same weight on each side.*" He begins again with large weights [for which there are no matched pairs]. "*I should have put one on each side. Since there aren't any, I had to put three on one side and two on the other. It stays straight because it's the same weight on each side.*" He predicts that equal distances are necessary for two unequal weights, but he does not find the law: heavier $\rightleftarrows$ nearer.—"If you put on C and E, where would you have to place them?"—"*I would say one hole and another hole* [= two different distances], *but they shouldn't go the same way* [at equal distances] *or it wouldn't make the same weight.*"

[2] From now on we will indicate objects of increasing weight by the letters *A, B, C,* etc. Increasing distances (which are measured for the child at three equidistant points where the hooks for the weights are attached) are indicated by the numbers 1, 2, 3, etc.

NEM (7 ; 4) discovers empirically that C on the left at a distance of 10 balances E on the right at a distance of 5. We ask him to place C on the right and E on the left, but he does not succeed in inverting the distance relationship. After the experiment, he exclaims, "*Ah! You have to do the same thing as before but in the opposite way!*"

Thus, from this point on the subject can order serially the weights he comes across as well as determine whether they are equal. He can add them in a reversible manner and correctly compare one pair of weights with another pair. What is more, he knows how to make use of the transitiveness of the relations of the equality or inequality of the weights. Moreover, all these operations reappear when he compares distances, but with the additional correspondence between distances oriented in opposite directions (symmetries relative to the axis).

Applied to the problems of the balance, these operations allow subjects to obtain the following results (by logical multiplication of relations):

Two equal weights B_1 and B_2 situated at equal distances L_x come into equilibrium by symmetry:

$$(B_1 \times L_x) = (B_2 \times L_x). \tag{1}$$

Thus, one of the weights is conceived of as compensating the other by reciprocity. Two equal weights B_1 and B_2 at unequal distances L_x and L_y do not balance each other:

$$(B_1 \times L_x) \gtrless (B_2 \times L_y) \text{ if } x \gtrless y. \tag{2}$$

Two equal weights B_1 and B_2 at unequal distances L_x and L_y do not come into equilibrium either:

$$(A_1 \times L_x) < (B_2 \times L_x). \tag{3}$$

Moreover, in each one of these relations the subject can substitute for one object an equivalent set of others through additive operations:

$$C_1 = (A_2 + A'_2 + B'_2) \tag{4}$$

and the same holds for distances.

On the other hand, in the case of unequal weights A_1 and B_2 and of unequal distances L_x and L_y, coordination is not yet possible at substage II-A. Even when the subject discovers by experi-

mentation that a large weight at a small distance to the right of
the axis balances a small weight at a large distance to the left, he
does not know how to invert these relations from one side to the
other and discovers too late that he should have "done the same
thing but in the opposite way" (NEM).

§   *Substage II-B. Inverse Correspondence of
     Weights and Distances*

The example just described (unequal weights and distances) is
resolved at substage II-B, not yet by metric proportions (with
the occasional exception of the relationship between 1 and 2) but
by qualitative correspondences bordering on the equilibrium
law: "The heavier it is, the closer to the middle."

FIS (10 ; 7) sees that P does not balance F "*because it's heavy: that
one* [F] *is too light.*"—"What should be done?"—"*Move it forward*
[he moves P toward the axis and attains equilibrium]. *I had to pull it
back from 16 holes* [arbitrary] *to see if it would lower twice* [arbitrary]
the weight."—"What do you mean by that?"—"*It raises the weight.*"—
"And if you put it back over there?" [moves P away].—"*It would make
the other one go up.*"—"And if you put it at the end?" [P].—"*It would
go up still more*" [F], etc. Conclusion: When you have two unequal
weights "*you move up the heaviest*" [toward the median axis]. But FIS
does not measure the lengths even for the relations of 1 to 2.

ROL (10 ; 10): "*You have to change the position of the sack because
at the end it makes more weight.*" He moves the lightest away from
the axis: "*No, it's heavier.*" He is presented with G at 2 and A at 14:
they balance "*because that one is there* [A at 14] *and it is less heavy
than the other one.*"

The difference between these reactions and those of substage
II-A is clear. At the earlier stage, when the subject comes across
two weights which do not come into equilibrium, he works mostly
with substitutions—additions or subtractions. In this way he
achieves certain equalizations by displacement, but only excep-
tionally and by groping about (regulations). On the other hand,
at the present stage the subject who comes to two unequal weights
tries to balance them by means of an oriented displacement on the

hypothesis that the same object "will weigh more" at a greater distance from the axis and less when brought closer to it. He is working toward the law, but without metrical proportions and by simple qualitative correspondences.

Thus, the new operation mediating the determination of the conditions of equilibrium is a double serial ordering of weights $A < B < C < \ldots$ and distances $L_1 > L_2 > L_3 > \ldots$ but with bi-univocal inverse correspondences:

$$
\begin{array}{ccc}
A & < B & < C & < \ldots \\
\updownarrow & \updownarrow & \updownarrow & \\
L_1 & > L_2 & > L_3 & > \ldots
\end{array}
\qquad (5)
$$

which can be translated into reciprocities (expressed in the language of relational multiplication):

$$(A \times L_1) = (B \times L_2) = (C \times L_3) = \ldots, \text{ etc.} \qquad (6)$$

But it is clear that such qualitative operations are inadequate to establish the law. The logical multiplications of type (6) allow some inferences but leave certain cases indeterminate:

$$
\begin{aligned}
&\text{heavier} \times \text{same distance} = \text{greater force,} \\
&\text{less heavy} \times \text{same distance} = \text{less great force,} \\
&\text{same weight} \times \text{further (from the axis)} = \text{greater force,} \\
&\text{same weight} \times \text{less far} = \text{less great force; but} \\
&\text{heavier} \times \text{further} = \text{indeterminate,} \\
&\text{less heavy} \times \text{less far} = \text{indeterminate; and} \\
&\text{heavier} \times \text{less far} = \text{less heavy} \times \text{further}
\end{aligned}
\qquad (7)
$$

(but only under certain metrical conditions).

However, at this level the subject can quantify the weights (he knows that $B = 2A$; etc.) as well as the distances (measurable by the number of holes). Given these facts, why must we await formal stage III before the schema of proportions is organized? We might say that it is a matter of book-learning, but, in contradiction, we are able to present some examples (analogous to those which we have already published elsewhere [3]) in which the

[3] See Piaget and Inhelder, *The Child's Conception of Space*, Chap. XII; No. 9; Piaget, *Les Notions de mouvement et de vitesse chez l'enfant*, Chap. IX, nos. 2 and 3; Piaget and Inhelder, *La Genèse de l'idée de hasard chez l'enfant*, Chap. VI, nos. 5 and 6.

proportionality schema is organized before any academic knowledge enters. Thus, it is probable that this schema requires, as a necessary and sufficient condition, a qualitative operational system that is both differentiated and unified, analogous to the I N R C group. This hypothesis is even more plausible when we consider that in this particular case a set of balanced actions and reactions is involved similar to that whose understanding we have analyzed in Chaps. 9 and 10.

§ *Stage III. Discovery and Explanation of the Law*

When the experimenter restricts himself to a procedure such as the foregoing, where the subject is allowed to hang the weights simultaneously on the two arms of the balance, subjects start to discover the law at substage III-A. It takes the form of the proposition $W/W' = L'/L$ (where W and W' are two unequal weights and L and L' the distances at which they are placed); this law is so immediately obvious that it does not give rise to a particular causal explanation even during substage III-B. ("It's a system of compensations," as CHAL will tell us.) But, when the experiment proceeds by successive and alternate suspensions of the weights, the subject's attention turns to the inclinations and the distances in height to be covered; this may lead him to an explanation in terms of equal amounts of work (displacement of forces). It is true that, although this explanation is already possible at substage III-A, it only rarely appears before substage III-B. Nevertheless we have observed it in several cases and think it worth analyzing.

First, we will present a case of the discovery of the law at substage III-A:

ROG (12 ; 11): for a weight P placed at the very tip of one arm [28 holes], he puts $C + E$ in the middle of the other arm, measures the distances, and says: "*That makes 14 holes. It's half the length. If the weight [C + E] is halved, that duplicates*" [P].—"How do you know that you have to bring the weight toward the center?" [to increase the weight].—"*The idea just came to me, I wanted to try. If I bring it in half way, the value of the weight is cut in half. I know, but I can't explain it. I haven't learned.*"—"Do you know other similar situations?"

—*"In the game of marbles, if five play against four, the last one of the four has the right to an extra marble."* He also discovers that for two distances of 1 and 1/4 you have to use weights 1 and 4; that for two distances of 1 and 1/3 you need weights 1 and 3, etc.: *"You put the heaviest weight on the portion that stands for the lightest weight* [which corresponds to the lightest weight], *going from the center."*

The rapidity with which the subject makes the transition from the qualitative correspondence to the metrical proportion seems at first to indicate the presence of an anticipatory schema. However, the analogy that the subject established between the balance and the game of marbles shows that this schema is taken from notions of reciprocity or of compensation. So we have to examine how, starting with substage III-B, the subjects proceed from the same conception to a search for an *explanation* in the strict sense of the term (with the apparatus using alternate suspensions):

CHAL (13 ; 6) quickly discovers that *"the greater the distance, the smaller the weight should be. It's staying up."*—"Why?"—*"It is compensated there and there."*—"What is compensated?"—*"The distances and the weights; it's a system of compensations. Each one rises in turn. For equal distances you need equal weights, and if it's inclined it rights itself and goes down on the other side."*—[We propose a test with two weight units at a distance of 5 and one at a distance of 10.] "What will the angles be?"—*"Larger on one side* [he points out the two-unit side] *and smaller on the other* [experiment]. *Oh! No, the same angles!"* He outlines them: *"The distance compensates for the weight."*—"What distances do they cover?" [heights H and H' are pointed out].—*"The smallest weight covers a greater distance and the large weight a shorter distance."*—"And what forces are required?" [strings which can be used to raise and lower the weights are pointed out].—*"For the smallest, there is more distance to pull, for the large one, less distance."*—"So where is more force required?"—*"Here* [two units]. *Oh! No, it's the same: the distance* [he is speaking of height] *is compensated by the weight."*

SAM (13 ; 8) discovers immediately that the horizontal distance is inversely related to weight.—"How do you explain that?"—*"You need more force to raise weights placed at the extremes than when it's closer to the center . . . because it has to cover a greater distance."*—"How do you know?"—*"If one weight on the balance is three times the other, you put it a third of the way out because the distance* [upward] *it goes is three times less.*"—"But once you referred to the distance [horizontal gesture] and once to the path covered?"—"*Oh, that depends on whether you have to calculate it or whether you really understand it. If you want to calculate, it's best to consider it horizontally; if you want to understand it, vertically is better. For the light one* [at the extremity] *it changes more quickly, for the heavy one less quickly.*"

TIS (13 ; 8) discovers the proportion 1 to 2 and shows the heights: "*If I replaced this weight* [one unit] *with that one* [two units], *it would only go halfway up* . . . [the distance in height] *is much longer when it is at the end of the arm than when it is in the middle.*"—"Does compensation take place?"—"*Yes, between the force and the height.*"— "How can you measure it?"—"*It's easier to measure the height, but it's really the same*" [as the horizontal distance].

These reactions, found at both substages of stage III, bring us back to the now familiar schemata of the I N R C group and in the same form that we found in Chaps. 9 and 10. But, above all, they show us how the general equilibrium schema is differentiated in the present case by constructing the proportions $W/W' = L'/L$ and $W/W' = H'/H$. Thus we have two questions to discuss— first, how is the proportional schema organized; second, how does it relate to the I N R C group?

In these responses the I N R C group first appears in a form which we could have described earlier when dealing with the problem of the oscillations of a liquid in communicating vessels (Chap. 9). One of the arms of a balance will lower when a weight is hung on it at a given distance from the axis; when an equal weight is placed on the other arm in a symmetrical position (= at the same distance from the axis as the first weight), this second arm will lower. "One goes up and then the other," says CHAL, "and if it's inclined [below the horizontal plane] it comes back to the middle and goes down on the other side."

In other words, a reciprocal relation operates in this case $(p \supset \bar{q}) = \mathrm{R}(q \supset \bar{p})$ in which p and q stand for the upward motion of the arms. But there is something new in the case of the balance: two factors are operative and they compensate each other; operating alone, a weight W at a distance L produces the same inclination as a weight $W' = nW$ at a distance $L' = L/n$. CHAL is astonished by this fact ("the same angles"), then finds it quite natural because "the distance is compensated by the weight."

The I N R C group reappears in the same form as in the problem of pressure and resistance in the equilibrium of liquids (prop. [8], Chap. 9 and prop. [1], Chap. 10). Two kinds of operations for reestablishing equilibrium can correspond to the operation in which a weight is placed on one of the arms at a given distance—the inverse N which consists of taking off this weight or the reciprocal R which consists of putting on equal weight at an equal distance on the other arm of the balance. Moreover, whereas the inverse N cancels the original operation, the reciprocal R compensates it without canceling it; still, N and R have the same final result—*i.e.*, they bring the arms back into the horizontal plane. It is not at all surprising that the transformations described in this connection (prop. [8], Chap. 9, and prop. [1], Chap. 10) reappear in the present context, for they are based on an extremely simple intuition already acquired at stage II through qualitative correspondences. But once again there is the additional fact specific to the balance—distances compensate weights.

Thus in both forms—*i.e.*, as it relates to pressures and resistances, or to oscillations and inclinations—the I N R C group doubles as a proportional schema: an inverse proportion of horizontal distances and weights $W/W' = L'/L$ in the case of pressures and resistances and an inverse proportion of heights and weights $W/W' = H'/H$ in the case of inclinations. There is a third proportion—it is direct rather than inverse ($L/L' = H/H'$), of a purely geometric character and obvious to our subjects (*cf.* TIS: "It's really the same" whether you measure horizontal distances or height). So our problem is to establish how our subjects construct the first two proportions. Is the construction done independently and by a direct structuring of the empirical data, or is it linked to the operational schema of equilibrium based on the I N R C group?

§   *The Proportional Schema and the I N R C Group*

First, we should remember that an understanding of proportions does not appear until substage III-A; this is true in all spheres and not only in the balance scale experiments. During substage

II-B it has often been noted that subjects search for a common denominator of the two relations that they compare, but this common relation is thought to be additive. Thus, instead of the proportion $W/W' = L'/L$ one would have an equality of differences $W - W' = L' - L$. Clearly, the formation of the notion of proportions presupposes that simple relations of difference be substituted for the notion of the equality of products $WL - W'L'$. But we must also note that the transition from the difference to the product rarely takes place from the start in a form that is metrical. The numerical quantification of the proportion is usually preceded by a qualitative schema based on a conception of logical product—*i.e.*, by the idea that two factors acting together are equivalent to the action of two other factors added together. "The larger the distance, the smaller the weight," says CHAL, using simple qualitative correspondence (*cf.* prop. [5]). But he adds, "They go together."

In other words a small weight combined with a great distance is equivalent to a large weight with a small distance. These logical multiplications are outlined at substage II-B (*cf.* props. [6] and [7]), but the subjects fail to generalize to all possible cases. Where does the generalization found at substages III-A and III-B come from? Without doubt, this is where the notions of compensation and reciprocity connected with the I N R C group come in.

It is clear that when the subject at stage III becomes able to understand transformations by inversion (N) and reciprocity (R) and to group them into a single system (I, N, R, and $NR = C$), by the same token he becomes able to make use of the equality of products in a more general form than in the multiplication of relations (6) and (7). Moreover, this form already implies the notions of compensation and cancellation. The possibility of reasoning in terms of a group structure—I N R C—indicates an understanding of the equalities $NR = IC$, $RC = IN$, $NC = IR$, etc., the equalities between the products of two transformations. The result is that the I N R C group is itself equivalent to a system of logical proportions:

$$\frac{Ix}{Cx} = \frac{Rx}{Nx} \text{ or } \frac{Rx}{Ix} = \frac{Cx}{Nx}$$

since $IN = RC$ (where $x =$ the operation transformed by I, N, R, or C).

For example, let us examine the subjects' reasoning on the changes of weight and horizontal distance (to simplify notation we shall disregard the constant weights and distances). Let p be the statement of a fixed increase of weight and q of a fixed increase of distance; let us call $\bar{p}$ and $\bar{q}$ the propositions stating a corresponding diminution of weight and distance on the same arm of the balance. Propositions p' and q' correspond to p and q, and $\bar{p}'$ and $\bar{q}'$ correspond to p' and q' on the other arm. By a process isomorphic to prop. (1) of Chap. 10, the subjects understand the following relations of inversion and reciprocity (the I N R C group but with $p.q$ chosen as the identical operation I):

I $(p.q) =$ to increase simultaneously the weight and the distance on one of the arms;

N $(\bar{p} \text{ v } \bar{q}) = (p.\bar{q}) \text{ v } (\bar{p}.q) \text{ v } (\bar{p}.\bar{q}) =$ to reduce the distance while increasing the weight or diminish the weight while increasing the distance or diminish both;   (8)

R $(p'.q')$ compensates I by increasing both weight and distance on the other arm of the balance;

C $(\bar{p}' \text{ v } \bar{q}') = (p'.\bar{q}') \text{ v } (\bar{p}'.q') \text{ v } (\bar{p}'.\bar{q}') =$ cancels R in the same way that N cancels I.

But, since R $(p'.q')$ is equivalent to compensating action I $(p.q)$ with a reaction (symmetry) on the other arm of the balance, we find that it can be written $\bar{p}.\bar{q}$; and since $(\bar{p}' \text{ v } \bar{q}')$ is also equivalent to compensating the action N by symmetry, we can write it $(p \text{ v } q)$. Therefore proposition (8) can be formulated as follows:

$$\begin{array}{l} \text{I } (p.q) \\ \text{N } (\bar{p} \text{ v } \bar{q}) \\ \text{R } (\bar{p}.\bar{q}) \\ \text{C } (p \text{ v } q). \end{array} \qquad (8a)$$

The system of these transformations, which states only the equilibrium of weights and distances, is in itself equivalent to the proportionality: [4]

---

[4] This logical proportion signifies the following:

$$\begin{array}{llll} (p.q).(\bar{p} \text{ v } q) = (\bar{p}.\bar{q}).(p \text{ v } q) & = \text{o} & \text{for I.N} & = \text{R.C} & \text{(a)} \\ (p.q) \text{ v } (\bar{p} \text{ v } \bar{q}) = (\bar{p}.\bar{q}) \text{ v } (p \text{ v } q) & = (p \circ q) & \text{for I v N} & = \text{R v C} & \text{(b)} \\ (p.q).(\overline{\bar{p}.\bar{q}}) = (p \text{ v } q).(\overline{\bar{p} \text{ v } \bar{q}}) & = (p.q) & \text{for I.(NR)} & = \text{C.(NN)} & \text{(c)} \\ (p.q).(\overline{p \text{ v } q}) = \bar{p}.\bar{q}.(\overline{\bar{p} \text{ v } \bar{q}}) & = \text{o} & \text{for I.(NC)} & = \text{R.(NN)} & \text{(d)} \end{array}$$

$$\frac{p.q}{\bar{p}.\bar{q}} = \frac{p \vee q}{\bar{p} \vee \bar{q}} \text{ thus } \frac{\text{I}x}{\text{R}x} = \frac{\text{C}x}{\text{N}x} \text{ (where } x = p.q\text{).} \qquad (9)$$

In other words, an understanding of the system of inversions and reciprocities (8) and (8a) follows directly from an understanding of this proportional relation; an increase of weight and distance on one arm of the balance is to the symmetrical increase on the other arm as an increase of weight or distance on one arm is to a reciprocal operation on the other.

Undoubtedly, this qualitative schema of logical proportions corresponds to the global intuition of proportionality with which the subject begins. And it is easy to pass on from this qualitative schema to more detailed logical proportions (involving a single proposition) and from there to numerical proportions.

In this respect, remember that, for a single proposition p, the correlative C is identical with I and the reciprocal R identical with N. From proportion (9) one can construct:

$$\frac{p}{q} = \frac{\bar{q}}{\bar{p}}, \text{ whence } p \vee \bar{p} = q \vee \bar{q} . \qquad (10)$$

In other words, the increase of weight is to the increase of distance as the decrease of distance is to the decrease of weight.

Secondly, beyond the direct proportions of types (9) and (10) the I N R C group includes what can be called reciprocal proportions, where one of the cross-products is the reciprocal R of the other:

$$\frac{p.q}{\bar{p}.\bar{q}} = \text{R} \frac{\bar{p} \vee \bar{q}}{p \vee q} , \text{ thus}$$
$$[(p.q).(p \vee q) = p.q] =$$
$$\text{R} [(\bar{p}.\bar{q}).(\bar{p} \vee \bar{q}) = \bar{p}.\bar{q}]. \qquad (11)$$

Hence, by virtue of (10) and (11), the reciprocal proportion:

$$\frac{p}{q} = \text{R} \frac{\bar{p}}{\bar{q}}, \text{ thus } (p.\bar{q}) = \text{R} (\bar{p}.q). \qquad (12)$$

The formulae demonstrate that the two logical proportions (11) and (12) are isomorphic to the numerical propositions which can be obtained by giving the same coefficient n either to an increase in weight (p) or to an increase in the distance (q). In other words, if $p = nW$ and $q = nL$, then:

$$\frac{p}{q} = \frac{\bar{q}}{\bar{p}} \text{ corresponds to } \frac{nW}{nL} = \frac{n:L}{n:W}, \text{ for example}$$

$$\frac{2 \times 4}{2 \times 8} = \frac{2:8}{2:4};$$

(13)

and

$$\frac{p}{q} = R\,\frac{\bar{p}}{\bar{q}} \text{ corresponds to } \frac{nW}{nL} = \frac{W:n}{L:n}, \text{ for example}$$

$$\frac{2 \times 4}{2 \times 8} = \frac{4:2}{8:2}.$$

(14)

Formulae (9) to (14) may seem much too abstract to account for the actual reasoning of our subjects. Actually, this is in part an independent result of the symbolism which we have introduced; nevertheless this is how proportions are discovered. Before introducing numbers as measurements for weight and distance, the subject usually begins by assuming:

$$p.\bar{q} = R\,(\bar{p}.q) \tag{15}$$

(increasing the weight and reducing the distance on one of the arms is the same as reducing the weight and increasing the distance on the other arm).

However, proposition (15) is none other than proportion (12), which then implies (10) and (9) and leads to metrical proportion (14). Thus we are justified in considering the preceding formulae symbolic expressions of the actual reasoning of our subjects.

As for the proportion between weight and height, as soon as they encounter alternating suspensions in the apparatus all the subjects understand that an increase in distance (q) implies a determinate increase in height (r), thus:

$$q \gtreqless r. \tag{16}$$

Consequently proportions (10) and (12) imply:

$$\frac{p}{r} = \frac{\bar{r}}{\bar{p}}, \tag{17}$$

and

$$\frac{p}{r} = R\,\frac{\bar{p}}{\bar{r}}. \tag{17a}$$

Finally, the transfer of a weight to a higher point constitutes work. This is expressed by our subjects in their own words, as they do not have the technical vocabulary of physics at their disposal: "There is more distance to pull" (CHAL) or "More force to raise the weight" (SAM). Actually, if a heavy weight hung at a small distance from the axis balances a weight n times smaller at a distance n times larger, it is because the same amount of work is needed to raise the first to a given height and to raise the second to a level n times higher than that height. As FIS says, there is compensation "between the force and the height." This idea of an equivalent amount of work, half-understood during stage III, provides the explanation of the phenomenon of equilibrium. However, since the reaction of these subjects is not completely spontaneous on this point, we must turn to the next experiment. There we replaced the overly-simple apparatus of the balance scale with one for hauling a weight on an inclined plane; we can see from this experiment how the concept of work is elaborated beginning with the concrete substage II-B; and we can see how it is used in the explanations of the formal stage III.

12

Hauling Weight
on an Inclined Plane[1]

ONCE AGAIN our subjects are given an equilibrium problem; one
not too different from the balance problem but especially designed to bring out work relationships. A toy dumping wagon is
drawn along a rail whose inclination can be varied. The task is to
predict the movements or equilibrium position of the wagon as a
function of three variables—the weight it carries, the counterweight suspended by a cable fastened to the wagon, and the inclination of the track. This last variable is calculated not in terms
of its angle measured in degrees, but in terms of its *sine—i.e.*, of
the (variable) height h. Thus, the law of equilibrium to be found
is $W/M = h/H$, where W is the (variable) counterweight, M the
weight of the toy wagon (which itself weighs 4 units, but which
can be loaded with varying amounts of weight), and H the total
height (the unvarying length of the track assuming it is held
vertically).

§ *Stage I. Failure to Distinguish Between One's Own*
   Actions and Objective Processes

At stage I the subject is most likely to explain the situation in
terms of the totality of actions which he can perform on the
apparatus:

[1] With the collaboration of A. Morf, H. Olivieri, former research assistant,
Institut des Sciences de l'Éducation, G. Mercier, and D. Royo.

BAC (6 years) pushes the wagon to make it descend: "It goes down?"—
"*No, it goes up.*"—"Can you do anything else?"—"*Push it by hand.*"
—"And to make it go up?"—"*You drive in the train.*"—"And with the
weights?"—He loads the counterweight. "*I put something on.*"—"Why
does it go up?"—"*I don't know. Because it's heavy.*"—"And to make it
go down?"—"*I don't know. You could push it.*"

HER (6 years): "What can you do?"—"*Make the wagon go.*"—"How?"—
"*With the chain*" [he pulls].—"And to make it go all by itself?"—"*Take
off the weights* [at W]; *put the weights on the wagon*" [he takes off
two units at W, replacing them at M].—"What else could you do?"—
"*Put the track up higher*" [he puts it at 45°].—"Will it go up?"—"*No,*

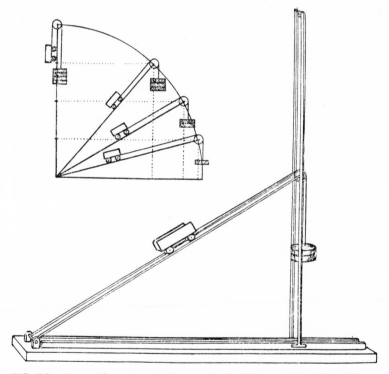

FIG. 10.   A toy dumping wagon, suspended by a cable, is hauled up
the inclined plane by the counterweights at the other end of the cable.
The counterweights can be varied and the angle of the plane is adjustable; weights placed in the wagon provide the third variable.

it will go down [it goes up]. *You have to push it.*"—"If you put it higher, will it go up?"—"*No, it will go down because it is slanted.*"— "What do you have to do to make it go up?"—"*Pull* [the rail is lowered]. *You have to push it*" [he pushes it by hand].

BEL (6 ; 6): *"That's* [W] *to pull with.*"—"How can you make it go?"— *"You have to lower that*" [he reduces the inclination].—"And what will the wagon do?"—"*It will go down* [experiment]. *It goes up!*"—"Why?" *"Because it's up high.*"—"And to make it go down?"—"*You have to raise the chain* [it is raised, and one unit of weight is put on the wagon and two units at W]. *It goes up even more! It can't go down. . . .*"

For these subjects, the apparatus is not yet seen as an independent set of causes and effects, but is still assimilated to the actions which they perform. There are two complementary senses in which this is true. First, the subject does not try to isolate relationships external to his actions, but locates his own roles in the same dimension as objective causes; second, the causes themselves are still conceptualized by assimilation to a motivational model. Thus, weight is conceived of as a force which can push or pull, etc. But it is also true that, at all levels of development, causality is an assimilation of transformations of reality to the subject's actions or operations with delegation of their power to the real world. In other words, when the subject has reached a certain operational stage of development, modifications of reality are conceived of as isomorphic to the operational transformations effected. But when the subject's activity consists of irreversible actions which are not as yet coordinated into systems of operations, then *reality* is represented as a set of equally uncoordinated forces which cannot possibly be differentiated from one's own actions.

§   *Substage II-A. Determination of the Role of the Weights Without Operational Coordination with the Inclinations*

The subject begins to relate the weights in the toy wagon to the counterweight because the two are homogeneous factors. He is also aware of the fact that the inclination of the track plays a role, but he cannot as yet coordinate it with that of the weight:

GOD (7 years): "What do you have to do to make the wagon go?"—
"*Pull.*"—"And if you don't pull?"—"*Well, you can push it.*"—"And if you
don't push."—"*Take off the load*" [he removes two units of weight from
the wagon which immediately goes up].—"Is there anything else?"—
"*Yes, you can put on a weight*" [he puts the two units at W].—"What
else?" [the subject is shown that the rail can be moved].—"*You have
to put it a little lower down.*"—"O.K. Now, what will happen if I add
here?" [two units at W].—"*The wagon will go up.*"—"And if I put the
weights in the wagon?"—"*It won't move.*"—[Experiment.] "So what do
you do to make it go up?"—"*You have to add another weight*" [he adds
one at W].—"And if I can't do that?"—"*Oh, you have to take some off
here*" [M].—The equilibrium is achieved for a given inclination: "And
if the track is lowered?"—"*I don't think it will move*" [the weights
are in equilibrium independently of inclination].—Experiment: "*It goes
up.*"—"Why?"—"*Because there were several weights here*" [W].

FER (7 ; 10): Same beginning; then, in order to make the wagon go up
one must "*take off some weights* [M] *and put on some weights*" [W].
—"And if I take off only one weight without adding any?"—"*It won't
move* [experiment]. *It moves!*"—"And to make it go down, what can
you do?"—"*Raise the chain: lift up the chain and take off a weight*"
[W]. But in making predictions FER takes only the weight relationships
into account, as if the equilibrium remained the same for given weights
independently of the inclination.

There is marked progress over stage I. A reversible composition
of weights appears; to add one unit to the counterweight W is
equivalent to removing a weight of the same value from the
wagon M, etc. But these are only simple compositions based on
the assumption that equilibrium between M and W is assured by
a simple weight equivalence, as in the case of a balance (when the
distances from the axis are equal). However, inclination is seen
as playing a partial role; the subject predicts that steepening the
slope works in favor of descent. However, he does not understand
that steepening the slope automatically reduces the effect of the
counterweight and that more weight is required at W in order to
raise the wagon on a steeper slope than on a gentler one. Thus
inclination is a secondary factor which operates in certain special
cases, but it is not yet taken as a general factor which can be
combined with the others.

§ Substage II-B. Discovery of the General Role of Inclination and Beginnings of the Concept of Work

From the start the subjects of this substage see that three factors are involved—the weight of the wagon, the counterweight, and inclination. But only gradually do they discover that the equilibrium between the weights involves more than simple equalities and varies according to inclination. As they discover this fact, they try to find the relationship. This leads them to the conclusion that more work is needed to pull a given weight along a steeper than along a more gently sloping track:

BOU (9 ; 6) notices the weights W and M from the start, then: *"Why do you have this curve?* [inclination]. *Can you change it?"*—"Would you like to change the slope?"—*"Yes, to see how it works. Could I put on another weight?"*—"Where?"—*"Here* [W]: *perhaps the wagon will go faster. If the weight is heavy enough, the wagon will go up. If it isn't heavy enough, it won't move."* He has several trials. *"If you put on more weight* [at W] *it goes up even faster* [he adds on up to 7 units]. *It can't go down because there is too much weight* [he takes off some at W; the wagon descends]. *It goes down if I put on less."* Next: *"I don't want to put any weight at all* [at W] *because I want to see if it stays down below or if it goes up more slowly."* The wagon goes down; then the subject varies the weights at M to see whether *"it makes a difference if you put several weights here* [he loads M]. *It still goes too fast* [he takes off weights at W and adds weights at M until they are equal]. *That makes equal weight because I put 4 weights here* [W] *and there are 4 weights on the wagon* [(which continues to move); he adds 2 and 2, 0.5 and 0.5, then 3 and 3]. *It doesn't move? No, it moves anyway!"* He has now discovered that equality of weight does not guarantee equilibrium.—"At the beginning you asked me what the curve was for; do you remember?"—*"Oh! Yes, you can lower the track, then it will go up."*—"Are you sure?"—*"It will go up* [he lowers the rail to 1 and the wagon goes up], *because now it doesn't slant as much so it's easier for it to go up."* Next, BOU varies the inclinations and realizes that when the track is vertical *"the weight* [W] *will pull the wagon"* because the counterweight is sufficient, but he does not vary it. At 45° and 1 unit at W he ascertains that the wagon descends and laughs: *"I thought it was going to go up because it goes up even when it stands up straight."* Then, realizing that it still goes down at 30°: *"You have to add or take off weight. You have to experiment when you see that it goes up or down too much."* But he does not do so systematically.

JAN (10 ; 8): *"To make it go up, you have to put a heavier weight here* [W].*"*—"What else could you do?"—*"Unload the wagon."*—"And for the wagon to stay at the same point?"—He puts 4 units on the wagon and 4 at W. *"The weights are equal. No, it doesn't move."*—"Can you do something with the rail?"—*"Maybe you could lower it; it's easier for the wagon to go forward because the track isn't as high."* —"If you lift the rail and add weight?"—*"It will stay poised because it's harder for it to go up."* Then he weighs the wagon and declares that it is equal to 4 weight units.—"So would it remain in equilibrium if you leave the wagon empty and put 4 weights here?" [W].—*"No, it would go up"* [thus he understands that the equilibrium depends on inclination].—"And if you raise the rail?"—*"It's harder for it to go up."* —"Why?"—*"Because the wagon gets heavier."*

Two main advances in thinking about the problem occur at this stage: (1) an understanding of the fact that the equilibrium is not due to a simple equality between weights, and (2) an understanding of the role of inclination—*i.e.*, more work is needed to pull a wagon up a steeper incline.

First the subject discovers that placing n units at counterweight W and n in wagon M does not guarantee an equilibrium; then he realizes that the wagon itself weighs 4 units, but that $p = 4 + M$ does not achieve equilibrium either (or does so only if the rail is vertical). This discovery leads him to focus on the problem of inclination.

Thus, the child discovers the role of the slope at this level either because of the preceding reason or because he varies the slope directly. "Now there is less slant," says BOU, "so it's easier for it to go up." "The wagon will go forward more easily," says JAN, "because the track isn't as high." Furthermore, he understands that equilibrium will be conserved if inclination and counterweight are increased simultaneously "because it's harder for it to go up." Finally, he predicts that at equal weights on W and M the wagon will go up and that if the slope is increased still more the wagon "gets heavier."

These latter protocols are instructive in that they show us how inclination gets to be thought of in the same terms as weight and is combined with weight in the form of work. Raising a large weight a little is equivalent to raising a lighter weight to a higher point; in other words, the amount of work is the same. JAN expresses this fact directly when he declares that "the weight gets heavier" if the slope is increased. Although he is not familiar with the parallelogram of forces,[2] the subject arrives at a pretty accurate intuitive understanding of the relationship between weight and inclination. Thus, even at the concrete level the concept of work is accessible in a qualitative form based on the multiplication of the relations between height and weight.

Now, it is remarkable that both the inverse relationship between weight and height found in the equilibrium of the wagon and the notion of work as the upward displacement of weight are structured at the same substage (II-B) as the discovery of the inverse correspondence between weights and distances in the balance. Both deal with the same physical law, but the child does not know this, since he thinks of neither work nor height in the balance problem unless the problem is presented in the form of an alternating suspension apparatus (see Chap. 11). Nor does he think of the balance in connection with the relations between weight and inclination dealt with in this chapter. Everything proceeds as if, at a certain level of development, the entire set of concrete operations applicable to a given subject arises simultaneously in the structuring of that delimited area (an example of such a delimited area would be the equilibrium of weight as a function of height and distance).

But there is a gap; the subject does not come to state the law in its entirety. He clearly takes the three factors into consideration (W, M, and inclination) and successively compares them two-by-two without changing the third. But it is not that he *intends* to hold one factor constant each time; he is *not trying* to apply the "all other things being equal" proof. Rather, in comparing any two factors he simply forgets the third, thus leaving it invariant without being aware of the fact. He does not get to formulate the law,

---

[2] The parallelogram of forces states that the portion of weight supported by the track (and making the wagon lighter by the same amount) increases as the inclination decreases, whereas the portion not supported by the track increases in direct proportion to the inclination.

for he lacks the means to coordinate the entire set of factors simultaneously.

It is easy to see why he has failed at total coordination—there are two main reasons and both are intrinsic to the nature of concrete operations. First, the relevant correspondences are too complex to be handled by proceeding in successive pairs or trios. It is true that the subject, since he possesses the required operations, could make a sufficiently complete inventory of factors. There is no need to describe the operations of serial ordering, equalization, and addition of weights at this time, for they have already been described in connection with the balance (Chap. 11). Obviously, the subject can also order the inclinations serially. Thus, if he wanted to, the subject could determine the correspondence between the weight of the wagon (M) and the counterweight (W) for each inclination so that he would know when the wagon would be in equilibrium and when it would go up or down. But one can see how complex such an empirically constructed triple-entry table would be. Besides, the idea does not occur to the subject, and he is content to deal with a few individual cases; hence, the first reason for his failure at total coordination of the relevant variables.

The second reason is that if the subject, instead of proceeding by successive correspondences, tries to utilize the form of logical calculus available to him—*i.e.*, multiplication of relationships (and he does in fact proceed in this way in determining the relationships of work)—then a certain number of products remain indeterminate. In other words, we find a parallel to the indeterminacy already encountered in the case of the logical multiplication of weights and distances in the balance scale (Chap. 11, prop. [7]). Starting from an equilibrium point:

$$\text{Inclination } x \leftrightarrow [Wy \leftrightarrow Mz],$$

the subject can certainly conclude that, if slope x is increased and y and z are left invariant, the wagon descends, whereas if x is diminished, it mounts, etc. But if slope x is increased at the same time as counterweight W ($> y$) is increased, or if slope x is increased while the weight on the wagon M ($< z$) is diminished,

the product is indeterminate: the double relation $\overset{x}{\rightarrow} \times \overset{y}{\rightarrow}$ or $\overset{x}{\rightarrow} \times \overset{z}{\leftarrow}$

can give rise to a product $>$, $<$, or $=$ as long as the given factors are not extensively quantified.

If an increase, lack of change and a decrease in the weight to be displaced upward are designated by $+m$, $=m$, and $-m$ respectively, and an increase, lack of change, and decrease in the height itself by $+h$, $=h$, and $-h$, we see that the double-entry table which characterizes the concept of work, worked out by means of logical multiplication at substage II-B, involves two indeterminate products out of nine (these products expressed in $+5$, -5, $=$, or $\pm$ signifying "more work," "less work," "same work," or "indeterminacy"):

	$-m$	$=m$	$+m$
$-h$		$-$	$\pm$ $=$
$=h$		$=$	$+$
$+h$	$\pm$ $=$	$+$	$+$

(1)

In summary, as long as the subject is limited to using concrete operations of classes and relations, he cannot determine the law (even in the form of implicit qualitative proportions found at substage III-A). The explanation is twofold; first, the correspondences which must be empirically established are too complex, and second, the products of the multiplications of relations are in part indeterminate.

§ Substage III-A. Qualitative Coordination of the Three Factors, but Without Proportion as a Function of Height

It is the nature of formal thought to consider an entire set of possibilities and to deduce from them what is real. With its appearance, subjects use a remarkably different approach to the problem. Instead of getting lost in the inventory of actual cases—an inventory which is in fact inexhaustible because the correspondences to be determined by successive experiments are much too complex—

in trying to cover all possible cases, the subject very quickly turns to a selection of crucial cases—*i.e.*, the extremes and the middle. It occurs to him to place the rail horizontally and vertically (an idea which is unexplainable without a preliminary explication of the possible transformations). Then he determines the demonstrative intermediate positions.

Thus, from the start the subject seeks to coordinate the three factors into a single law, itself a qualitative proportional schema. But since the subjects think in terms of the angle expressed in degrees and not in terms of its sine—*i.e.*, height—they still fall a little short of discovering the law they are groping for.

LAV (10 ; 6 advanced): *"To make it go up you have to put on more weight* [at W], *to make it go down, less."*—"What else can you do to make it go up?"—*"Lower the rail."*—"And to make it go down?"—*"Take off all the weights* [W]. *You can also put some weights in the wagon. You can put the track higher up, too, because the wagon comes down"* [with more force].—"What do you have to do for the wagon to stay in place?"—*"That depends on how you place the track and whether you put more or less weight on the wagon"* [he puts the track at about 45° and finds that 3 units in W balance the wagon].—"Are there other places where the wagon rests without moving?"—*"Put the track horizontal and take off the weights here"* [W]. Then he measures the weights needed when the track is vertical after having announced *"4 or 5* [units] *because the wagon has a weight of 4."* And he discovers the point of equilibrium for W = 4.—"Tell me what you have to do so the wagon won't move?"—*"You have to put the track way down without any weight or way up with 4 weights and also put the rail halfway with 2 weights"* [he has not tried it at the midheight; *i.e.*, 33°].—"Are there other points?"—*"Yes, everywhere"* [he finds one unit for 15°].— "Did you understand everything?"—*"Yes, the higher up you go* [inclination of the rail], *the more weight you have to put on for the wagon to stay where it is; the more you go down, the less you need."*

END (11 ; 6): *"You can take off a weight to make it go down, put on one to make it go up or raise the track more."* Then he experiments by himself; he starts by lowering the track to the horizontal point, then lifts it to the vertical and says, *"If you want to put the track straight up* [vertical], *you have to put on more weight; you need 4 weights."*— "Why 4?"—*"Because with 4 it doesn't go up. You can compare it with a balance-scale: on one side 400 grams and on the other 400 too."*

Next: "*I want to see how many weights it takes at 45°*" [he finds 2.5]. Then he concludes: "*The more you lower it, the more weights you have to take off. The more you go up, the more you have to add.*"

SCU (11 ; 12): "*To make it go up you can lower the track or take off weight in the wagon or put on some at*" [W]. He seeks the equilibrium positions by varying the slope: "*When it is low, there isn't enough weight in the wagon and the wagon goes up; the counterweight pulls harder.*" He weighs the wagon [$M = 4$], then tries at 33° and establishes that $W = 2$ at the equilibrium point. "*That's funny: a minute ago I saw that the wagon weight was 4 and now it's 2 [$W = 2$] and it pulls a wagon that weighs 4. You have to raise it more to make it equal [$4 = 4$] and calculate its relationship to the inclination.*" He raises the track higher and higher: "*Still not enough [he has reached 80°]. That's almost it. It has to be straight [90°]. When there is an inclination the equilibrium changes, and when it is straight the relationship is one to one.*" But for half of the inclination he tries at 45°, although he had already established 2 units for 33° and does not understand that height alone plays a role.

CLA (11 ; 6): "*To make it go down, you can either pull up the line or take off some weight from the counterweight [W] or add some in the wagon*" [M].

Unlike the subjects at the advanced concrete substage (II-B), subjects of the first formal substage (III-A) immediately or very rapidly coordinate the three factors into a single relationship. At first this integration is the simple statement of factors in the form of a ternary disjunction. If we call p the increase in weight at W (and $\bar{p}$ its decrease), q the increase of weight at M (and $\bar{q}$ its decrease); r the increase (or $\bar{r}$ the decrease) of the inclination, and t the rise of the wagon (or $\bar{t}$ its descent), we see that SCU and CLA start out with reciprocities:

$$t \supset (p \text{ v } \bar{q} \text{ v } \bar{r}) \text{ and } \bar{t} \supset (\bar{p} \text{ v } q \text{ v } r). \tag{2}$$

In practice these ternary disjunctions can be distinguished by the fact that the subject no longer modifies two of the factors without thinking of the third but looks for covariations. From then on he is quickly convinced that the equilibrium of the wagon and the counterweight varies according to the inclination, and the subject makes several tries. Most often the subject tests the extreme positions almost at once—the horizontal position where the wagon is at equilibrium without any counterweight ($W = 0$) and the vertical, where it is in equilibrium as if on a balance when the counterweight is equal to its own weight ($W = M = 4$). Hence, the qualitative law: the more the inclination is increased, the greater the counterweight required to bring the wagon into equilibrium—until the upper limit (vertical inclination) where the counterweight is equal to the weight of the wagon.

Since he possesses disjunctive operations (2) from the formal standpoint, the subject also possesses the I N R C group in the form:

$$
\begin{aligned}
&\text{I}(p \vee \bar{q} \vee \bar{r}) \\
&\text{N}(\bar{p}.q.r) \\
&\text{R}(\bar{p} \vee q \vee r) \\
&\text{C}(p.\bar{q}.\bar{r}),
\end{aligned}
\tag{3}
$$

and the utilization of the structure naturally furnishes the schema of proportionality evident in several cases (in particular LAV):

$$
\frac{p \vee \bar{q} \vee \bar{r}}{\bar{p}.q.r} = \text{R} \frac{\bar{p} \vee q \vee r}{p.\bar{q}.\bar{r}}.
\tag{4}
$$

In other words, all the subjects at this level understand the possible compensations between p and $\bar{r}$ and between q and r.

If this group (for this proportionality) and these equivalences are to result in the formulation of the law $h/H = W/M$, a fourth variable corresponding to absolute elevation H (the length of the rail measured when it is held in a vertical position) must come into play. Only then would the *propositional* reciprocities and inversions express the reciprocities and inversions *operant* in the equilibrated system being analyzed. But the subject calculates the inclinations in degrees, which results in one constant for H ($90°$) whatever the apparatus chosen and which leads him to look for half of the inclination at $45°$, where the counterweight does not have the value of 2 but an intermediate value between 2 and 3. Either the subject generalizes falsely or he is prevented from discovering a simple law.

In spite of the appearance of formal implications and disjunctions with the consequences that they imply—(3) and (4)—at this stage the subject does not manage to exclude the angle (in degrees) in favor of height. This fact may seem curious, since even at substage II-B the child formulates the concept of work as a function of the lifting of a weight. But this is because at this stage the subject is limited to qualitative reasoning and is not yet able to separate the concepts of angle and height. On the other hand, when the subject at substage III-A wants to go beyond this qualitative relation of inclination to find a metrical expression, he thinks of the angle rather than the height, doubtlessly because the apparatus governing the inclination of the track describes a rotating movement.

§ *Substage III-B. Discovery of the Law*

As soon as the angle measured in degrees has been excluded in favor of the height (sine), the subject discovers the proportionality of heights and weights. But, curiously enough, this exclusion is not easy (we suggested it to the second of the three following subjects):

GIL (12 ; 7) is asked to find the equilibrium points and to extract the law. He finds $W = 4$ at the vertical, then looks for the midpoint [$W = 2$] at 45° and then at 60°: "Why?"—"*I count half the distance* [horizontal]. *No, that doesn't work. You have to find the*" [halfway point]. He does the experiment for $W = 2$. "*It's about 30°. But there is also that* [height]. *Here, with 2 weights, it's 32°. For 3 weights, you have to put it at 3* [in height]. *Anyway, I think so* [experiment]. *Yes, for 1 weight* [$W = 1$] *you have to put the rail at 1, for 2 weights at 2, for 3 weights at 3 and for 4 way up at the top.*"—"Can you give a single rule?"—"*Yes, for 2 weights you put it halfway up. For the halfway height it's half of the weight of the wagon, for one-fourth it's a quarter of the total weight,*" etc.

DIZ (14 ; 3): "*If the rail is vertical, you have to put enough units here* [W] *to make them equal* [the weights and M: he finds 4]. *The 2 weights are the same, one on each side. For the half, you have to put on half the weight.*"—"For half of what?"—"*Of the inclination: 45°* [experiment: 2.5]. *No! Maybe there is friction.*"—"It's possible, but it doesn't play an important role."—He finds that $W = 2$ corresponds to 33°. "*33°, that makes about two-thirds. Look here* [height]. *Oh! The height! It isn't the angle that does it, but the height* [he tries elevations 1/2, 3/4, and 1/4]. *The weight pulling the wagon* [W] *has to*

*be equal to the height; for example, if you have a height of 2 you need
2 over 4, if it's 1, then 1 over 4"* [1 and 2 in elevation are the fourth
and the half].

VUL (15 ; 6) determines 4W for M when the rail is vertical; then: *"At
33° I find 2; at 15° I find 1; at 60° it should be 4 but it isn't. If it
isn't proportional to the angle, then. . . ."*—"Is there something else
you might consider?"—*"The height corresponds to the angle. If I take
twice the height: height 2 corresponds to 2 weights. Let's see: elevation 3 gives 3 in weight. Good, it's in proportion to the height. Each
time you increase the height by a certain amount, you have to add a
proportionate amount of weight."* Summary: *"The height is proportional to the weight."*

There are two complementary types of response at this stage.
For the first type (DIZ), the discovery of height is a result of suggestion, but the subject immediately formulates the law of proportionality once the suggestion is made. But, in the second type,
height is discovered as a factor because of the search for proportionality—i.e., the simple proportionality which is still beyond the
III-A subjects.

In both cases the law is discovered: $h/H = W/M$. If we let s
stand for an increase in the total height (vertical length of the
rail, which in fact, does not vary in our experiment), the logical
proportion is as follows:

$$\frac{r}{s} = \mathrm{R}\,\frac{p}{q},\tag{5}$$

where $q.r$ stands for an increase in the work to be accomplished
and $p.s$ for an increase in the work furnished by the counterweight as a function of the total height.

Given the already established equivalences between p and $\bar{r}$ or
between q and r and given the equivalence between $\bar{q}$ and s (the
result is the same when the weight of the wagon is increased or
when the value of H is decreased), proportion (5) can be deduced
from:

$$\frac{\bar{p}}{\bar{q}} = \mathrm{R}\,\frac{p}{q}.\tag{6}$$

Thus, in the present case, the forms of the equilibrium schema
(I N R C group) and the proportionality schema are the same,

mutatis mutandis, as in the balance-scale problem (Chap. 11). This isomorphism raises an interesting problem for the psychology of formal thinking. It is true that the structure of the two laws is the same. If we call B the heavier of the two weights in equilibrium on the balance and A the lighter, B corresponds here to the weight of the wagon (M) and A to that of the counterweight (W). If we call L the horizontal distance (from the axis) corresponding to weight A (thus the greater of the two distances for the smaller of the two weights) and l the distance that corresponds to weight B (the smallest distance for the greatest weight), then in this case L corresponds to H and l to h. The formula is:

$$l/L = A/B, \text{ as } h/H = W/M .$$

However, the intuitive content of the two laws is quite different (so different that many psychology students take a great deal of time trying to understand their identity). In the balance problem, the relations between weight and lengths (l and L measured horizontally on the two arms) are crucial, whereas the heights are potentially effective factors only if the system is in equilibrium, unless an alternate suspension apparatus is used (as in Chap. 11); thus, the concept of work is not immediately elicited. But, in the case of the wagon, the horizontal distances do not influence the system; the inclination is intuitively given and the concept of work is elicited as soon as the system actions are observed. Then the psychological question becomes: is this difference in intuitive content the determining factor in the development of operations, or, on the contrary, is the underlying operational structure its determinant? To answer the question, we must compare the results for the two problems stage by stage.

In both cases the system processes throughout stage I are explained by an assimilation to the subject's own action, pulling and pushing, etc. But, since the balance is noticeably symmetrical, there is a more rapid equalization of distances and weights (as intuitive regulations without operations) in that experiment.

At substage II-A operational equalization of weights occurs in both cases, and the subject understands that distance plays a role in the case of the balance and that inclination is relevant in the wagon problem. But the subject cannot combine these heterogeneous factors with weight (except in certain special cases).

At substage II-B the subject discovers the inverse correspondence between the weight and distance for the case of the equilibrated balance (lighter corresponds to further from the axis and heavier to closer to it) and discovers as well the fact that the larger the inclination, the heavier the counterweight needed to balance the wagon (in the second case). In the latter case, the coordination between weight and inclination gives rise to the structuring of the concept of work—more work is required to raise the same weight to a greater than to a lesser height. But in both cases the coordination remains qualitative. Certain compensations are understood (heavier = less far, and heavier = less high), but there is no possibility of solving the more general problem because of the indeterminacy of logical multiplication.

At substage III-A the subject discovers the metrical proportion in the balance problem and looks for the same proportion in the case of the wagon, discovering the qualitative law coordinating the three factors (weight, counterweight, and inclination). The time lag in the second case is due to the fact that the height has to be dissociated from the angle (measured in degrees); half of the height is not 45° but 33°.

At substage III-B the metrical proportion is finally discovered in the case of the wagon and is explained directly in terms of work. On the other hand, in the balance problem the metrical law found at substage III-A appears as a compensation system that is self-sufficient as long as the two weights are hung up at the same time; it does not occur to the subject to invoke either height or work in his explanation. But when the weights are presented successively with a suspension apparatus that brings out the alternating differences in height, the subject discovers the inverse proportion between the weights and the height attained. In the wagon problem he explains the equilibrium in terms of the equal amounts of work needed.

Thus, it is clear that between the two lines of development there are a set of intuitive differences which result from the nature of the apparatus and from the questions asked of our subjects. So it is all the more striking that we find the same operational mechanism underlying the apparent divergences. In both cases, after the same preoperational representations of stage I and the same initial operations at substage II-A, the inverse correspondence is discovered at substage II-B. In both cases the operational schema of equilibrium is established only when the I N R C group comes into play at the level of formal or propositional operations. And in both cases this leads to the schema for proportions and compensations in their general form. Thus the differences at the intuitive level only give rise to slight differences in timing within stages II and III, while the over-all progression of organization is the same.

13

The Projection of Shadows[1]

In ADDITION to the usual problem of the formal operations needed to establish the table of possibilities that allows the discovery and verification of a law, the present research raises a question about the formal operational schema relative to proportionality. But we are dealing with a new type of proportionality. Whereas the proportions in the problems of the balance and of hauling a weight on an inclined plane derive from a model of physical equilibrium, the proportions we shall study in connection with the projection of shadows are of an essentially geometrical nature. They denote relationships between distances and diameters in a physical phenomenon that can be explained in terms of simple projective geometry.

The problem we have set for ourselves is to discover whether the proportions involved in the present experiments will be discovered at stage III, as in our previous experiments, and whether or not this discovery is a function of the I N R C group. If it is, one must think of the I N R C group in a more general sense than in the earlier problems.

The law to be discovered in this experiment is extremely

[1] With the collaboration of Vinh Bang, research assistant, Institut des Sciences de l'Éducation; B. Reymond-Rivier, research assistant, Institut des Sciences de l'Éducation; and F. Marchand.

simple. Rings of varying diameters are placed between a light source and a screen. The size of their shadows is directly proportional to the diameters and inversely proportional to the distance between them and the light source. Specifically, we ask the subject to find two shadows which cover each other exactly, using two unequal rings. To do so he need only place the larger one further from the light, in proportion to its size, and there will be compensation between distances and diameters.

The stage I reactions need not be presented for this problem. The preoperational subjects do not understand the formation of shadows and in another work we have described the representations of shadow typical of 2–7-year-old children in sufficient detail to make it unnecessary to take up the question here.[2]

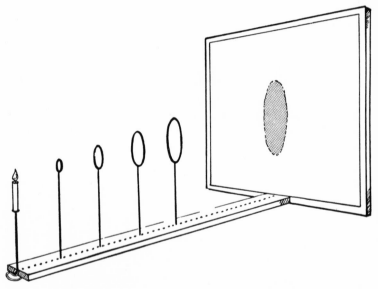

FIG. 11. The projection of shadows involves a baseboard, a screen attached to one end of this, a light source, and four rings of varying diameters. The light source and the rings can be moved along the baseboard. The subject is asked to produce two shadows of the same size, using different-sized rings.

[2] See Piaget, *The Child's Conception of Physical Causality*, Chap. VIII, nos. 1 and 2, and *Play, Dreams, and Imitation in Childhood* (Norton, 1951), Chap. IX.

§ Stage II. Discovery of the Role of the Size (II-A), then of Distance (II-B)

The II-A child knows that the size of shadows depends on the size of the object, but his knowledge goes no further:

PEL (7 ; 10) predicts correctly that a ring 10 cm. in diameter will produce a larger shadow than a ring of 5 cm., etc. "If I move it to this side, where will the shadow be?"—"*There*" [accurate]. "Does it stay the same size or does it get bigger or smaller?"—"*It's the same.*"

BAR (8 ; 8) starts with the same reactions. Then, through experiment, he discovers that the shadow of the same ring varies in size with the distance. He is then asked to produce a single shadow using 4 unequal rings: he places the 20 cm. circle at 70 cm. distance, the 10 cm. at 41 cm., the 5 cm. at 23 cm., and the 1 cm. at 11 cm.

It is possible to order serially the sizes of the rings and the sizes of the shadows and to formulate accurate correspondences at equal distances. In fact we find an accurate serial ordering of distances, but the subject does not relate this to the size of the shadows. He starts out with the assumption that the distance does not modify size (PEL). When corrected by the experiment, he expects haphazard transformations and does not find regular correspondences (BAR).

On the other hand, at substage II-B the subjects no longer think of light as being "everywhere" without rays having a determinate direction, and they begin to predict the effect of divergent rays. At least, they establish an empirical correspondence between the decreasing sizes of the shadow thrown by the same object and the increasing distances from the light source. In other words, they understand that the closer the object is to the screen, the smaller the shadow.

MAND (9 ; 6): "*As it advances* [toward the light], *it* [the shadow] *always becomes bigger, because when it is closer* [to the screen] *it gets smaller, and when it's further away* [from the screen] *it gets bigger.*"

NOV (10 ; 5): "*You have to put the smallest ring in front* [toward the light source], *because it keeps getting bigger*" [cf. the cone of light rays]. Then, in order to obtain equal shadows he puts the 5 cm. ring at 44 cm., the 10 cm. at 55, the 15 cm. at 56, the 20 cm. at 57.

OLI (10 ; 2) puts the 5 cm. ring at 10 and the 10 cm. at 19, then 15 at 38 and 20 at 50. "Why did you put them that way?"—*"Because with those* [5 and 10 diameters] *it's bigger* [because closer to the light] *and those* [15 and 20 diameters] *get smaller."*

CHRI (10 ; 0) puts the 5 cm. ring at 2 cm., 10 at 15, 15 at 29, and 20 at 42 cm., *"because the smallest have to be further back to make the same size as the big ones: you have to put them at almost equal distances apart."*

DEL (11 ; 11) puts the 5 cm. ring at 11 cm., 10 at 21 cm., 15 at 32, and 20 at 43. Then 5 at 41, 10 at 52, 15 at 63, and 20 at 74 cm.

MAU (11 ; 10) puts the 5 cm. ring at 55, 10 at 63, 15 at 71, and 20 at 80 cm.

The qualitative correspondence between the shadow sizes and distance is clearly formulated, but with two peculiarities which are extremely instructive for analysis of the opposition between concrete level compensations obtained through logical multiplication and the true proportions based on multiplicative compensations found at the formal level.

The first of these peculiarities is that, although the stage II-B subjects know how to construct inverse correspondences, they prefer direct ones. Consequently they tend to calculate the distances by starting from the screen rather than from the light source (*cf.* "it advances" for MAND, "further back" for CHRI; these expressions are relative to the screen). Nevertheless, certain subjects reach an intuitive understanding of the light cone (*cf.* NOV, "it keeps getting bigger"), but when they measure the distances they do so from the screen so that they can make a direct correspondence—the larger the distance (from the screen) the larger the shadow (the 5 and 10 cm. rings are "bigger" than those of 15 and 20, says OLI).

Nevertheless, they clearly understand the compensation between the distance and the size of the ring. However, the second and most important peculiarity of these reactions is that at this stage the attempts at metrical quantification to which this compensation gives rise cannot be interpreted as derived from a true proportion—*i.e.*, multiplicative relationships. Rather, they derive from any constant additive differences whatsoever in the serial orders and correspondences. For example, after a gap of 11 cm. between the first two rings, NOV places the following rings 1 cm. apart (54, 55, 56, 57 cm.). Likewise MAU puts 8 to 9 cm. between rings at distances of 55 to 80 cm., and CHRI calculates constant differences of 13 to 14 cm. OLI seems to approach an understanding of proportionality, but he makes a simple dichotomy between the large and small circles. Then he distinguishes additive differences of 9 and 12 cm. within each set, but a one-to-two ratio between the two. As for DEL, whose chronological age would put him at stage II, his initial proportion is more or less accurate, but on the second trial he regresses to an arbitrary additive difference of 11 with 41 cm. as the starting point.

§ *Stage III. Proportionality in the Correspondences (III-A), then Generalization and Formulation of the Law (III-B)*

At substage III-A an inverse metrical proportionality between distances and diameters first appears, but it is not yet generalized to all possible cases. The subject measures the diameters and the distances and looks for a metrical hypothesis based on the divergent structure of light rays, taking into account the distance between the light source and the first ring (the smallest or the largest):

CHE (12 ; 8) measures the rings and finds that their diameters differ by 5 cm. He concludes that one must *"find a distance between them which is a multiple of 5."* He places them correctly in proportion to size.

DUC (12 ; 1), after having placed the 20 cm. ring at 83, says: *"Now you have to count from here to there* [to the light source] *and divide by 4."* Then he puts the 5 cm. ring at 21, the 15 cm. at 61, but the 10 cm. at 51. "And if I only give you three rings?" [5, 10, and 15].—*"You have to count from the largest and divide by three."*

WAL (12 ; 4) divides the shadow of the large ring by the distance, obtaining 50/40 = 1.25, and looks for multiples of 1.25 as differences. Then he places the 5 cm. ring at 25 and the 10 cm. at 50: *"It's half!"* Next, he places the 15 cm. at 75 and the 20 cm. at 100: *"It works. It's always 5 cm. more [for the rings] and the lengths are always 25 cm. more. It's the same scale."*—"Can you find me another distance with the shadow the same for all rings?"—He places them with 26-cm. distances between each ring and moves the light back 4 cm.—"I mean without moving the light."—*"I don't think you can. You can't enlarge the scale."* —The experimenter places the 15 cm. ring at 46; the subject then puts the 10 cm. at 23, then the 5 cm. at 7, the 10 cm. at 30, the 15 cm. at 53 and the 20 cm. at 76.

It is clear that at this level the subject assumes proportionality from the start. But the proportion is only found in one or two instances and is not yet generalized to all cases. All of the previously developed relationships of concrete serial ordering and correspondence are coordinated in an organized view of the whole; all of the relationships are subordinated to the geometrical representation of divergent rays (in the experiment the subject's goal is to control his placement of the rings) and the representation is correctly given the property of proportionality.

In sum, at stage III-A the subject begins to calculate distances from the light source rather than from the screen, and in his calculation he takes the distance between the light source and the first ring into account and not simply the distances between the rings (two new operations not present at stage II).

But he is satisfied when he has verified his hypothesis on a single case and does not yet conceive of the relationship as changeable and as capable of taking a series of equivalent forms. In other words, he does not yet look for the general law, defined as a system of necessary relations which are adequate to account for the obtained result.

However, at substage III-B the law is generalized and made explicit:

WAH (14 ; 1): *"You can take any distance as long as the ratio is the same."*

MIC (14 ; 6): *"Since the diameters all have regular differences, the differences between the distances have to be the same."* Then he places the rings of 5, 10, 15, and 20 cm. at distances of 8, 16, 24, and 32 cm., respectively; next, he takes another arbitrary distance and finds the proportion in the same way: *"The distances have to have the same relation to each other as the rings."*

FAU (15 ; 6): *"For the shadow to be equal, the same with two rings, the fraction of the axes [distances] has to be equal to the fraction of the two rings."* Then: *"The shadow is never smaller than the actual ring."*

HUE (15 ; 6): *"The angle made by the light rays gets wider and wider. For the light to make twice the size [of the shadow], it takes twice the distance,"* etc.

MART (16 ; 2) begins getting the rings to coincide: *"You have to put the largest the furthest away, and the ratio between the diameters of the rings and the distances has to be the same."* He is successful in discovering the proportion.

GUY (16 ; 6): *"It should send out a ray like this [he shows us a conic form] from the small ring. I think that the first ring will give a shadow whose outline will depend on a kind of ray that increases in size. . . ."*

Thus there is a difference between the set of these children and the set of the substage III-A subjects. From the start their formulations are dependent upon a hypothesis that is both explanatory and general, and at this stage the hypothesis no longer deals only with the divergent light rays but includes a conception of the cone itself. Thus, proportionality is implied by the explanatory schema itself and holds, as WAH says, at "any distance at all."

But we cannot forget that proportionality was anticipated before this final view of the whole was constructed. The proportions are deduced from the whole figure only after the child understands the divergence of light rays, but at substage III-A the proportionality was discovered without having first projected this figure. (Moreover, this may often happen at substage III-B.) What, then, is the nature of this proportionality, which does not stem from a mechanical schema like the proportions in the earlier chapters but is accompanied from the start by geometrical representations (the divergence of rays of light, then the shape of the cone)? In a sense, the stage III subjects discover proportionality because they have access to propositional logic and, therefore, are able to understand and transform the equality of two products. This is possible only when the subject can state that a given increase in the diameters of the rings combined with a given increase in distance can give the same results as other combinations of increases or decreases. On the other hand, it is clear that this equality of products is understood only as an instrument which enables the subject to express a multiplicative compensation between changes in the diameters and the distances.

Let us designate increases in diameter and distance by p and q respectively, and decreases in diameter and distance by $\bar{p}$ and $\bar{q}$. Let r_0 be the conservation of the size of a shadow and $\bar{r}_0$ its modification; let r be an increase in shadow size and $\bar{r}$ a decrease (thus $\bar{r}_0 = r \vee \bar{r}$). Then the subject will state the following propositions. First, in the case of modification of diameters and distances, conservation r presupposes either the simultaneous increase or the simultaneous decrease of both:

$$r_0 \supset [(p.q)] \vee (\bar{p}.\bar{q}). \tag{1}$$

Second, combinations $p.\bar{q}$ and $\bar{p}.q$ *always* correspond to modifications of the shadow:

$$(p.\bar{q}) \vee (\bar{p}.q) \supset \bar{r}_0 . \tag{2}$$

But in two opposite senses:

$$(p.\bar{q}) \supset r \text{, and} \tag{3}$$
$$(\bar{p}.q) \supset \bar{r} . \tag{3a}$$

Finally, the same results can be obtained either by increasing the diameter or by diminishing the distance and vice versa:

$$r \supset (p \vee \bar{q}) - i.e., \ r \supset [(p.\bar{q}) \vee (p.q) \vee (\bar{p}.\bar{q})] \tag{4}$$

with exclusion of $\bar{p}.q$, and

$$\bar{r} \supset (\bar{p} \vee q) - i.e., \ r \supset [(\bar{p}.q) \vee (\bar{p}.\bar{q}) \vee (p.q)] \tag{5}$$

with exclusion of $p.\bar{q}$.

Actually, if r_0 implies $(p.q)$ or $(\bar{p}.\bar{q})$, the reciprocal is not true and $(p.q)$ or $(\bar{p}.\bar{q})$ can imply either r or $\bar{r}$ when the diameters and the distances are modified.

In this way, reasoning by implication reveals to the subject that the same products can result from either of two different causes;

thus, he discovers the qualitative schema of proportionality. From (1) he concludes:

$$(p.q) = \mathrm{R}(\bar{p}.\bar{q}), \tag{6}$$

from which, by reciprocity of cross-products:

$$\frac{p}{\bar{p}} = \mathrm{R}\frac{\bar{q}}{q}. \tag{7}$$

And, from (2), he concludes:

$$(p.\bar{q}) = \mathrm{R}(\bar{p}.q), \tag{8}$$

from which, by reciprocity of cross-products:

$$\frac{p}{q} = \mathrm{R}\frac{\bar{p}}{\bar{q}}. \tag{9}$$

But we remember that this proportion (9) corresponds to the numerical proportion $\dfrac{nx}{ny} = \dfrac{x : n}{y : n}$.

In a general way, the discovery of proportionality in this particular case results from an understanding of multiplicative compensations. Even at substage II-B, the child is aware of the fact that a change in the diameter of the circles can be compensated by a change in distances, but he is unable to interpret this compensation except by an additive formula (equality of differences). If he is to assign the true multiplicative form to the compensation, the child must simultaneously distinguish and coordinate two kinds of general transformations: transformations by inversion, which cancel the modification in question, and transformations by reciprocity, which compensate it without canceling it. But this is exactly what propositional operations enable the subjects to do at stage III. In distinguishing two independent variables, each of whose modifications can be canceled (inversion) but which also can compensate each other without cancellation (reciprocity), subjects get to make effective use of a group of four transformations (of course, they are not aware of this), and the discovery of proportionality is a direct consequence.

It is striking to note that expression $(p.\bar{q})$ in proposition (3)—increasing the diameter and decreasing the distance—is the correlative of expression $(p \vee \bar{q})$ in proposition (4), just as expression $(p.\bar{q})$ in proposition (3a) is the correlative of expression $(\bar{p} \vee q)$ in proposition (5). Each of these pairs of propositions—(3) and (4) or (3a) and (5)—is linked to an inverse result of the other (r or $\bar{r}$).

Thus, we have the group:

$$I(p \vee \bar{q})$$
$$N(\bar{p}.q)$$
$$R(\bar{p} \vee q) \tag{10}$$
$$C(p.\bar{q}).$$

Hence, the possible proportion:

$$\frac{p \vee \bar{q}}{p.\bar{q}} = \frac{\bar{p} \vee q}{\bar{p}.q} \text{ thus } \frac{Ix}{Cx} = \frac{Rx}{Nx} \tag{11}$$

where x is $(p \vee \bar{q})$.

But it is unlikely that propositions (10) and (11) actually play a part in the subjects' reasoning, although they involve nothing more than propositions (3), (3a), (4), and (5), which are themselves direct expressions of the stage III statements. Thus we are dealing with an example of a structure which is merely potential but which is implied by the actual reasoning we have observed. On the other hand, one can say that the I N R C group does play a part in the subjects' thought processes in a simpler "unary" form, and that this accounts for proportions (7) and (9). If the increase in diameter is set forth as the identical transformation ($I = p$), then its decrease is the inverse transformation ($N = \bar{p}$). But the increase in distance from the light source compensates for the increase in diameter without canceling it. Consequently, it plays the part of the reciprocal transformation ($R = q$). Finally, the negation of the reciprocal produces the same effect as the increase in diameter and thus plays the role of the correlative ($C = \bar{q}$). The following proportions result:

$$\frac{p}{q} = \frac{\bar{q}}{\bar{p}} \text{ or } R \frac{\bar{p}}{\bar{q}} \text{ (proposition [9])}.$$

The foregoing analysis demonstrates that the I N R C group has a more general function than that of explaining mechanical equilibrium. It comes into play when two distinct reference systems (as in relative motions: see Chap. 17) have to be coordinated, as we shall see in Part III of the present work. As we see in this chapter, it even operates in the coordination of changes in two independent variables when multiplicative compensation of their effects is possible (as opposed to the additive compensations first formulated at the concrete operational stage).

NOTE. It should be noted that logical proportionality is not tied up only with the I N R C group but can also be derived from the general structure of proportionality found in the lattice (in a way which seems to us to involve the I N R C group as well; see J. Piaget, *Essai sur les transformations des operations logiques*, pp. 166–68). A lattice is a partially ordered set of inclusions (and in logic, of propositions and implications as well), such that for any two elements of the set, x and y, there is always a least upper bound UB (= the smallest element which includes both x and y) and a greatest lower bound LB (= their intersection). Now, there is a proportional relationship in any lattice such that $\frac{LB}{x} = \frac{y}{UB}$. For example, in logic, the proportional relationship is $\frac{p.q}{p} = \frac{q}{p \vee q}$. Passing from the lattice in propositional logic to the lattice in whole numbers, we know that LB = the greatest common divisor (GCD) and UB = the least common multiple (LCM). We then have:

$$\frac{GCD}{x} = \frac{y}{LCM}, \text{ for example } \frac{2}{4} = \frac{6}{12} \text{ or } \frac{1}{20} = \frac{3}{60}.$$

But, in the shadow problem, the subject could find the logical proportion directly:

$$\frac{p.q}{p} = \frac{q}{p \vee q}, \tag{12}$$

where $p = (p.q \vee p.\bar{q})$ and $q = (p.q \vee \bar{p}.q)$; and (p, q, and their negations carry the same meaning that they do in propositions [1] to [11]).

However, we have no evidence that the stage III subjects actually resort to the lattice structure in the solution of the shadow problem, for this structure is psychologically manifested in qualitative reasoning by an explicit utilization of combinatorial operations, which is not the case here. Moreover, from the formal standpoint, proportion (12) can be reduced to proportions which derive from the I N R C group, whereas the converse of this statement is not true.[3]

[3] See Piaget, *Les Transformations des opérations logiques*, p. 225. (Not transl.)

14

Centrifugal Force
and Compensations[1]

THE SCHEMA of proportionality has been examined in several
forms, both in connection with the equilibrium schema (Chaps.
11 and 12) and independently of it (Chap. 13). We now have to
examine one more case in order to define the relationship between
the proportionality schema and the schema of multiplicative compensation. Here we are not speaking of compensation in the most
general sense of the term, in which it is synonymous with reversibility. Rather, we are referring to compensation between heterogeneous factors x and y, such that an increase in the value of one
gives the same result as an increase or decrease in the value of
the other. We have already come across compensations of this
type: in the flexibility problem (Chap. 3); in the balance problem,
where distances and weights compensate each other; in the problem of traction on an inclined plane, where inclination and
weights are involved; and finally, in the case of shadow projection,
where diameters and distances compensate each other. Still, we
thought it worth while to analyze a new example, one in which
two possibilities are open to the subject. He can construct metrical
proportions (which he could not in the flexibility problem), and
he can isolate the factors that determine equilibrium in terms of

[1] With the collaboration of M. Meyer-Gantenbein and L. Vergopoulo,
Institut des Sciences de l'Éducation.

the "all other things being equal" method (which he could not do in the traction problem). Our aim is to discover whether, psychologically, proportions carry with them the idea of compensation or whether it is the other way around.

Three metal balls of different weights are placed on a disc at three different distances from its center. The disc is rotated faster and faster until the balls roll off the disc because of centrifugal force. The problem is to predict in what order they will leave their initial positions and why. Obviously, the law of centrifugal force is a complex one—i.e., $f = mv^2/r$ where $m =$ mass, $r =$ the radius (distance from the center), and $v^2 = r^2w^2$ (where $w =$ the angular speed). When v^2 is replaced by r^2w^2, $f = mw^2r$ is obtained. But, since the speed of the disc is constant with the initial acceleration, the subject need isolate only factors m and r—i.e., need understand only the following two relationships: a ball is displaced *sooner* in direct proportion to its weight and *later* in inverse proportion to the distance from the center. Consequently, a problem of compensation arises. A heavy ball placed at a point nearer the center may move at the same time as a lighter one closer to the periphery. (The three weights are calculated in such a way as to compensate exactly for the three distances.)

§ *Stage I. Preoperational Interpretations*

Subjects under 7 years refer to all possible factors to explain the order of succession of movements, including among others the size and the distance:

coq (6 ; 11): *"One takes off on one side, the other on the other side, because they don't want to go on the same side."*

pav (6 ; 2): *"The ball on the third circle will take off first because it's nearer the edge.* [Large and medium on the same circle.] *Together. It doesn't make any difference if they're bigger* [experiment]. *The big one took off first because it's heavier. The heavy ones always go first."*

cum (6 years): *"They rolled because it turns. I want to put on the little ones because the big ones roll* [experiment]. *The big one takes off first."*

Thus, even at this level the subject is able to account for the two factors, for he accepts anything he sees and lacks any operational caution (*cf.* PAV who goes from "never" to "always" when he has observed a single case!). More accurately, at this stage the child does observe the facts but does not have a sufficiently developed set of inclusions or relationships between "all" and "some" to establish laws. Moreover, he tends to believe that everything that occurs has to be as it is. This assumption is both the principle from which he generalizes and a failure to distinguish between the moral and the physical.

§ Substage II-A. Partial Correspondences

When concrete operations appear, the child can correctly order serially the sizes (or weights) of the balls and formulate the correspondence with the take-off order, but only in those cases in which the distances are equal (independently of the subject's manipulations). He also discovers the correspondence between the take-off order and the distances when the weights are equal. But the multiplication of the two relationships appears only in exceptional cases. Moreover, it never occurs when compensation is involved—*i.e.*, when the two factors do not vary in the same direction.

GUI (7 ; 4): *"They all moved . . . the big ones moved away and pushed the little ones."* Prediction: *"They won't move because I put a big one on* [he turns the disc]. *They took off anyway* [new trial]. *That one stayed because it's too light"*—"Why?"—*"The light balls stay; sometimes they take off because it turns fast."*—The experimenter places large balls of equal weights at distances D_3 and D_2. [2]—*"[D_3] was first, [D_2] second."*—"Why?"—*"Because it's always first when it is at"* [1].—"Why?"—*"It doesn't have far to go."*—The experimenter puts W_2 at D_3 and W_1 at D_2.—*"W_2 will be first* [He turns the disk]. *Both at the same time; the small one [W_2] a little before."*—"Does that surprise you?"—*"No. The big ones go faster* [another trial]. *The little one [W_2] was first and the big one [W_1] next, but they got there at the same time. The big one went first and pushed the little one* [actu-

[2] Distances D_1, D_2, and D_3 are numbered from the center in that order; balls P_1, P_2, and P_3 are numbered in order of increasing weight.

ally both took off at about the same time on opposite sides]. *The big one was first because it is bigger* [another trial: more or less simultaneous]. *The little one goes first because it only has a little way to go.*"

MON (8 ; 0) prediction: "*They will go all over the place if I turn it: it's the same for all of them* [experiment]. *The nearer ones go first . . . those there* [near the edge: D 3] *go first.*" He is given two different weights: "*The middle ones go off first because they are lighter* [experiment]. *No, it's the other way around.*"—"Can you do it again?" —"*If it's like that once, it will be always!*"—We put W 1 at D 3 and W 2 at D 1.—"*No, I'm pretty dumb. It* [W 1] *took off before because* [W 2] *is nearer the pivot.*"

MOR (8 ; 0) after observations of this sort arrives at the following logical multiplication: "How are you going to place the balls [W 1, W 2, and W 3] if you want them to take off one after the other?"—[He puts W 3 at D 3, W 2 at D 2, and W 1 at D 1.] "*Because the big one is heavier and it's further forward, the second is smaller and further back, and the third is smaller and still further back.*"

These facts are relevant to the problem of compensation. When the experimenter dissociates the factors in order to make the child's task easier, the child discovers what part each plays. The balls take off according to weight, and those closest to the periphery are displaced before the most distant. But the subject could not have found the two laws by himself, since the "all other things being equal" method does not come into play before stage III. Also, when, after his independent discovery of the two factors, the subject is asked to compare balls of different weights and at different distances, he runs into a number of difficulties.

First, the subject is unable to gauge the simultaneity of take-off when balls are displaced in opposite directions. (Also, the experimental apparatus causes some minor discrepancies along this line.) In other words, the subject is not yet able to eliminate deviations due to uncontrolled factors.

Secondly, logical multiplication is inadequate for a solution to the problem. At this level subjects are able to use logical multiplication effectively when two factors reinforce each other (*cf.* MOR). But when the two factors work against each other simultaneously, the child brings one and then the other into his explanation, but he fails to understand that they can compensate each other. Nor does he look further for a multiplicative product. This fact raises a problem—must we assume that multiplication of relations remains incomplete and operates only in the intuitively favorable cases, or do we have to consider the possibility that the child regresses when faced with the indeterminate result of the multiplication "heavier $\times$ closer" and "lighter $\times$ further away"? One could argue against the second explanation on the grounds that the product is just in front of his eyes, since, in fact, there is compensation. On the other hand, the child fails to perceive it because he does not understand it, or he sees it without understanding that compensation is involved. Thus, in this particular case, it is likely that the operational mechanisms governing the multiplication of relations are already complete for those situations in which the product is determinate, but his operational mechanisms do not yet appear in a form that can be generalized to products that admit of three possibilities (heavier $\times$ closer $=$ the heavier ball takes off after the lighter ball or before it or at the same time because of compensation).

§   *Substage II-B. The Beginning of Concrete Compensation*

Unless the problem is simplified by the experimenter, at this level the subject cannot always explain why two balls of different weights can take off at the same time if they are placed at different distances from the center. He cannot isolate the variables without help; thus, he cannot conceive of the compensation of two opposing factors based on possible combinations. But when the experimenter simplifies the task by varying the factors one at a time, the child can discover both the role of distance and the part played by weight. Then, he begins to understand compensation and in some cases even predicts it:

JOL (9 ; 3). W 3 at D 2 and W 3 at D 3: *"They didn't take off at the same time because one is far away and the other up close."*—"And when they are like that?" [W 3 at D 1 and W 2 at D 3].—*"That one [W 3] will go first because it's bigger and the other is smaller* [experiment: contrary result]. *It's because it [W 2] is closer to the edge."*—"What makes it change?"—*"The weight and the size."*—"At that?" [W 1 at D 3 and W 2 at D 2].—"*Together, because if you had put this one* [W 1] *with a big one* [W 3] *the big one would go first, but with* [W 2] *it goes at the same time because it is little and it has a little distance and the other is big and it has a longer distance.*" "And?" [W 3 at D 3, W 2 at D 2, and W 1 at D 1].—"*That one* [W 3] *will go first because it's closer to the edge and it is bigger;* [W 2] *is further from the edge and smaller and* [W 1] *is less close* [than W 2] *and less big.*"—"What do you have to do to make them both go at the same time?"—He puts W 3 at D 1; W 2 at D 2 and W 1 at D 3 "*because the smallest one has a small distance*" [to cover].

CRO (10 ; 2). Experiment: W 3 at D 2 and W 3 at D 2. "*They will go together because they are both the same size.*"—"And?" [W 3 at D 1 and W 3 at D 2].—"*The furthest from the center went off first, because it's nearer the edge.*"—[W 3 at D 1 and W 2 at D 1.] "Which one will take off first?"—[W 3] "*first because it's heavier.*"—[W 3 at D 2 and W 2 at D 3?]—"*Both together because that one is smallest but closer to the edge, the other one is bigger but further from the edge, and the biggest is furthest from the edge.*"

Thus the main advance over substage II-A is that, once the subject knows the respective roles of the two factors, he completes the coordination (or logical multiplication) of weights and distances. Of course, the multiplication is performed correctly when the results are cumulative ($W_1 \times D_1$, $W_2 \times D_2$, and $W_3 \times D_3$). But some subjects explain or even predict (CRO) for case W_1D_3, W_2D_2, and W_3D_1, in which the three balls take off at the same time because the three factors compensate each other. In these cases the operation is:

$$
\begin{aligned}
&(+W) \times (+D) = \text{before} &\quad &(=W) \times (+D) = \text{before} \\
&(-W) \times (-D) = \text{after} &\quad &(=W) \times (-D) = \text{after} \\
&(+W) \times (-D) = \left. \begin{array}{l} \\ \end{array} \right\} \begin{array}{l} \text{compen-} \\ \text{sation} \end{array} &\quad &(+W) \times (=D) = \text{before} \quad (1) \\
&(-W) \times (+D) = } \text{possible} &\quad &(-W) \times (=D) = \text{after} \\
& (=W) \times (=D) = \text{equality} & &
\end{aligned}
$$

where $\pm W = \pm$ heavy, $\pm D =$ more or less distance from the center.

So once again we encounter the same double-entry table we found for work relationships (Chap. 12), density, etc., and again we encounter it at substage II-B.

One may be tempted to believe that the concept of multiplicative compensation is acquired at this substage. But two circumstances contradict this assumption. First, as we have stated above, the subject does not yet succeed in isolating the factors by himself. He does perform the operation correctly when the data are prepared in advance, but this is not the same as a spontaneous organization of deductive proof based on his previous structuring of the relevant elements (the latter is the general procedure of experimental method and requires the formal combinatorial system). Second, the operation of logical multiplication—the only one available to the concrete level subject—remains indeterminate in the cases $(+ W) \times (- D)$ and $(- W) \times (+ D)$. There may be compensations in these cases, but they do not follow from these products. The result cannot be determined completely without making use of proportions, and proportions necessitate formal thought.

§   *Stage III. Spontaneous Isolation of Variables and Compensation by Proportionality*

At substage III-A the subject can organize the experiment without outside help, and he can anticipate compensations by using a system of propositional operations. Still, his deductions are incomplete:

CHAM (10 ; 7, advanced) begins haphazardly by placing W 3 at D 3, W 2 at D 2 and W 1 at D 1: "*They left one after the other.*"—"Does that surprise you?"—"*No. The biggest should move after* [the others] *because it turns more slowly. . . . Oh! No. It's the other way around. The closer it is to the edge, the faster it turns.*" To test his hypothesis, he puts W 3 at D 3, W 3 at D 2, and W 3 at D 1; then, on a new trial, W 1 at D 3 and W 1 at D 2. He varies the distances at equal weights and uses the extreme weights for the counterproof. "*Those* [D 3] *took off first. The middle ones go next*" [D 2]. Then, after several new trials for distance, he discovers that when he puts W 3 at D 1 and W 1 at D 3, they take off at the same time, but that when he puts W 3 and W 2 at D 3, "*the big one goes first.*" Next, he compares the problem to one of equilibrium: "*It's a little like a balance scale* [in the sense of movements required for balancing]; *it's a question of equilibrium.*

If [W 3] and [W 1] are at [D 3] they have no equilibrium; if they are at [D 2] there is a little more; if it's at [D 1] it's still better in equilibrium; and if it's in the center it's completely in equilibrium. The closer it [the ball] *is to the center, the slower it goes. . . . But if [W 3] is at [D 1] and [W 1] at the center, the big one goes first.”—*“What determines the result?”—*“The size and the holes a little; no, only the size because the holes are all the same; and the force with which the ball is thrown off.”—*“What determines the equilibrium?”—*“The size of the balls and their places. If you put the same size [W 1] at [D 3] it moves before [W 1] at the center. If you want them to go at the same time, you have to put a big one here”* [at D 1 when W 1 is at D 3].

DEF (11 ; 2) also discovers the role of the weight and distance. He is asked to predict the result when W 3 is at D 1 and W 2 is at D 2: *“They will go at the same time. The big one has a larger distance to cover, but it is heavier and heavier things go faster.”—*“And?” [W 3 at D 1, W 2 at D 2, and W 1 at D 3].—*“They will go at the same time. The big one is heavier but has a larger distance to cover, so it comes out the same.”*

VIS (12 ; 9), after the same train of reasoning: *“That's a compensation.”*

DUB (13 ; 4), after he has discovered the two factors by himself: “Can you make the balls go at the same time?”—*“You have to form a proportion: the weight and distance* [he puts W 3 at D 1 and W 2 at D 3]. *They didn't go together because of the difference in weight and the difference in distance. They have to counterbalance each other exactly.”*

It is worth noting that the subjects adopt a new attitude toward compensation at exactly the point when they become able to isolate the variables (without help by the experimenter) and to experiment spontaneously; they conceive of it both as the product of an equilibrium and as a proportion. For example, instead of the former linear representation of the distances as measured from the center to the edge of the disc, we find that CHAM, viewing them from a kinematic standpoint, considers them in relation to the center. The further the ball is from the center, the greater its motion, whereas the closer it is, the more it approaches a state of equilibrium. From this stems the inverse proportion between weight and distances to which DUB explicitly resorts and that VIS terms compensation.

It is not by chance that these experimental methods and conceptions have converged. For one, the method of isolating variables results from the combinatorial system that appears at the beginning of formal thinking. Secondly, the structural integration of formal thinking requires the I N R C group, which is the source of both the equilibrium and the proportionality schemata; both of these are necessary for the appearance of the concept of compensation.

Thus, the compensation schema which appears at substage III-A is quite different from the one that we observed at stage II. At substage II-B compensation is still seen in additive terms, even though it is a result of multiplications. Compensation is explained as the logical multiplication of "heavier" or "more weight" (in the sense of an additive difference) by "closer" (also seen as an additive difference). But at substage III-A, compensation stems from a definite feeling for proportionality. For example, when DUB says that the "difference in weight" does not compensate the "difference in distance" in case $W_3D_1W_2D_3$, and that it does not "counterbalance," he uses the terms "difference" and "counterbalance" to mean a relationship between two propositional relations which can compensate each other exactly for determinate values.

It is true that the III-A subjects do not make specific mention of metrical proportions. But, first, they were not asked to do so, and secondly, they show signs of possessing a qualitative schema for proportionality that automatically becomes quantitative during substage III-B:

VIR (13 ; 0) produces combinations W_3-D_3 and W_3-D_2 by himself, then W_2-D_1 and W_3-D_1, saying: "It's a law. The lightest one goes latest in relation to the heaviest." Then W_3-D_1 and W_3-D_2, and finally W_1-D_1 and W_1-D_2: "The furthest [from the center] went first." Next he places W_2-D_3 and W_3-D_2: "They will go at the same time because there's an equivalence of weight and an equivalence in the amount of force. It's as if I put [W_2] at [D_2] and [W_1] at" [D_3].

CAI (14 years): "It depends on how they are placed. If we want them to take off together, the weight has to be reduced in proportion to length."—"How do you prove it?"—"A ball of ⅓, another of ⅔, and *the third of* 1" [for distances $D\,3$, $D\,2$, and $D\,1$].—"And if the distance is greater?"—"*You have to get the same proportional combination.*"

The proportion foreshadowed at substage III-A now becomes an actual proportion to which numbers can be assigned (CAI); the quantitative "equivalences" (VIR) assure compensation. Keeping these protocols in mind, let us return to the problem stated at the beginning of this chapter. Psychologically speaking, does the idea of compensation lead to the development of proportions, or is it the other way around? Or, as a third possibility, are they interdependent from the start?

§ *Proportions and Compensation*

The foregoing analyses (Chaps. 11 to 13) allow us to say that proportionality is a general schema linked to the double reversibility of reciprocals and inverses (I N R C group); sometimes it takes the form that we saw in the case of mechanical equilibrium; sometimes it takes other forms (geometric, etc.). But the diverse forms of the I N R C group have a common property—the operation of compensatory processes expressed through reciprocity R, as opposed to simple inversion N. The intervention of judgments of compensation are common to all the forms of proportionality discovered by the subjects. Given two independent variables, the subject constructs the qualitative proportionality schema when he understands that an increase in one gives the same result as a decrease in the other. In all cases the structure of proportions requires an element of compensation. When one has $\dfrac{x}{y} = \dfrac{x'}{y'}$ the products $xy' = yx'$ constitute a system of compensations such that any change in the value of x must be compensated by a modification of at least one of these terms if the equality is to be preserved.

However, the intuitive evidence for the idea of compensation may differ greatly from one case to the next. First of all, it is much more obvious when compensation guarantees the conservation of a visible effect (size of a shadow, etc.) than when purely relational features are involved—*e.g.*, conservation of parallelism in Thales' theorem. Compensation is particularly clear when the reciprocal factors assure equilibrium in a mechanical system, etc.

What determines the ease with which proportionality can be isolated? It depends on whether direct or inverse proportions are involved and the ease with which factors can be compared from the standpoint of their units of measurement. For the centrifugal force problem, the compensations to be analyzed are perceptually clear. We reduced the weights to but three unit differences—W_1, W_2, and W_3—and the distances D to three also—as a result, the proportion $W_1/W_3 = D_1/D_3$ is so intuitively obvious as to eliminate any need for calculation.

It is remarkable that, even in so simple a case, the schema of compensation still precedes the proportionality schema. In other words, the subject first wants to isolate the potential conservation for the same result (*i.e.*, simultaneous take-offs) so that he can find the proportions, whereas he could have started from the operant relationships and their proportions in order to come to the idea of a potential compensation.

Said differently, the subject becomes able to find the inverse proportion of weights and differences when he discovers that an increase in weight leads to the same result as an increase in distance and, therefore, can be compensated by a decrease in distance. But after having seen that the balls took off sooner in proportion as they were heavy and later in proportion as they were close to the center, the subject could have set up the inverse proportion and deduced its possible equivalences and compensations.

If we let p stand for an increase in weight (and thus $\bar{p}$ its decrease) and let q stand for an increase in distance from the center (and thus $\bar{q}$ a corresponding decrease if p, $\bar{p}$, q, and $\bar{q}$ apply only to changes), and if we let r and $\bar{r}$ stand for the propositions asserting that one ball takes off before or after another and r_0 represent the statement that they take off simultaneously, we see that the subject begins by assuming:

$$p \supset r \qquad \text{and} \qquad q \supset r, \qquad (1)$$

from which we get:

$$r_0 \supset (p.\bar{q}) \vee (\bar{p}.q), \qquad (2)$$

which expresses the compensation: A ball has to be both heavier and closer to the center or lighter and further from the center if it is to take off at the same time as another when weight or distance is modified.

The proportion as stated below is derived from this statement:

$$\frac{p}{\bar{q}} = \frac{q}{\bar{p}}, \text{ for example } \frac{W_3}{D_1} = \frac{D_3}{W_1}, \text{ or} \tag{3}$$

$$\frac{\bar{p}}{p} = R\frac{\bar{q}}{q}, \text{ for example } \frac{W_1}{W_3} = \frac{D_1}{D_3}. \tag{4}$$

But he could have derived the qualitative propositions below from (1):

$$\frac{\bar{r}}{r} = R\frac{\bar{p}}{p}, \text{ and} \tag{5}$$

$$\frac{\bar{r}}{r} = R\frac{\bar{q}}{q}, \tag{5a}$$

from which he could have derived proportion (4) by transitiveness. From (4) he then could have derived the products:

$$p\bar{q} = R\,\bar{p}q \tag{6}$$

which results in the possibility of a compensation for r_0 (prop. [2]) which can be verified experimentally. But he does not actually proceed in this manner; for he discovers the compensation by concrete operations starting with substage II-B. Moreover, the subject begins with compensation and works through to proportions not only in the centrifugal force problem but in all of the cases which we have studied before this one (Chaps. 11 to 13). Of course, the fact that we usually asked the subjects to find compensations or equivalences may have influenced our findings, but this was done because after experimenting we found that this was the easiest path for the adolescent to follow since the compensations were discovered in a concrete mode at substage II-B in each of these experiments.

Given the consistency of these findings, we have to look for an explanation. First we have to go back to the general characteristics of the compensation schema. Let us take two factors, for which an increase is stated by p or q and a decrease by $\bar{p}$ or $\bar{q}$, which lead to a final result whose increase is stated by r, decrease by $\bar{r}$, and conservation by r_0. Two clearly different cases may then occur:

First, one can have:

$$p \supset r \quad \text{and} \quad q \supset \bar{r} \tag{7}$$

(for example, an increase in the diameter of the circles implies an increase in the size of the shadow, whereas an increase in distance reduces the size of the shadow: Chap. 13).

In this case, compensation takes the form:

$$r_0 \supset (p.q) \vee (\bar{p}.\bar{q}) \tag{8}$$

where there is a direct proportion between p and q. Let us call p_1 and q_1 (or $\bar{p}_1$ and $\bar{q}_1$) the changes conjointly effected in these elements and p_2 and q_2 (or $\bar{p}_2$ and $\bar{q}_2$) the changes in the second pair. Then we have for (8):

$$\frac{p_1}{q_1} = \frac{\bar{q}_2}{\bar{p}_2}, \text{ or} \tag{9}$$

$$\frac{p_1}{q_1} = R \frac{\bar{p}_2}{\bar{q}_2} \tag{9a}$$

(for example, the diameter of the larger of the two circles is to its distance from the light source (also the larger distance) as the diameter of the small circle is to its distance from the light source).

But, secondly, one can also have (as in the centrifugal force and balance problems):

$$p \supset r \quad \text{and} \quad q \supset r \tag{10}$$

$$(cf. \text{ prop. } [1]). \tag{10}$$

In this case, the compensation is:

$$r_0 \supset (p.\bar{q}) \vee (\bar{p}.q) \quad (cf. \text{ prop. } [2]). \tag{11}$$

Consequently, there is an inverse proportion—*i.e.*,

$$\frac{p_1}{q_2} = \frac{\bar{q}_1}{\bar{p}_2}, \text{ or } \frac{p_1}{q_2} = R \frac{\bar{p}_2}{\bar{q}^1} \tag{12}$$

(for example the weight of the large ball is to the greater distance of the small ball as the weight of the small ball is to the small[er] distance of the large ball).

Given the above formulations, one can understand why the compensation schema is more directly accessible than the proportionality schema. First, compensation is based directly on qualitative logical relationships such as (7)–(8) or (10)–(11), whereas the proportions acquire an experimentally verifiable structure only when they are quantified. Consequently, there is a kind of logical anticipation of proportions (as the respective equality of two products and two sums) before they are put into metrical form. This fact enables us to grasp the import of the anticipatory schema, since it is always derived from the compensation schema. Secondly, the compensation may be additive or multiplicative as is the case for logical proportions; this is not true for metrical proportions. This accounts for the initial tendency of the child to look for proportionality in the equality of additive differences. Finally, compensation derives directly from the idea of reciprocity, since (prop. [8]) $\bar{p}.\bar{q}.$ is the R of $p.q$ and (prop. [11]) $\bar{p}.q$ is the R of $p.\bar{q}.$

Moreover, for relations, reciprocity is at the core of reversibility, and reciprocity cannot be combined with inversion N except through the mediation of the I N R C group, the basis of proportions.

15

Random Variations
and Correlations[1]

PROBLEMS OF CHANCE are relevant to the study of formal thought from two standpoints. In a general way, formal thought has the property of dealing with what is possible and not only with what is real. But the probability that an event will occur is nothing more than the relationship between the possible instances of an event and those which actually occur. Moreover, a probability estimate of relations or laws presupposes certain special operational instruments such as the calculation of "correlations" or "associations." In its simplest form, the notion of correlation is a formal operational schema related to those we have just studied—particularly the proportionality schema. The aim of this chapter is to analyze the two-sided problem of how subjects from 5 to 15 years react to chance fluctuations that occur during the experiments and how they construct the correlation schema.

RANDOM VARIATIONS

Nearly all the phenomena studied in the foregoing experiments involve chance fluctuations. We have emphasized elsewhere [2] that one of the essential tasks of experimental reasoning *or* induction

[1] With the collaboration of Vinh Bang and S. Taponier, research assistant, Institut des Sciences de l'Éducation.
[2] See Piaget, *Introduction a l'epistemologie génétique*, Vol. II, Chap. 6.

is that of separating the deductible from the random. The earlier chapters have shown how the child and adolescent organize the deductible; now we shall examine how they react to chance and how they assimilate it to the deductible, though they do so indirectly through probability. For this purpose, we are not going to devise new experiments (we have already devoted an entire volume to the child's handling of chance).[3] In the present work we shall simply analyze some of the protocols already collected. Moreover we shall limit this analysis to two of the experiments which especially touch on this problem—the experiment that involved launching a ball in a horizontal plane (Chap. 8) and the experiment that involved equilibrium between the pressure of a piston and liquid resistance (Chap. 10).

Both the stopping points of the balls and the equilibrium levels reached by the liquid and the piston entail notions of probability, as neither is strictly constant under any given set of conditions. So the subject must first fit a probability law to the fluctuations. Secondly, he has to isolate the laws or causes of the phenomenon under study (motion or equilibrium) in spite of the fluctuations. But the first task is precisely the problem of the probability of random variations, and the second that of correlations.

§ Stage I. Neither Conservation nor
Law of Distribution

At the preoperational level, the subjects' attitudes toward chance are paradoxical. They expect that under similar conditions given phenomena will be repeated either identically or in terms of a definite progression, etc. When they do realize that there are small fluctuations, they first deny the conservation of the relevant quantities (matter, etc.) and then conclude that the stopping points are completely arbitrary:

MEY (6 ; 8), in the stopping-point problem, sees that the large aluminum ball stopped at a given point [20 cm.]: "And if we throw it again, where will it go?"—"*Further*" [experiment: 21 cm.].—"And if you throw it again?"—"*A little further because it already went further.*"

3 Piaget and Inhelder, *La Genèse de l'idée de hasard chez l'enfant.*

[experiment: 19 cm.].—"Why did it stop there?"—*Because the little flag* [that marks the stopping points] *is still there.*"—"And if you throw it again?"—"*A little further because it goes a little further every time.*"

CROS (5 ; 9) for liquids, predicts that if the red box [500 grams] is put back on, the water will return *"to the same place,"* pointing out the red flag. The water goes a little further; CROS explains this *"because the box is heavier, . . . it goes faster,"* etc. He is then shown a series of positions and asked whether or not they are possible for the same red box: they are all possible. The same for the other boxes.

TAC (6 ; 0): "Where will the water go if this box is put back on? [1,500 grams].—*"There. It will go where it was before."*—"Why?" *"It's the same."*—[Experiment: the water level is higher.] "Why?"— *"Because it's heavier"* [the water].

These responses are familiar ones in the protocols of the young children. They deny chance, but when faced with fluctuations they believe anything is possible or look for a hidden order (effect of the flag for MEY) or a temporary disorder masked by invisible reasons which have to be divined. In both cases, the subject's attitude is reinforced by his lack of notions of conservation; the box can become heavier in moving more quickly, the water can increase in quantity or weight, etc.

§ *Stage II. Diffuse Probabilistic Responses (II-A) then Determination of a Zone of Distribution (II-B)*

After 7–8 years of age, the subject's responses are quite different. Not only does he cease to be surprised by variability, but his predictions often take it into account ("it will go about the same place"). As usual, the appearance of the notion of chance is at first characterized by a generally negative attitude based on caution and a feeling that it is hard to make predictions:

BOUT (7 ; 6): "If you throw it [the same ball on the horizontal plane] 10 or 20 times?"—"[It can get] *there* [1.60 m.], *there* [1.79], *or there* [1.80].—"Do you think it will ever go all the way to the end?"—*"No."*

DUB (7 ; 5): "[The ball will reach] *about the same place.*"—"Can it go all the way to the end?"—*"No."*

And in the liquid problem:

GUI (8 ; 0): "Why didn't the water go up as high this time?"—*"Because a little water went out"* [into the piston tube].—"And now, why is it a little higher?"—. . . [no explanation for two trials].

DES (8 ; 0), in order to explain the deviations in an upward direction, first hypothesizes that the clamps are tighter, then says *"because I put it* [the box] *down harder."*

Above all, these diffuse responses show how the subject is disturbed by chance—*i.e.*, by that which resists his budding operations. But by about 9 years, the subjects are no longer satisfied with characterizing fluctuations in terms of the essentially negative notion of "about" and try to find systematic causes for them. When they are asked to predict the results of 10 to 50 successive trials, they may even come to delimit true zones of variation.

In the liquid experiment (Chap. 10), the search for causes was particularly stressed:

BUC (8 ; 5): *"Sometimes it goes down faster, sometimes slower."*

ZBI (9 ; 4): *"Maybe I let it go harder. . . . Maybe I let the water fall to here* [different starting point] *and let the box fall harder."*

In the horizontal plane problem, we asked for predictions:

COR (9 years): "Where will it go now?" [the large aluminum ball, 36 cm.].—*"The same as before, around there."*—"And if you try 10 times in a row, can it go here?" [1 m.].—*"No."*—"And there?" [65 cm.].— *"Yes."*—[Experiment: 42, 36, 37, and 38 cm.] "Can you say now where it will go?"—*"Yes, around there"* [indicates between 27 and 47 cm.]. —"Could it go all the way up here?" [60 cm.].—*"Yes, sometimes."*— "And there?" [1 m.].—*"No, between here and there* [from 10 to 65 cm.].

WIN (9 ; 8). Small aluminum ball to 1.80, then to 1.62 m.: "Does the difference surprise you?"—"No."—"And if you throw it again?"—*"Not as far as the first time"* [experiment: 1.36, 1.51, and 1.48 m.].—"And if you threw it 50 times?"—*"About there"* [between 1.27 and 1.95 m.].

JOS (10 ; 2). Small brass ball to 27 and 32 cm. "And if you sent it off 50 times?"—*"There will be bigger spaces between the flags."*—"Will it go everywhere?"—*"No, around here"* [20 to 40 cm.].

LUC (11 ; 0). Same ball: *"No further than that* [37 cm.] *and no closer than that"* [20 cm.].—"If you throw this ball 10 times, where is it most likely to go?"—*"Around here"* [28–29 cm.].—"Why?"—*"Because it's about the middle."*

FRA (11 ; 8): "If you played 50 times?"—*"Between 20 and 50 cm."*— "How can you tell?"—*"You see about where it goes."*—"Will there be more flags in some places?"—*"Yes, here* [35–40 cm.]. *because if it's thrown regularly there will be more in the middle than at the edges."*

Whereas, at first, chance was seen in opposition to operations, during substage II-B it begins to be assimilated to them through the search for the causes of fluctuations and a determination of their amplitudes. The result is that subjects set the boundaries for a zone of variation and, toward the end of the substage, they understand that the deviations comprise a curve with a higher frequency in the median region and a lower frequency at the extremities. In this way FRA intuits the Gaussian curve. In another work we studied the representation of this curve at different stages of development.[4]

Moreover, we are struck by the fact that those substage II-B subjects who spontaneously point out a zone of variation in the way seen above are the same ones who from the start perceive at least two factors as causes of stopping—*e.g.,* weight and volume, volume and matter, etc. (see Chap. 7). But, since the variables are not yet systematically isolated at this stage, the subjects' assertions about causality and the variation zone cannot be taken to imply a latent correlation structure. Rather, they see the variation simply as the result of a multiplicity of causes and make no effort to isolate the respective parts.

§   *Stage III. Explanation of the Distribution and Determination of the Law which Underlies the Chance Fluctuations*

At stage III the subjects begin active experimentation. They also begin to make allowance for random variations (whose form they now try to discover) and to isolate the law which underlies them.

    [4] *Ibid.,* Chap. II.

Moreover, isolation of the law sooner or later requires the formation of a correlation schema.

First, a case from the liquid problem:

BOI (14 ; 6): "Where will the water go?"—"*To the same level as before.*" —[Higher.] "What do you have to say about that?"—"*It depends on the jiggling of the apparatus. Maybe it jiggled more.*"—"Sure?"—"*No. It could have been clogged up at the top. No, it would be lower; the air would be compressed.*"—"If we do it again?"—"*If the experiment is precise, the water will go to the same level when the piston slides easiest.*"—"How high will it go when it slides easiest?"—"*If you perform the experiment several times, you'll find the spot.*"—"If you do it 5 times?"—"*Take the average of 5.*"—"When it slides easiest, you get the average?"—"*It will be the best reading we'll get.*"

The following examples refer to the stopping point experiment:

CHAP (13 ; 3): "*It will go about there*" [0.9 to 1 m.].—"Can it go all the way up there?" [1.60 m.]—"*No, because it's too heavy.*"—"And that one?" [wider and lighter].—"*Around here*" [1.50 to 1.79 m.].— [Experiment: 1.60 m.] "Why?"—"*Because it's not as heavy as the other one.*"

RAY (14 ; 4) describes a range of 1.10 to 1.55 m.: "*Maybe it's launched in different ways* [he tries to throw the ball with a constant force]. *I see; it's the launching force that varies. Theoretically it should go to the same place. The friction has to be reduced.*"—"Can you show that the friction has an effect?"—"*Theoretically the small one should go further. The air resistance varies with the volume* [he has now confused friction with air resistance]. *If you take two balls of the same weight with different volumes, you can prove that the friction has an effect.*" He performs the experiment and proves his point in spite of the fluctuation; the number of confirming cases seems sufficient to him.

LEV (15 ; 5) wants to demonstrate the role of volume: "*These two balls have the same weight but not the same volume* [experiment]. *They roll* [approximately] *the same distance, so volume doesn't play the main role.*"—"A small role or none at all?"—"*It plays a small role because the small wooden one goes further than the big one.*" Thus he is aware that there are confirming and nonconfirming cases (without calculating them) in spite of the small deviation. As for the fluctuation, he sees it as caused by the launcher: "*Sometimes I launch it harder than others.*"—"What zone?"—"*Here and there* [9 to 20 cm.]. *That's*

about the path it takes with a normal launching. The exceptions go to-
ward the edge. It scrapes a bit; you see [traces]; *that proves it acts*
as a brake."

NIC (15 years) wants to show that the light ball goes further than two
heavier balls of the same volume, but he realizes that there are fluctuations: "*My hypothesis should be right but only with small pushes.*"
—"Then the hypothesis is wrong?"—"*If my hypothesis is wrong, I don't
know what to think. I didn't think of the fact that the launchings don't
all have the same force.*" Thus he maintains the variable of lightness
in spite of the interference of variations in results.

It is easy to see what is new in these responses as compared
with those of stage II. The stage II subjects are limited to describing the raw experiment with concrete operations, without
separating the variables. In this way they can discover the fluctuations just as they discover everything given in the raw experiment.
They even construct laws of distribution (bell-shaped curve with
maximum frequency at the center). But since they try to arrive at
conclusions without separating the empirically given variables
according to the possible combinations, they can do no more. In
contrast, the stage III subjects want to find laws by accurately
separating out variables in terms of all possible combinations.
But they run up against the obstacle—one that must also be systematically analyzed—of the fluctuation of results. Thus we must
ask what method they use in their analysis.

Sticking to spontaneous reactions and leaving aside for the
moment the subjects' enumeration of the instances which they feel
confirm and those which disprove a particular hypothesis, we find
that stage III subjects make use of a method that brings us to the
problem of correlations.

First, we see that a simple statistical enumeration of cases
(which determines the zone of variation with its high frequency
"middle") is not enough to solve the problem. Thus BOI proposes
to take the mean of the water levels to estimate a normal level
for a given weight, but he realizes, in the case in which the piston
slides most easily, that the mean differs from the "best point
obtained." NIC, especially, who wants to prove that a heavy ball
does not go as far as a light one, sees that he has to account for
the interference of scatter and realizes that his hypothesis is

proved only for "small launches"—*i.e.*, for the shortest paths obtained with the light and heavy balls.

Secondly, this results in the formation of a new intellectual attitude specific to stage III, one in which confirming instances are distinguished from nonconfirming instances. The subject sees that he must decide which are more frequent. We saw that RAY and NIC held to their hypotheses in spite of the scatter they noted (and, in NIC's case, in spite of the exceptions due to "forceful launchings"). And, because of the number of nonconfirming cases, LEV reinstates the importance of volume, which at first he wanted to prove was but a minor factor.

This is why we must try to reconstruct the reasoning of the adolescent as it bears on the question of confirming and nonconfirming instances. We have seen constantly that the *qualitative* operational schema for proportions (based on interpropositional links) precedes a *metrical* treatment of these same proportions. In the same way, we shall now see how a simple *qualitative estimate* of the range of scatter develops in the direction of an operational schema for *correlations*.

As an example, let us assume that the subject wants to prove that the smallest balls go the furthest. Let p be the proposition that the ball under consideration is smaller than the standard ball and $\bar{p}$ the statement that it is larger; let q stand for the fact that the small ball goes further than the other and $\bar{q}$ that it does not go so far. Observation of the fluctuations then leads the subject to assume the truth of the four possible combinations (where * is the sign for "complete affirmation"):

$$(p.q) \text{ v } (p.\bar{q}) \text{ v } (\bar{p}.q) \text{ v } (\bar{p}.\bar{q}) = (p * q). \tag{1}$$

Now one can see immediately that these four possibilities contain the four cells of the type of association table used in calculat-

	small (p)	large ($\bar{p}$)
far (q)	$a = p.q$	$c = \bar{p}.q$
near ($\bar{q}$)	$b = p.\bar{q}$	$d = \bar{p}.\bar{q}$

ing simplified forms of correlation called "coefficients of association" (Yule or Bravais-Pearson formula). In the present instance, the subjects do not break down the cases so that they correspond to the four possibilities, nor do they calculate ratios between the numbers in the table. Instead of this numerical quantification, they attempt an intensive quantification (using $>$ or $<$) and are content to stop there. But the intensive quantification seems to involve estimating both the number of a and d cases in the table (*i.e.*, $p.q$ v $\bar{p}.\bar{q}$) and the number of b and c cases (*i.e.*, $p.\bar{q}$ v $\bar{p}.q$) and comparing the two sets. The a and d cases confirm the hypothesis being tested and the b and c represent the nonconfirming instances, but our subjects seem to make exactly this comparison. Thus their reasoning can be represented as follows (E stands for the sum that verifies the propositions as to confirming and nonconfirming instances):

$$E\ [(p.q)\ \text{v}\ (\bar{p}.\bar{q})] \gtrless E\ [(p.\bar{q})\ \text{v}\ (\bar{p}.q)]\ ; \ i.e.,$$
$$(a + d) \gtrless (b + c). \tag{2}$$

Consequently, if the subjects determine numerically the difference $(a + d) - (b + c)$ and its relationship to the whole $(a + d) + (b + c)$, instead of being satisfied with the comparison based on $>$ or $<$, one could say they are explicitly using a notion of correlation. Does this mean that correlations are structured during stage III? A more direct approach will help to answer this question.

CORRELATIONS

Thus, this latent correlation schema, which may be present at stage III, needs a more detailed analysis. To study it, we have devised an apparatus such that the subject can easily count the cases that confirm and those that fail to confirm a hypothesized relationship between two variables. It will allow us to see what relationships will be established between confirming and nonconfirming cases and if these relationships resemble any of the "association" formulae used in the calculation of correlations.

The problem set for the subjects involves simply a correlation between eye and hair color. Subjects are shown 40 cards, each with a face drawn on it. The eyes and the hair are colored according to the following four associations:

$a =$ blue eyes and blond hair $\quad(= p.q),$
$b =$ blue eyes and brown hair $\quad(= p.\bar{q}),$
$c =$ brown eyes and blond hair $\quad(= \bar{p}.q),$
$d =$ brown eyes and brown hair $(= \bar{p}.\bar{q}).$

The subject is then given a set number of cards and asked whether he thinks there is a relationship between eye color and hair color (*i.e.*, not whether there is such a relationship in real life, but whether one can be discovered in the given data). At the start, it is possible to proceed in either one of two different ways; one can let the subject form his own classification (construct the four boxes of a double-entry table) or give him the cards already classified according to the four possibilities. The latter method puts more emphasis on the possible numerical combinations. For example, we might apply combinations *abcd* to 4, 0, 0, and 4 faces respectively; or to 4, 4, 4, 4; 6, 6, 2, 2; or 13, 8, 3, 8; etc., asking the subject in each case to estimate the relevant relationships. In addition, the subject can be shown two different sets (*i.e.*, 6, 4, 2, 4, and 4, 4, 4, 4) and asked which shows the clearest correlation. Finally, the subject can be asked to remove cards in such a way as to strengthen the correlation; then he can be asked to discuss which of the four associations he used as a basis for eliminating cards.

This is the experimental situation; it provides us with a number of interesting responses to examine from the standpoint of propositional logic—the only standpoint we are concerned with now— as opposed to questions of calculation or induction in the narrow sense of the term. If we let p stand for the presence of blue eyes and q for the presence of blond hair, cases favorable to the correlation will correspond to equivalence $(p.q) \text{ v } (\bar{p}.\bar{q})$ and nonconfirming cases will correspond to reciprocal exclusion $(p.\bar{q}) \text{ v } (\bar{p}.q).$ If one is to show a correlation one must first establish two classes of individuals, each class corresponding to the conditions stated in one of the two kinds of links we discussed, $(p = q)$ and $(p \text{ vv } q);$ then, one has to determine the relationship between these two classes.

But, from the start, the organization of these classes or these links raises a number of difficulties. Even at substage III-A the subject often begins by considering association a (*i.e.*, $p.q$) independently, without understanding that the d cases (corresponding to $\bar{p}.\bar{q}$) are just as crucial. And when he tries to relate the a cases ($p.q$) to another association, he may at first think of case b ($p.\bar{q}$) rather than case d. In so doing, he proceeds vertically (eyes) or horizontally (hair) in the double-entry table before he understands that the diagonal comparison has to be made.

Once the initial difficulties have been overcome, the problem is to discover that correlation is not a simple probability—*i.e.*, an elementary ratio between the confirming cases ($a + d$) and the total number of possible cases ($a + b + c + d$). The basic difficulty in correlations is certainly found here. And herein lies an explanation of the fact that correlation is not acquired before substage III-B, whereas simple probability in its multiplicative form (ratios) is found at substage III-A.[5]

In its most elementary form—*i.e.*, the additive—the association coefficient R is derived from the following formula:

$$R = \frac{(a + d) - (b + c)}{(a + d) + (b + c)}.$$

If we let E stand for all the items which confirm a particular interpropositional link, the expression can be written as follows:

$$\left\{ \begin{array}{l} E\,[(p.q) \vee (\bar{p}.\bar{q})] - E\,[(p.\bar{q}) \vee (\bar{p}.q)] \\ E\,[(p.q) \vee (\bar{p}.\bar{q})] + E\,[(p.\bar{q}) \vee (\bar{p}.q)] \end{array} \right\}. \tag{3}$$

Thus the discovery of the correlation takes place as follows: After having found the probabilities $(a + d) / (a + b + c + d)$ and $(b + c) / (a + b + c + d)$, the subject still must learn that correlation is a function of the difference $(a + d) - (b + c)$ divided by the total number of cases.

Of course, we do not expect our subjects to invent these formulae anew or to perform a complete calculation amounting to the same thing. But from the standpoint of the qualitative reasoning performed on the numerical combinations, the difference $(a + d) - (b + c)$ can be said to be present when the subject has constructed the two classes $(a + d)$ and $(b + c)$, which correspond to the two links, $(p.q \vee \bar{p}.\bar{q}) = (p = q)$, and $(p.\bar{q} \vee \bar{p}.q) =$

[5] *Ibid.*, Chap. V, No. 3, and Chap. VI, No. 6.

$(p \vee\vee q)$, and realizes that, if they are of equal probability, the correlation is zero. Conversely, the correlation is strengthened as inequality $(a + d) > (b + c)$ increases.

But if it is interpreted in this way, independently of any explicit formula, we shall see that the idea of correlation is certainly discovered during substage III-B as a consequence of the utilization of propositional logic. In this context it is not worth returning to the nonprobabilistic responses of stage I or to the first probabilistic schemata of stage II. Rather, we shall begin with the analysis of substage III-A, for it is only at this point that the subject is able to reason about the sets of cards that he is given without having to appeal to the empirical world (hypothetico-deductive reasoning).

§ *Substage III-A. Probabilistic Interpretation of Frequencies Considered in Isolation but Without Relating the $(a + d)$ Cases and the $(b + c)$ Cases*

As we would expect, the substage III-A subject can estimate probabilities as relationships between positive confirming cases and the cases which are possible relative to the characteristic under consideration. Consequently, he knows how to judge the chance that a given individual has blue eyes if he has blond hair by comparing the number of a cases to the number of b cases or the sum $a + b$. But in spite of this, he cannot yet add up the set of positive and negative confirming cases $(a + d)$ and relate them to the nonconfirming cases $(b + c)$ or to the set of all possible cases. Below are some examples:

LYN (12 ; 4): "Can you find a relationship between hair color and eye color in the cards?" [6, 0, 0, 6].[6]—"*Yes. These [d] have the same color eyes as hair.*"—"But for the group as a whole [all of the cards are shown], is there a relationship between the color of the eyes and the hair?"—"*No.*"—"Here?" [d].—"*Here it's only brown.*"—"And here?" [a].—"*It's blue. They are all blue.*"—"And here [6, 2, 4, 4], is there a relationship?"—"*No. Yes, the four [sub-sets] separately, but not when they are together.*"—"Why?"—"*Because some are yellow* [blond] *and*

---

[6] The number of cards in each set corresponding to the four possible combinations are indicated in the order, a, b, c, d.

blue, and some yellow and brown."—"And like that?" [4, 4, 4, 4].—
"*You have more chances here; they are all 4. There if you go wrong, there are 2 [b] while here there are 4 and 4*" [indicates *a* and *b*].—
"How many chances do you have of finding blond hair if you only see blue eyes?"—"*Four and 4; the chance is the same*" [*a* and *b*].

She is given all the cards to classify; she does it immediately according to the four associations. Then she is asked to form two groups so that the chances are higher of finding a relationship between eye color and hair color in one than in the other. She gives 3, 3, 4, 4 and 3, 6, 6, 4: "*The chances are higher here (3, 3, 4, 4) because you have 3 and 3, and 4 and 4, while there, 6 and 4, and 6 and 3.*" In other words, although LYN has organized the sets correctly, she organized them and justified her view by reasoning about relations *a/b* and *c/d* and not in terms of the diagonal relations $(a + d) / (b + c)$.

MOR (13 ; 6), when he is given the set [10, 2, 3, 9], answers that "*there's a chance of being wrong.*" In attempting to predict eye color from hair color or vice versa he points out the nonconfirming cases, *b* and *c*, but does not calculate the ratios. "And here?" [12, 0, 0, 12].
—"*They are the same colors; you're not likely to be wrong.*"—"And this set?" [8, 4, 6, 5].—"*You could be wrong.*"—"And what can you do to be sure?"—[He takes off the *b* and *c* cases.]—"And if you have these two groups [8, 4, 4, 8 and 11, 1, 7, 5], where are you most sure?"—
"*Here* [11, 1, 7, 5] *you have less chance of being wrong because there are fewer exceptions*" [he points out the ratio of *a* to *b*—i.e., 11 to 1—but without paying any attention to the 7 *c* cases].

BON (14 ; 3) claims that "*there is no relation between eye color and hair*" for the set [5, 1, 2, 4].—"Why?"—"*Precisely because there are different cards*" [he shows *b* and *c*].—"And there?" [3, 0, 0, 3].—"*Yes, because the brown eyes are with the brown hair and the blond hair with the blue eyes. . . . There is a maximum relationship.*"—"And in these two sets?" [4, 2, 2, 4 and 3, 3, 1, 5].—"*There is no relationship at all because you have a different number of people in each part.*"—
"But is it the same in both groups?"—"*No. Here* [4, 2, 2, 4] *you have 4 and 4 and 2 and 2* [beginnings of the diagonal relationship], *but there* [3, 3, 1, 5] *you have 3 and 3* (= *a* and *b*); *it's the same number on each side*" [from which he concludes that no relationship between the variables can be demonstrated].

These cases suffice to bring out the two main difficulties that stand in the way of the substage III-A subjects. The first, shown clearly in LYN's case, is that when she encounters the four combinations $a\,(=p.q)$, $b\,(=p.\bar{q})$, $c\,(=\bar{p}.q)$ and $d\,(=\bar{p}.\bar{q})$, she has no difficulty in understanding the relationship between brown hair and brown eyes or between blond hair and blue eyes, but she does not understand that it is the same or the reciprocal relationship ($p.q$ and $\bar{p}.\bar{q}$). She places the a cases in opposition to the b cases and the d cases in opposition to the c cases without seeing that the a and d cases reinforce each other and form a single whole—i.e., composed of the cases which are favorable to a *general* relationship between eye and hair color.

A second difficulty stems from the same source as the first but its effect lasts longer (*cf.* MOR and BON who get around the first difficulty but are stopped by the second). Once the subject sees that cases a and d confirm the relationship he seeks and that cases b and c oppose it, he does not calculate the ratio of confirming cases to nonconfirming or to possible cases by comparing the sum $(a+d)$ to the sum $(b+c)$. Instead, unless he is limited to one of the pairs (either ab or cd, neglecting the other), he compares only a to b and d to c. For example, when MOR compares the two equal ratios $(8+8/4+4$ and $11+5/1+7)$ he finds the second more favorable because he compares $1/11$ to $4/8$. LYN does the same when she limits herself to the relationships between a and b and d and c.

One of the interesting facets of the reasoning from relations ab or cd rather than from diagonals ad and bc is the reaction of the subjects to the cases in which total cases and nonconfirming instances are equal: $ad = bc$. The subjects are a long way from understanding that the correlation is exactly zero in such cases. On the contrary, at this level they tend to assign them special importance, because when they reason about a and b only (or d and c only), the positive and negative chances are equal. Thus LYN prefers set 4, 4, 4, 4 to set 6, 2, 4, 4 because there are "4 and 4; the chances are the same." She also concludes that 3, 3, 4, 4 gives a higher correlation than 3, 6, 6, 4. Although this is correct, she explains it by saying that there is an equality $a = b$ and $c = d$ "because you have 3 and 3 and 4 and 4." Of the above subjects, only BON understands that, in case 3, 3, 1, 5, $a = 3$ and

$b = 3$ proves nothing ("because it's the same number on each side"), but even he has used reasoning having to do with a and b only.

In contrast, toward the end of substage III-A, we find a level intermediate between III-A and III-B in which the subject gradually comes to the diagonal relationship and begins to consider combined probabilities $(a + d)$ and $(b + c)$ even though he starts off with the same kind of reasoning found at substage III-A. Below are some examples:

BAB (14 ; 3) when he is given the set [5, 2, 1, 4] says: *"There are several with blond hair and blue eyes, but there are others too"* [he shows cases *b, d,* and *c* in order].—"Is there a relationship?"—*"There's a relationship anyway; most of the ones who have blue eyes have blond hair and most of the ones who have brown eyes have brown hair."*— "How many chances do you have of being right?"—*"In this group [a] 5 chances and here [d and c] 4 chances of being right and 1 of being wrong."*—"And on the whole?"—*"Three chances out of 12 of being wrong* [thus $(b + c)/(a + b + c + d)$] . . . *3 chances out of 12."*

"And here?" [6, 0, 0, 6].—*"You have an equal number of chances. . . . No, you will always be right."*—"And there?" [5, 1, 3, 3].—*"One chance in 12 of being wrong; no, there are those too [c]; no, 4 chances in 12."*—"In which of these two groups [5, 2, 1, 4 and 5, 1, 3, 3] are you most likely to be right?"—*"The same: 5 and 5* [he counts the *a* cases]. *It doesn't make any difference."*—"And how many chances of being wrong?"—*"Three chances in 12 and 4 chances in 12."*

"And in these two groups?" [4, 2, 2, 4 and 3, 3, 1, 5].—He classifies the cards, then compares the *a* cases with each other and the *d* cases with each other, then counts the whole: *"Four out of 12 come out wrong here and there!"* Finally he is asked to form a set such that no prediction can be made. He chooses 1, 1, 1, 1, then 2, 2, 2, 2, etc.; he must understand that the correlation is zero in such cases.

VEC (14 ; 6) classifies set 5, 1, 2, 4 correctly: "Is there a relationship?" —*"Not really. There are those"* [he puts the *a* and the *d* aside].—"So there is a relationship that has to be eliminated?"—He shows *b* and *c*. *"There is a half—no, 2/6 and 4/6* [he calculates *c* over *c* + *d*, and *d* over *c* + *d*] *and in the blue eyes 5/6 and 1/6"* [he calculates *a* over *a* + *b*, and *b* over *a* + *b*].—"But if you take the whole group into account?"—*"Approximately 9/10—no, 9/12—which are pretty much covered by the law and 3/12 which are not covered."*

"And with these?" [6, 0, 0, 6].—*"The law is exact; there are no exceptions at all."*—"And here?" [4, 2, 3, 3].—*"There the proportion is pretty weak; it's about half and half."*—"Exactly?"—*"7/12 in the rule and 5/12 exceptions."*—"Can you say it's a law?"—*"Less. For the brown hair you can't say it's a law; it's half and half. For the blond hair, it's better."*—"And together?"—*"You could say there's a law but it's not very regular."*

He is asked to form a group in which there is no relationship; he immediately produces 3, 3, 3, 3: *"It cancels out. You can't say it's a law; 3/6 are covered by rule and 3/6 aren't for the brown hair and for the blond hair."*

Finally he is asked to compare sets 4, 2, 2, 4 and 3, 3, 1, 5: First he compares *a* to (*a* + *b*) and *d* to (*c* + *d*).—"But for the whole set?"—*"There are 4/12 outside the law and 8/12 covered by it. It would be the same for the whole set."*

One can see how much progress has been made since the beginning of substage III-A. Although BAB begins by reasoning about the blue eyes (*a*) independently of the brown (*d* cases), he soon realizes that the *a* and *d* cases make up a single whole confirming the same law, as opposed to the *b* and *c* cases which oppose it (VEC grasps this fact almost immediately). But there is still a tendency, of varying strength, to reason about either the *a* and *b* or the *c* and *d* cases in isolation (see VEC's calculation for the first set, 5, 1, 2, 4). But as soon as they are reminded of the totality of possible cases, the subjects of this intermediate level start to compare the (*a* + *d*) cases added to the (*b* + *c*) cases or (*a* + *b* + *c* + *d*). This addition marks the appearance of the idea of correlation in the strict sense of the word. The distinctive feature of substage III-B is that the subjects succeed in performing it at the start and spontaneously.

§ Substage III-B. Spontaneous Relating of Confirming Cases to Nonconfirming Cases and to the Sum of the Possible Cases

After the age of 10 ; 4, we have found exceptional subjects who respond in terms of the final schema of substage III-B, but it is usually toward 14–15 years that the frequency of these cases is high enough to define a stage. Below are two examples:

DAN (14 ; 0) classifies the set [5, 1, 2, 4]: "If you look at the hair, can you find the color of the eyes?"—"*It's safe. . . . There are exceptions, but they are rare; 3 exceptions out of 9. In a case like that you can say that there's no absolute law, but some kind of law.*"—"And here?" [3, 0, 0, 3].—"*Here you have an absolute law.*"—"And there?" [4, 2, 3, 3].—"*The exceptions are rare compared to the number*" [of confirming cases].—"Is it the same as in the first group?"—"*No, the exceptions aren't so rare: before there were 3 out of 9, now there are 5 out of 7.*" He is asked for a zero correlation: he gives 1, 1, 1, 1. Then he is asked to compare 4, 2, 2, 4 with 3, 3, 1, 5: "*There are more chances of being right here* [second group]. *No, they are exactly the same.*" He is finally asked to form two sets, one containing more irregularities than the other: he constructs an inverse correlation [1, 2, 2, 1 and 1, 1, 1, 3]: "*There there are 2* [confirming] *and 4* [nonconfirming]; *here there are 4* [confirming] *and 2*" [nonconfirming].

COG (15 ; 2). Set [5, 1, 2, 4]: "*Most of the people who have brown hair have brown eyes and most of those with blond hair have blue eyes.*"—"What is the relationship?"—"*Not maximum, but not weak . . . 9 people out of 12 have hair the same color*" [as eyes]. "And?" [6, 0, 0, 6].—"*It's the maximum.*"—"And a group where there is no relationship?"—"*You have to mix them up*" [he makes up 1, 1, 1, 1].—"And compare these two groups" [4, 2, 2, 4 and 3, 3, 1, 5].—"*The relationships are the same; there are the same number of cards*" [!].—"Did you count them?"—"*Yes. In both groups there are 8/12* [confirming] *and 4/12*" [nonconfirming].—"What is the best way of seeing whether or not there is a relationship?"—"*You have to compare* [a] *and* [d] *with* [b] *and* [c]." He describes the four combinations by grouping them by diagonals.

The qualitative notion of correlation of which we spoke at the beginning of this section is easily discerned in the substage III-B responses. First, as soon as he has classified the four possibilities according to the schema $(p.q)$ v $(p.\bar{q})$ v $(\bar{p}.q)$ v $(\bar{p}.\bar{q})$ (prop. [1]), the subject realizes that the a ($= p.q$) and d ($= \bar{p}.\bar{q}$) cases are linked together (hair color the same as eye color), but by a link of reciprocity and not of identity (reciprocity here, as always, being the same relationship but in the negative: blond $\times$ blue, and not-blond $\times$ not-blue):

$$\bar{p}.\bar{q} = \text{R} \, (p.q). \qquad (4)$$

Whence the equivalence (which no longer stands for identity but for necessary reciprocal correspondence):

$$(p = q) = (p.q) \, \text{v} \, (\bar{p}.\bar{q}). \qquad (5)$$

Propositions (4) and (5) define the relevant characteristics of the cases confirming the law. Proposition (5) states the law (correspondence between p which asserts presence of blue eyes and q that of blond hair, or between $\bar{p}$ which asserts the presence of brown eyes and $\bar{q}$ that of brown hair), whereas proposition (4) states the relationship adding the two pairs of characteristics which operate in the law.

In the same way, the subject understands the reciprocity between the $b \, (= p.\bar{q})$ and $c \, (= \bar{p}.q)$ cases:

$$\bar{p}.q = \text{R} \, (p.\bar{q}). \qquad (6)$$

Whence the negative equivalence [7] characteristic of the b and c cases:

$$(p = \bar{q}) = (p.\bar{q}) \, \text{v} \, (\bar{p}.q), \qquad (7)$$

where $(p = \bar{q}) = (p \, \text{vv} \, q)$ (reciprocal exclusion of p and of q).

But the subject sees immediately that the b and c cases constitute the inverse of the a and d cases—*i.e.*, the sum of the nonconfirming cases:

$$(p = \bar{q}) = (p \, \text{vv} \, q) = [(p.\bar{q}) \, \text{v} \, (\bar{p}.q)], \text{ thus} \atop (p \, \text{vv} \, q) = \text{N} \, (p = q). \qquad (8)$$

In other words, as soon as he understands the reciprocity between the confirming cases a and d (*i.e.*, $p.q \, \text{v} \, \bar{p}.\bar{q}$), the subject also understands the reciprocity between the nonconfirming cases b and c and the relationship of inversion (prop. [8]) existing between the two classes so characterized. Thus, from the start the substage III-B subject reasons according to diagonals and looks

[7] We know that the negative equivalence or reciprocal exclusion $(p \, \text{vv} \, q)$ can also be written $p = \bar{p}$.

for the numerical relationships between the sum $(a + d)$, verifying proposition (4), the sum $(b + c)$, verifying proposition (6), and the sum of all possible cases. But from this standpoint, what is new in these cases in equilibrium at substage III-B compared with the intermediate cases (BAB and VEC) is that, instead of looking for the ratios $(a + b) / (a + b + c + d)$ and $(b + c) /$ $(a + b + c + d)$, the subject relates the $(a + d)$ cases directly to the $(b + c)$ cases in order to estimate the degree of correlation. Thus DAN compares $3(b + c)$ to $9(a + d)$ and $5(b + c)$ to $7(a + d)$ and estimates that the first correlation is higher; COG says explicitly: "You have to compare $(a + d)$ with $(b + c)$." The consequence of this formulation is that the three following cases are discovered (if E is the sum verifying the propositions considered):

$$E\,[(p.q) \text{ v } (\bar{p}.\bar{q})] = E\,[(p.\bar{q} \text{ v } \bar{p}.q)] = \text{zero correlation.} \quad (9)$$

DAN and COG understand this when they form the sets 1, 1, 1, 1, etc., as showing no relationship at all between the variables.

$$E\,[(p.q) \text{ v } (\bar{p}.\bar{q})] > E\,[(p.\bar{q}) \text{ v } (\bar{p}.q)] = \text{positive correlation} \quad (10)$$

(which is strengthened as the inequality increases).

This is what the subjects express when they compare two distributions to estimate their respective equal or unequal correlations.

$$E\,[(p.q) \text{ v } (\bar{p}.\bar{q})] < E\,[(p.\bar{q}) \text{ v } (\bar{p}.q)] = \text{negative correlation.} \quad (11)$$

This is what DAN sees imperfectly at the end of his interrogation. Since these differences between the two sets $E\,[(p.q) \text{ v } (\bar{p}.\bar{q})]$ and $E\,[(p.\bar{q}) \text{ v } (\bar{p}.q)]$ are evaluated in relation to the total number of possible cases, we are not going too far in stating that these subjects make use of the various links operating in proposition (3); that is, they hold an implicit conception of correlation.

Part III

THE STRUCTURAL INTEGRATION

OF FORMAL THOUGHT

THE GROWTH of thinking may be studied from either of two distinct and complementary points of view: the first is that of equilibrium conditions and the second that of structure formation. From the first perspective, thought processes seem to tend toward states of increasingly stable equilibrium throughout a variety of specific forms.[1] Here the problem is to isolate the reasons for which an equilibrium state is more or less stable. There seem to be two main factors which can account for the difference between the stable and unstable forms: the first concerns the relative degree of extension of the cognitive field included in a given equilibrium, and the second, the instruments of coordination—*i.e.*, the level of development of the cognitive structures which are available at any particular age. From the second perspective, the problem is to determine how the structures arise and how they follow one another along the genetic series. However, this structure formation depends on three principal factors: maturation of the nervous system, experience acquired in interaction with the physical environment, and the influence of the social milieu. But it is also true that the respective and concurrent operations of these factors are limited by laws of equilibrium,

---

[1] Of course equilibrium is not stable in the sense that it is a state of rest in a closed system (such as a balance of forces in mechanics). From the psychological standpoint, a system is in equilibrium when a perturbation which modifies the state of the system has its counterpart in a spontaneous action which compensates it. Consequently equilibrium is a function of the actor's behavior. The equilibrium laws express the probability of the occurrence of various possible forms of compensation in function of neuro-physiological conditions, physical environment, *and* social milieu.

which determine the best forms of adaptation compatible with the sum of these operant social and physical conditions. Thus equilibrium and structure are really two complementary aspects of any organization of thought.

The first of the following chapters will take up the problem of formal thinking from the standpoint of equilibrium, and the second from the structural point of view. The concluding chapter sums up what we know about the differences between the child's thinking, which is preoperational or limited to the use of concrete operations, and the adolescent's thinking, which follows directions unknown to the child as a result of the growth of formal logic.

16

Formal Thought from the Equilibrium Standpoint

THERE IS no doubt that the most distinctive feature of formal thought stems from the role played by statements about possibility relative to statements about empirical reality. Thus, compared to concrete thought, formal thought constitutes a new equilibrium, one that can now be analyzed from the two perspectives just outlined: the extension of the cognitive field that it entails, and the coordinating instruments operative in its functioning. However, we will first analyze briefly the equilibrium characteristics of the stages prior to formal thought.

§ Preoperational Thinking, Concrete Operational Thinking, and Formal Thinking

I. At the end of the sensori-motor period, during which intelligent behavior is limited to coordinating actions, the appearance of symbolic processes enables the child to organize elementary representations. Thus a distinctive form of thinking is worked out between 2 and 7–8 years; signs of it can be found throughout the empirical data presented in this work (see stage I).

This type of thinking is preoperational, *i.e.*, prelogical,[1] and it differs from concrete operational thought on three points: (1) When the child considers static situations, he is more likely to explain them in terms of the characteristics of their configurations at a given moment than in terms of the changes leading from one situation to another; (2) When he does consider transformations, he assimilates them to his own actions and not as yet to reversible operations. We can reduce these two differences to one by saying that at the level of preoperational thought the static states of a given system and its modifications do not yet form a single system, whereas at the level of concrete operations the static states will be conceived as resulting from transformations in the same way as the results of various operations are also subordinated to the operations. (3) Nevertheless, even at this level we find tendencies toward the organization of integrated systems. And we can discern in them an orientation toward certain forms of equilibrium. However, the only instruments available to the subject for the organization of such systems are perceptual or representational regulations, in contrast to actual operations.[2] The difference between these two sorts of mechanisms is that reversibility remains incomplete in the first case but is achieved in the second.

The primacy of static situations over transformations is not too clearly shown by the facts described in this book because the problems given our subjects have to do with transformations for the most part. However, all we have seen in our earlier research on the absence of notions of conservation before 7–8 years illus-

---

[1] Wallon, who has adopted the same point of view with modifications of our vocabulary, speaks of "precategorical" thinking, meaning thinking which cannot deal with the links between subject and predicate. But this aspect of preoperational thought is not sufficient to cover all of its characteristics. "Precategorical" is a term which remains relative to the linkages characterizing class logic. Although it is true that preoperational thinking does not come to control the most elementary "groupings" of class logic, it is no less able to deal with the elementary groupings of the logic of relations. Thus it is "prerelational" to as great an extent as precategorical. The term "preoperational" has the advantage of covering both aspects at the same time, and in addition it emphasizes operations—*i.e.*, action more than the verbal side of thinking. Verbal productions can by no means fully account for the structures of intelligence.

[2] Regulations are incomplete or approximate compensations, in contrast to operations which entail complete compensations.

trates this aspect of preoperational thinking. When, after having transferred a given quantity of liquid into a beaker which is more elongated than the initial container, the 4–6-year-old child believes that the quantity has increased because the form of the container is different, he has based his assertion on the static perceptual configuration; he has perceived each state of the system individually instead of envisaging the situation in terms of a reversible transformation which leaves the quantity constant. In the present experiments, analogous static perceptions are found in connection with the problem of equilibrium between the pressure of a weight and the resistance of a liquid, a case in which neither the quantity of liquid nor the weights involved are conserved (Chap. 10). Likewise, in the balance-scale problem (Chap. 11), the substage I-A subjects do not know how to take away a weight when they want to reestablish the equilibrium (inverse operation) and are limited to continually adding weight to one side or the other in an irreversible succession of states, without returning to the earlier situations.

On the other hand, the second difference is clearly shown by the data collected in this study on the stage I subjects. Instead of considering the experimental apparatus as a system of autonomous causal relations, these subjects fail to distinguish between the physical processes which they observe and the effects of their own actions; often they assimilate the objective relationships to a conception of causality based on the model of these actions (using the impetus, "push," etc., as explanations; see, for example, Chap. 12).

This means that the system transformations are assimilated to the child's own action and static situations are explained in terms of characteristics of the perceptual configuration; as a result there is no homogeneity between the two areas and the only instruments of coordination between them are regulations. On the static side, it is simply a matter of perceptual regulations, this being the case where in the child's eyes the configuration takes the place of an explanation (for example, the elementary equalizations of the weights on the balance are due to a simple perceptual symmetry; Chap. 11). On the dynamic side, these regulations take the form of the corrections and adjustments inherent in the child's action. These latter regulations foretell the advent of operational processes in orienting the child's actions toward reversibility. Although they do not result in complete compensations, such regulations do thrust a small wedge of potential transformations (*i.e.*, elementary processes based on "possibilities" as distinct from "reality") into a type of cognition which is still almost completely bound to reality (either in the sense of external perceptual reality or in the sense of imagined actions).

II. With the appearance of concrete thought, the system of regulations, though maintained in an unstable state until this point, attains an elementary form of stable equilibrium. As it reaches the level of complete reversibility, the concrete operations issued from the earlier regulations are coordinated into definite structures (classifications, serial orders, correspondences, etc.) which will be conserved for the remainder of the life span. Of course this is no bar to the organization of higher systems; but even when higher systems emerge, the present system remains active in the limited area of the organization of immediately given data. What then is the nature of this form of equilibrium?

First, the dichotomy between static situations and transformations no longer obtains in those cognitive spheres organized by concrete thinking; from this point on static situations are subordinated to transformations in that every state is conceived of as the result of a transformation. For example, the subject regards each position of the balance-scale (Chap. 11) as the result of previous additions and subtractions of weight or of equalities and inequalities introduced between the weights on the two arms of the apparatus and between the distances from the center, etc.

Secondly, to say that the system of transformations is in equilibrium means that these transformations have acquired a reversible form and the potentiality for coordination according to fixed laws of composition. Thus, from this point on transformations are assimilated to operations. Operations result from the internalization of actions and preoperational regulations from the earlier stage.

Thirdly, compared to preoperational or intuitive thought, concrete operational thought is characterized by an extension of the *actual* in the direction of the *potential*. For example, to classify a set of objects means that one constructs a set of class inclusions such that at a later point new objects can be included in systematic relationship with those already classified and such that in this way new class inclusions are continually possible. Similarly, to order objects serially permits further possible subdivisions, etc. But at the beginning of the reasoning process these "possibilities" inherent in concrete operations do not at all expand into a set of hypotheses, as will be the case for formal possibilities.

But the equilibrium attained by concrete thought covers only a relatively narrow field. Consequently, the boundaries of this field remain unstable. These two circumstances make the elaboration of formal thought necessary.

The equilibrium field of concrete operations is limited in the sense that, like any equilibrium, the characteristics of such operations are determined by the compensation of potential work (operations) compatible with the system links; however, these links are limited both by the form of the operations involved and by the actual content of the notions to which they are applied.

From the standpoint of form, concrete operations consist of nothing more than a direct organization of immediately given data. The operations of classification, serial ordering, equalization, correspondence, etc., are means for inserting a set of class inclusions or relations into a particular content (for example: lengths, weights, etc.), means which are limited to organizing this content in the same form in which it is presented to the subject. The role of *possibility* is reduced to a simple potential prolongation of the actions or operations applied to the given content (as, for example, when, after having ordered several objects in a series, the subject knows that he could do the same with others, this by virtue of the same schema of expectation for serial ordering that enabled him to perform his actual serial ordering).

From the standpoint of content, concrete thought has the limiting characteristic that it cannot be immediately generalized to all physical properties. Instead, it proceeds from one factor to another, sometimes with a time lag of several years between the organization of one (for example, lengths) and the next (for example, weights). This is because it is more difficult to order serially, to equalize, etc., objects whose properties are less easy to dissociate from one's own action, such as weight, than to apply the same operations to properties which can be objectified more readily, such as length. Thus, from the standpoint of content, the "potential transformations compatible with the system links" which determine the boundary line between real and possible operations are still more limited than is implied by the form of the operations involved; this is a second reason why the form of possibility characterizing concrete operations is nothing more than a limited extension of empirical reality.

Aside from these two kinds of limitation of the equilibrium field of concrete operations, it should be noted that, although the equilibrium is stable at the interior of a given field, it becomes unstable at its boundaries. In other words, although each field of concrete organization easily attains stable forms of equilibrium, instability reappears when fields have to be coordinated. There is no general concrete composition; that is to say, after the subject has classified or has ordered a set of properties serially, etc., or after he has found the correspondences between series with two or more different properties, concrete thought does not solve all the problems raised by the interference of heterogeneous operations or by the intersection of different properties. Thus, from the concrete point of view, the comparison of a weight and an upward displacement may be sufficient to determine the amount of "work" required in some cases (heavier × higher = more work). But the arrangement of factors may remain indeterminate and alio-transitive in others (heavier × less high = more or less work or the same work). In the latter cases the problem cannot be solved with concrete thought.

In sum, concrete thought remains essentially attached to empirical reality. The system of concrete operations—the final equilibrium attained by preoperational thought—can handle only a limited set of potential transformations. Therefore, it attains no more than a concept of "what is possible," which is a simple (and not very great) extension of the empirical situation. This characteristic of concrete thought is particularly clear in the present research, where the child's concrete thought and the formal thought of the adolescent or preadolescent may be compared at every point. The adolescent and preadolescent start by formulating sets of hypotheses and then choose experimentally the ones which are compatible with the facts. But, strictly speaking, at the concrete level, the child does not formulate any hypotheses. He begins by acting; although in the course of his action he tries to coordinate the sequence of recordings of the results he obtains, he structures only the reality on which he acts. But if the reader objects that these cognitive organizations are in fact hypotheses, we would answer that in any case they are hypotheses that do no more than outline plans for possible actions; they do not consist of imagining what the real situation would be if this or that hypothetical condition were fulfilled, as they do in the case of the adolescent.

III. Finally, in formal thought there is a reversal of the direction of thinking between *reality* and *possibility* in the subjects' method of approach. *Possibility* no longer appears merely as an extension of an empirical situation or of actions actually performed. Instead, it is *reality* that is now secondary to *possibility*. Henceforth, they conceive of the given facts as that sector of a set of possible transformations that has actually come about; for they are neither explained nor even regarded as facts until the subject undertakes verifying procedures that pertain to the entire set of possible hypotheses compatible with a given situation.

In other words, formal thinking is essentially hypotheticodeductive. By this we mean that the deduction no longer refers directly to perceived realities but to hypothetical statements— *i.e.*, it refers to propositions which are formulations of hypotheses or which postulate facts or events independently of whether or not they actually occur. Thus, the deductive process consists of linking up these assumptions and drawing out the necessary consequences even when their validity, from the standpoint of experimental verification, is only provisional. The most distinctive property of formal thought is this reversal of direction between *reality* and *possibility;* instead of deriving a rudimentary type of theory from the empirical data as is done in concrete inferences,[3] formal thought begins with a theoretical synthesis implying that certain relations are necessary and thus proceeds in the opposite direction. Hence, conclusions are rigorously deduced from premises whose truth status is regarded only as hypothetical at first; only later are they empirically verified. This type of thinking proceeds *from* what is possible *to* what is empirically real.

We may ask whether there is a simpler way to characterize formal thought than by referring to the notion of *hypothesis* or

[3] Based on the transitiveness of inclusions of classes or relations.

possibility. Now, the most prominent feature of formal thought is that it no longer deals with objects directly but with verbal elements; and at first we tried to contrast formal and concrete thought in terms of this distinguishing factor. In fact, it is often sufficient to translate a concrete operation into simple propositions and deny the subject the use of manipulatable objects for working out the operation in question for the problem to become insoluble before the formal level. This is the case for the verbal problem: "Edith is lighter than Suzanne and Edith is darker than Lily; which is the darkest of the three?" It seems that no factors are required for the solution other than a coordination of relations "more x" and "less x," as in any serial ordering of asymmetrical transitive relations. However, starting with the age of 7, the child knows how to arrange small strips of wood in a series according to length; he places the smallest of those which are left over each time next to the strip he had just set in the series. Thus, for any element E he coordinates the two relations $E > D, C, B, A$ (already placed in the series) and $E < F, G, H$, etc. (remaining elements). Thus, he can handle the two relations "more x" and "less x." Nevertheless, the verbal problem concerning complexion is not solved until 11–12 years, although the concrete serial ordering of shades is no more difficult than setting lengths in series.

However, this is not the whole problem, for all verbal thought is not formal and it is possible to get correct reasoning about simple propositions as early as the 7–8-year level, provided that these propositions correspond to sufficiently concrete representations. Even if the content of the complexion problem requires nothing more than serial order operations, the fact that it cannot be solved in exclusively verbal terms until several years after the child can solve it with the aid of physical props shows us that some other factor is at work here. If we consider the mental images involved in the problem we see how difficult it is for the subject to set up the data in his own mind (because only the relations are given). The result is that the subject is unable to translate the data into representational imagery and has to formulate them in exclusively hypothetical terms if he is to see the necessary consequences. This conjunction between what is possible and what is theoretically necessary makes it indispensable that the serial ordering operations used be inserted into a set of implications, made up of the relations which are to be ordered serially, which serve as an interpropositional form for the intrapropositional content itself.

When verbal statements are substituted for objects, a new type of thinking—propositional logic—is imposed on the logic of classes and relations. Here we see a second fundamental property of formal thinking. However, propositional logic offers a much greater number of operational possibilities than simple groupings of classes and relations. So in the first instance and from the functional standpoint, formal thought differs from concrete thought in that it offers a greater number of operations, aside from the question of operational structures (to which we shall turn presently). The proof of this is that propositional logic appeared in its most characteristic forms as readily when the subject dealt with our experimental apparatus as when he was confronted with a purely verbal task. In the former case, rather than have the subject apply his reasoning to data that were already formulated, he was urged to set his own problems and work out his own methods for solving them. Thus, one can see that the role of formal thought is *not* simply to translate into words and propositions concrete operations that could have been performed without its benefit anyway. On the contrary, it is *in the course of experimental manipulations* that a new series of operational possibilities arise during the early stages of formal thought; it consists of disjunctions, implications, exclusions, etc. These operate as soon as the subject starts to take in the factual data and organize the experiment. They are even imposed on simple groupings of classes and relations. Again, this is because, from the moment of contact with factual problems, formal thought starts off with hypotheses— *i.e.*, possibilities—instead of limiting itself to a direct organization of perceived data. Thus, in spite of appearances and current opinion, the essential characteristic of propositional logic is not that it is a verbal logic. First and foremost, it is a logic of all possible combinations, whether these combinations arise in relation to experimental problems or purely verbal questions. Of course, combinations which go beyond a simple registration of data presuppose an inner verbal support; however, the real power of propositional logic lies not in this support but rather in the combinatorial power which makes it possible for reality to be fed into the set of possible hypotheses compatible with the data.

When we saw that a purely verbal criterion was inadequate to define formal thought, we sought to define it in terms of a third property; formal thought as a system of second degree operations. Concrete operations may be called first degree operations in that they refer to objects directly. For example, this is the case in the structuring of relations between given elements. But it is also possible to structure relations between relations, as for example in the case of proportions (our experiments show that proportions are not mastered before the formal level). In this sense proportions presuppose second degree operations, and the same may be said of propositional logic itself, since interpropositional operations are performed on statements whose intrapropositional content consists of class and relational operations. And obviously this notion of second degree operations also expresses the general characteristic of formal thought—it goes beyond the framework of transformations bearing directly on empirical reality (first degree operations) and subordinates it to a system of hypothetico-deductive operations—*i.e.*, operations which are *possible*.

In the light of the foregoing, it seems to be the case that the most general property in terms of which we can characterize formal thought is that it constitutes a combinatorial system, in the strict sense of the term. Although it seems more restricted than the characteristics we mentioned earlier, this property implies all the others and thus is more general than they are. So the main feature of propositional logic is not that it is verbal logic but that it requires a combinatorial system. Secondly, the combinatorial operations are second degree operations: permutations are serial orders of serial orders; combinations are multiplications of multiplications, etc. They do not in fact appear until the formal level, either in the form of explicit [4] mathematical operations or in the implicit form that they take in our present research (Chap. 7, etc.).

But even this opposition between combinatorial and noncombinatorial operations can be derived from the difference between *possibility* and *reality;* only a formal combinatorial system furnishes the total number of possibilities, and on the experimental level the search for new combinations is exactly what characterizes hypotheses. So however general it may be, the combinatorial

[4] See *La Genèse de l'idée de hasard chez l'enfant,* Part III.

nature of formal thought is secondary to a still more general property—i.e., the subordination of *reality* to *possibility*.

Thus, aside from the structural consequences it implies, the most fundamental property of formal thought is this reversal of direction between *reality* and *possibility*. And so we must use this essential difference between concrete and formal thought as a starting point in accounting for the form of equilibrium found at this last stage in the development of thought.

§ *Reality and Possibility in Formal Thought*

We have just seen that concrete thought is the form of equilibrium toward which preoperational thought tends. When it appears, static situations and transformations are integrated into a single system such that the former are subordinated to the latter and such that the latter form an operations structure attaining reversibility by mutual compensation of the transformations. But we have also seen that the scope of this equilibrium is still limited, as much because of the form of the operations as because of the resistance of the content. Since the instruments of general coordination between the concrete operational groupings are lacking, the subject still regards what is possible as but a direct extension of empirical reality. With the more complex instruments of coordination found in formal thinking, a new form of equilibrium appears, encompassing all the partial fields covered by concrete thought and coordinating them into a general system. The links of this system are at once the second order operations and the combinatorial system by means of which propositional logic comes to assign reality a place within a structured set of possible transformations. But before we can understand the nature of this new equilibrating process, we must first spell out the meaning of *possibility* and *reality* in formal thought.

I. The possibilities entertained in formal thought are by no means arbitrary or equivalent to imagination freed of all control and objectivity. Quite to the contrary, the advent of possibilities must be viewed from the dual perspective of logic and physics; this is the indispensable condition for the attainment of a general form of equilibrium; and it is no less indispensable for the establishment of the requisite connections utilized in formal thought.

A. From the point of view of physics, we know that a state of equilibrium is characterized by compensation between all of the potential modifications compatible with the links of the system. Even in experimental physics—*i.e.*, in the science relating to reality in its most material aspect—the notion of possibility plays an essential positive role in the determination of the conditions of a state of equilibrium. To the extent that psychology feels the need to consider states of equilibrium as well, it is thus *a fortiori* indispensable to resort to the notion of *possibility* expanded into the notion of "potential" actions or mental transformations (in the sense that the physicist uses when he speaks of potential acceleration or potential energy). But, since the concept of equilibrium even plays a role of some importance in perceptual theory, as Gestalt psychology has shown us, it is even more useful in the analysis of intellectual operations; for once an operational system is established it may remain unchanged for the remainder of a person's life.

So we are not being vague in stating that at the formal level the subject subordinates *reality* to *possibility* and that this assertion can be related directly to the theory of equilibrium states. First, it means that when faced with a determinate situation the subject is not limited to noting the relations (between the given elements) that seem to thrust themselves on him. Rather, in order to avoid inconsistencies as new facts emerge, from the start he seeks to encompass what appear to be the actual relations in a set of relations which he regards as possible. In other words, to equilibrate his successive assertions (which is equivalent to avoiding contradiction by subsequent facts), the subject tends to insert links, which in the first instance he assumes as real, in the totality of those which he recognizes as possible; he does this in a way that allows him later to choose the true ones by the examination of certain transformations performed accurately within the set of the possible links. *Possibility* plays its role in this way from the moment of the first spontaneous attitude taken by the formal level subjects; and from their subjective point of view, it is the very condition of the equilibrium of their thought.

But the following point arises as soon as we shift from the subject's to the observer's point of view: if the subject is to conceive of the total set of possible links or transformations in a given situation, he has to deduce them by means of adequate logical operations. But this set of operations is still a system of potential transformations with respect to physical equilibrium. (Of course, only a part of the set is utilized in a particular situation or problem.) Only the operations actually at work in a given situation are for the moment "real"; the others are merely potential. There are two reasons why these potential operations (or logical transformations) are as necessary to functional equilibrium as the ongoing or "real" operations: (1) they insure reversibility, and (2) they can be developed as they are needed for the acquisition of possible new links.

Summarizing, in order to conceive of what is possible, formal thought must have a wide range of operations ready for any particular situation, operations in addition to those which are being made use of at any given moment. These *potential* operations are a necessary condition of equilibrium for two reasons: (1) they correspond to what equilibrium theory terms "potential transformations," and (2) equilibrium actually exists to the extent that these potential transformations compensate each other exactly. Or, in operational language, it exists to the extent that these possible operations form a system that is strictly reversible from the logical standpoint.

B. There is a second sense in which *possibility* plays a role in formal thinking, namely the logical sense. In a new perspective it will show us once more that the role of possibility is indispensable to hypothetico-deductive or formal thinking. Logically, formal possibility is the required correlate of the notion of deductive necessity. An assertion that refers to empirical reality only, such as an existential or predicative judgment, could not be considered deductively necessary; it is true or false insofar as it corresponds or does not correspond to a factual datum. But a deduction logically derived from a hypothesis (or from a factual datum assumed hypothetically) is necessarily true from the formal point of view independently of the value of the assumed hypothesis. The connection indicated by the words "if . . . then" (inferential implication) links a required logical consequence to an assertion whose truth is merely a possibility. This synthesis of deductive necessity and *possibility* characterizes the use of possibility in formal thought, as opposed to possibility-as-an-extension-of-the-actual-situation in concrete thought and to unregulated possibilities in imaginative fictions. But what is the nature of formal possibility? Anything which is not contradictory may be called possible. But the non-contradictory, strictly speaking, is the totality of reversible transformations performed in such a way that the composition of an operation and its inverse result in a product termed "identical" or null: $p.\bar{p} = o$. Thus, whereas from the standpoint of physics operational reversibility signifies the exact compensation of potential transformations (or operations), from the logical standpoint it refers to deductive necessity.

In the end, these two aspects of the concept of *possibility*, stemming from physics and logic, are psychologically equivalent. In feeding an actual set of conditions into a set of possible transformations, formal thought guarantees its own equilibrium by insuring that the structures it elaborates will be psychologically conserved. At the same time it assures its value as a necessary logical instrument in that it makes use of these structures as deductive instruments. The mental equilibrium being determined by a general operational structure, the notion of *possibility* acts in two capacities—as an equilibrium factor and as a logical factor. Which of these is brought into play depends on whether the observer deals with a problem from the standpoint of *explanation*, which is for the most part the point of view of the observing psychologist, or from the standpoint of *understanding*, which is above all the subject's own point of view. This is why the reversal of direction between *reality* and *possibility*, which marks the advent of formal thought, is a turning point in the development of intelligence, at least insofar as intelligence may be thought of as an organization tending toward a state of equilibrium that is both stable and mobile.

II. But we must go one step further, for psychological equilibrium cannot be reduced entirely to physical equilibrium. The comparison between physical and psychological explanation of the equilibrium state is as instructive for pointing up their differences as for pointing up their common aspects.

In physics a system such as a balance is said to be in equilibrium when all of the potential energies (elevations of the weights at one side or the other as expressed as angular displacements) completely compensate each other. From the standpoint of *possibility*, on the other hand, the balance could be displaced in a particular direction and consequently an opposing force would appear which would reestablish the equilibrium. But as long as these transformations do not actually take place, the "potential energy" which defines the intervention of possibility in equilibrium theory does not, strictly speaking, exist outside of the mind of the physicist. Certainly it is possible to maintain that it plays a role in reality, but it is an essentially negative role; it explains the fact that nothing changes if the system is not modified from without. Thus, in a state of physical equilibrium, one must distinguish a "reality" which is causal and temporal and a set of "possibilities" which are deductive, extemporaneous, and which reside in the brain of the physicist who constructs the theory about reality.

Psychologically, a system in equilibrium, such as the total number of relations understood by a subject who has succeeded in explaining the mechanism of the balance, has two aspects to it, one referring to *reality* and the other to *possibility*. On one hand the subject really performs certain mental operations and organizes certain relationships which really apply to the object which he has before his eyes at a given moment—for example, because he perceives and conceives it effectively, the subject asserts that the balance is horizontal and that a 2-kg. weight at a distance of 10 cm. from the axis compensates a weight of 1 kg. at 20 cm. from the axis suspended from the other arm. But these real operations and relations, real in the sense that they are used at a given moment, are not sufficient to account for the equilibrium attained in the act of understanding, since in the latter an entire set of possible or potential operations and relations intervenes. However, the line of demarcation between possibility and reality is much more difficult to trace in psychology than in physics. But it is just this difference between the two forms of equilibrium which is most instructive for the study of formal thought and which we may now analyze more closely.

First of all, we must distinguish carefully between two meanings of the word *possibility* as it is used here. First, one may speak of possible operations and relations in designating those which the subject himself regards as possible—*i.e.*, those which he knows he is able to perform or construct, even without actually trying them out. We shall call this *instrumental possibility*. The reader will see immediately that this is identical with our earlier description of possibility from the subject's point of view. But one could also mean by *possibility* the operations and relations that the subject would be capable of performing or constructing without his thinking of doing so—*i.e.*, without the subject's becoming aware of the contingency or even of his own capacity with regard to it. We shall call this *structural possibility*. Thus, it is possibility defined by the observer rather than the subject.

Let us return to our example concerning "instrumental possibility." In ascertaining the compensation between a 2-kg. weight at 10 cm. and a 1-kg. weight situated at 20 cm. from the axis, the subject, although not performing the physical actions, may assume that he might move the 2-kg. weight by 5 cm. away from the axis and likewise the 1-kg. weight outward by 10 cm. In so doing, he does not go beyond the realm of *potential* actions (actually, he does not move anything); still, the subject will deduce that these two hypothetical displacements will maintain the initial equilibrium because the inverse proportion of the weights and the distances will remain constant, etc. Returning to *reality* (the actual apparatus), he could verify his assertion. But, as he is deductively certain of the result, he could dispense with this verification. In both cases, these possible actions will help him and will even be indispensable to him if he is to understand the empirical data (horizontal position of the balance for 2 kg. at 10 cm. and 1 kg. at 20 cm.); thus, he will interpret the real relations as a function of a set of potential actions and explain them as necessary consequences of these possible operations or actions which are combined among themselves.

But one can see immediately that these "instrumentally possible" operations or relations are, from the psychological point of view, as real as the initial "real" operations or relations. It is only from the subject's point of view that they appear as possibilities. In other words, the subject distinguishes what is physically present from the transformations which it would be possible to introduce in the perceived system. But even when the subject's performance is based purely on potential actions, these possible transformations are thought over and thus, psychologically, may consist either in representations or in real operations. Hence, "the instrumentally possible" is but a particular modality of the real thinking of the subject. This modality gains a special importance at the level of formal thinking when, as a result, the modality of "reality" can be subordinated. Although this may contribute to the stabilization of operational equilibrium to the extent that the transformations effectively realized are further conditioned by the transformations which the subject conceives of as possible, on the psychological level this realm of the "instrumentally possible" is not wholly comparable with potential transformations in the theory of equilibrium in physics. To be more exact, it does not correspond to anything in the physical realm, since it is intermediate between pure reality (immediately present relations) and pure possibility (the "structurally possible") whereas in physics no *tertium* intervenes between the immediate and the potential.

It is an entirely different situation in the case of "structural possibility." For at the level of psychological equilibrium, "structural possibility" is wholly equivalent to the "potential transformations" of a physical system in equilibrium. Alongside the operations which he has actually performed (either as actions as such or as effective perceptions or as simple hypotheses "instrumentally possible"), the subject could have performed others, which he has not either in action or thought. Instead of imagining a displacement of 5 and 10 cm. for the 2- and 1-kg. weights respectively, he could have modified the weights themselves (in act or mentally), or he could have diminished a distance on one side and a proportional weight on the other, etc. It may even happen that, without explicitly referring to it, the subject acts as if he had utilized the operational schema of proportions; the same question which for another subject would not have provoked a conscious reaction for lack of an underlying structure would be for the subject in question a starting point for an immediate organization of new, well-formulated relationships. Thus, such observations, which are common, will furnish the proof—but only after the fact—that a determinate operation was "structurally possible" for one subject while it was not for the other. In short, in any act of intelligence one must always consider both real operations, in the sense of those actually made use of in the subject's conscious thought, and "structurally possible" operations, in the sense of those which the subject does not perform, but could perform. As we have seen above, it is this possibility, relative to the operational structures which are available to the subject, which constitutes *possibility* from the observer's point of view and which thus corresponds in the field of physics to the potential transformations not actually realized.

But if we are to construct a theory of the psychological mechanisms involved in operational equilibrium, we must give separate consideration to each one of these two distinct meanings of the notion of possibility even though, in this case, we are faced with three analytic levels (the real, the instrumentally possible, and the structurally possible) and not two (the real and the potential) as is the case in physics. Indeed, the instrumentally possible is closely related to the structurally possible of which it constitutes a first stage of realization—*i.e.*, it is insofar as the subject has access to a large enough number of structurally possible operations that he is enabled to imagine the instrumentally possible transformations. Without a certain set of structural possibilities (and the consequences of this lack are clearly evident at the preoperational and even the concrete operational levels), he could not do more than ascertain the state of the facts he perceives in reality, whether static states or transformations in process, and would not succeed in representing to himself the hypothetical transformations which serve for devising new experiments. Thus instrumental possibility depends on structural possibility, but the first is more impoverished than the second. For this reason, even if the former derives from the latter at each instant, the two varieties must be carefully distinguished, for the second is both theoretically and practically of a much greater importance.

In these terms we can return to the comparison between the physical and psychological concepts of equilibrium. From the physical point of view, only reality is of a causal nature, for the potential or the possible plays a positive role only in the mind of the physicist; in other words, possibility is only an instrument of calculation or deduction, whereas reality alone is causal. But the situation is entirely different in the case of operational equilibrium. For, in this realm, reality and possibility are both of a psychological nature—*i.e.*, both have their locus in the subject's mind and causally intersect only within the mechanisms which make up the subject's mental life. It is true that it could be maintained that only reality and the "instrumentally possible" (which in this case is still part of psychological reality) play a role in the subjects' minds, whereas the "structurally possible" is a concept that refers not to the subject himself but to the psychologist who attempts to analyze and explain the subject's behavior. In this case there would be complete parallelism between the physical and psychological points of view. However, to the extent that the "instrumentally possible" is conditioned by the "structurally possible"—and all of the research described in this work leads to this conclusion—we must exclude the notion of not attributing the structurally possible to the subject as such.

But then we are led to the paradoxical consequence that, in a state of psychological equilibrium, possibility (both structural and instrumental) plays as important a causal role as real operations. It could even be maintained that the whole of mental life is dominated by this sort of causality of the possible.

As for "instrumental possibility," this assertion does not contain anything that is especially surprising and it corresponds to the universally acknowledged fact that "hypothesis" plays a role in the functioning of thought. The causal function of instrumental possibility is actually hypothesis behavior, behavior which permits the subject to go beyond what he perceives or conceives while admitting the validity of the present situation and to involve himself in that which can be conceived without an immediate decision concerning its verification. From sensori-motor trial-and-error transfer to the most sophisticated experimental hypothesis, adaptation to present reality is complemented by a progressive adaptation to future reality. Thus possibility enters the adaptation process, in the field of adaptations indispensable to action, in the form of the potential future, since, when an internalized action becomes an operation, possibility intervenes at each bifurcation—i.e., in every case in which the subject, after having imagined where each of two or several possible courses of action leads, must make a choice. Finally, at the formal level, hypothesis intervenes from the moment of contact with reality insofar as immediate fact is thought to admit of giving rise to several interpretations.

But this causality of the "instrumentally possible" has nothing mysterious about it, since to think about a possible event by virtue of thinking itself is (and we repeat) to perform one or several real operations.

On the other hand, the potential causality related to the "structurally possible" raises a completely different problem but one which is much more important—how can the operations actually performed be causally affected by operations which are not accessible to the subject, at least not at the conscious level, and which sometimes remain latent to such a degree that they never belong in an explicit form to the subject's realm of available knowledge?

Let us first note that such a problem is by no means limited to the psychology of logical operations or even to psychology in general. In embryological theory for example, note is taken of partial or total "potentialities" whose appearance is linked with a determinate level of development and only a portion of which is actually realized. So it is difficult to accept the notion that these limited realizations are independent of the entire system of "potentialities" from which they carry a given segment of *possibility* over into *reality*. If, in psychology, we accept the view that the development of mental functions is linked to the maturation of the nervous system (a hypothesis which is a simple extension of the embryological point of view just mentioned) it follows as a matter of course that a coordination could appear in a potentially general form although it would first give rise to certain specific applications only. These latter, though the only ones realized, depend on the system of possible coordinations which appear as innate possibilities and which are more or less retarded in maturation.

Without committing ourselves to such hypotheses, we may note that the causality of possibility is resorted to in an implicit way more often than one would think by those studying mental processes and even general biology. But as regards our present problem, the frequency of certain synchronisms bears witness to the causal role of possibility; for these synchronisms would seem to be inexplicable if one or more integrated structures were not organized at a given moment as sources or reservoirs of possible operations. Only some of these come to be realized, but their realization is a function of the total system and consequently a function of the *potential* as much as the *actual*.

Beginning at the level of concrete operations, we have observed the striking fact that although they do not see the relationship between two objectively analogous problems and do not know that they are applying similar operations to them, on the average, the subjects react synchronously in the same manner to the two problems at a given level. For example, if the substage II-B subjects' responses to the balance scale problem are compared to their responses for the toy-dumping-wagon problem (Chap. 12), it is clear that at 9–10 years the subjects begin to establish an inverse correspondence between weights and distances (from the axis) in the first case and begin to understand in the second that the more the rail is inclined, the more the wagon weighs. But in the second case they bring in the concept of work while in the first case they have no conception of it; nevertheless, the operations of relating weights to distances from the center or to the height is the same in both cases. In either situation, the child understands that the force exerted by the weight changes as a function of the spatial relationships. Likewise, weight is related to volume at the same levels in the most varied problems without the subject's appealing explicitly to analogies. A great many examples of the same type could be furnished from the concrete level.

At the formal level, the synchronization of like reactions in the face of analogous problems is still more striking. For, in contrast with the preceding level, the operational form is entirely dissociated from thought content. Beyond the possibility of reasoning formally—*i.e.*, by implications, exclusions, disjunctions, etc.—we see concepts of proportions and especially combinatory considerations [5] appearing at the same level in the most diverse areas. It is as if the system of possible operations were an internal network along which a given thought content, once it had engaged the network, spread out immediately in all directions at the same time. Thus the causal role of possibility is manifested as a kind of action of implicit schemata on explicit operations, the latter being determined not only by the cognitive acts actually performed just prior to the new operation but by the totality of the operational field constituted by the possible operations.

Certainly, nothing is more dangerous than recourse to the

[5] Not taught in class. The schema of proportions also appears in many cases before its scholastic introduction.

implicit—*i.e.*, to the potential. But there are safeguards permitting the dissociation of the abusive usage of the concept (such as the passage from the potentiality to the act) and its legitimate usage. The true potential differs from the false in that it is calculable and is simply a response to the exigencies of the conservation of the total system (as the potential in physics). But in the case of "structurally possible" operations we are faced with a legitimate potential, since there are algebraic instruments which enable us to uncover the role of general structures and to calculate their extension as well as the elements. The following chapter will demonstrate this point.

Meanwhile, let us limit ourselves to the following conclusion: in a state of *physical* equilibrium reality alone is causal and possibility relates only to the mind of the physicist who deduces this reality; in contrast, in a state of *mental* equilibrium the succession of mental acts is affected not only by the operations actually performed but also by the entire set of possible operations insofar as they orient the subject's searching toward deductive closure. For in this case it is the subject who deduces, and the possible operations are part of the same deductive system as the real operations he performs.[6]

§   The Problem of Structures

Despite this difference between physical and mental processes, it remains true that in both cases a system is in equilibrium when all of the potential transformations compatible with the system links compensate each other. This physical definition of equilibrium corresponds in the mental field to the following considerations: the state of fact—or reality—corresponds to the operations explicitly performed by the subject, whereas the potential transformations correspond to the possible operations that the subject could perform and that he may explicitly perform later on but which he has not or has not yet performed at the moment con-

---

[6] From the standpoint of applied psychology, this is equivalent to saying that a subject should be evaluated not only by what he actually does but also by what he could do in other situations—*i.e.*, by his "potential" or his aptitudes.

sidered; the system links correspond to the givens of the problem posed—*i.e.*, to the content on which his operations are exercised. Our problem is to find the point at which it can be said that equilibrium is attained for such a mental system.

In the first place, it is not attained as long as the problem is not solved—*i.e.*, as long as it is still necessary to perform explicit operations. But here we must distinguish between two cases: (1) where the subject possesses all of the methods and all of the operations required for the solution; and (2) where he has not yet acquired them. Naturally, in the second case it would not be possible to speak of equilibrium, since a more or less considerable effort must still be provided before the question posed may be considered as resolved. In the first case, if there is still disequilibrium, it is only momentary and partial—*i.e.*, relative to the single new problem whose solution is not immediately visible. For the rest—*i.e.*, for the entire set of methods and operations which the subject utilizes in the solution of problems of this class—it can be said that even in a permanent sense equilibrium is attained, since the subject has become capable of solving all similar problems.

Thus, as in a physical equilibrium, equilibrium of this type is characterized by the compensation of the total number of potential transformations. This is equivalent to saying that once the data are given the subject can submit them to an indefinite number of operational transformations beyond those which he chooses in trying to answer the question posed but that these transformations are relative to a structure (the integrated structure of the operations available to the subject) and that this structure is reversible. There is then an equilibrium, because to each transformation that the subject could perform (as a function of the operational structure considered) there is a corresponding inverse possible transformation that could also be realized. Or, stated in simpler fashion, the system is in equilibrium when the operations which the subject is capable of constitute a structure such that these operations can be performed in either one of two directions (either by strict inversion or negation or reciprocity). Thus, it is because the total set of possible operations constitutes a system of potential transformations which compensate each other—and which compensate each other insofar as they conform to laws of reversibility—that the system is in equilibrium. The operational reversibility and the system equilibrium constitute, after all, a single unified property, and it is because the possible operations are reversible and mobile (*i.e.*, can be combined in all ways, but with a complete liberty for a return to the initial starting point) that possibility acts in a continuous manner on the choice of new operations to be performed.

If such is the nature of operational equilibrium, the solution of the problem of formal thought is to be sought in an analysis of the structural integration which characterizes formal operations in contrast to concrete operations. On the one hand, the functioning of the equilibrium such as we have just attempted to describe implies the existence of an integrated structure, since only an integrated structure can explain the presence and the extent of the possible operations as well as their influence on the performed operations. On the other hand, as we have seen, formal thought is characterized by reversal of direction between reality and possibility in that the first is subordinated to the second and the second acquires an importance unknown up to this point. But here as well, if we are to determine the proportions between real operations and possible operations at the concrete and formal levels as well as the indefinite extension of the second to this latter level, we must compare the structural integrations of the operations involved in the two cases. This comparison will be undertaken in the next chapter.

But first, several further remarks are in order on the choice of the logico-mathematical instruments which we shall make use of in this analysis. When the problem is to determine the "factors" involved in a mental effort or simply the correlations between various returns, it seems obvious to psychologists that the mathematical methods of factorial analysis or probability calculation should be used, since they do not bias the results. But we must ask ourselves whether there are methods which are as precise as these in their calculational technique (and as objective as analytic instruments) which can resolve the question of the structural integration of operations.

The answer is *yes*. For several decades, mathematicians have striven to isolate the integrated structures which are found in the most varied areas but whose structural laws are independent of any application to a particular realm. Synthesizing the results already acquired in this field and adding the original contribution that their efforts at revising the principles of mathematics have made possible, the Bourbaki in their remarkable works have reached the conclusion that there are three kinds of basic structures of which the multiple combinations explain all of the others: (1) algebraic structures whose prototype is the "group"; (2) the order structures one of whose principal forms is the "lattice"; and (3) the topological structures relative to the continuum.

But, leaving aside the topological structures, since they are not relevant here,[7] we see that group and lattice structures are common to both mathematical and logical operations. In other words, the general analytic instrument forged by mathematicians is as valuable in the qualitative study of the structures found in thinking as in any other structural research. It is easy to discern the *lattice* structure in the propositional calculus of symbolic logic, and we have shown elsewhere the multiple forms under which the *group* of four transformations (*Vierergruppe* or "Klein group")[8] reappears.

Thus it is not as logic (for logic has no more place in psychology than psychology does in logic), but as a calculus or an algebra that we are here using symbolic logic. Considered in this perspective, symbolic logic is badly needed at this point as an analytic tool for at least two reasons.

First, such an analytic instrument is generally the only possible one for determining the exact extent of possible operations. We know, for example, that with 1, 2, 3, 4, . . . propositions, it is possible to organize respectively 4, 16, 256, 65,536, etc., operations, and it is even easy to enumerate them one by one—an interesting exercise—up to the 256 ternary operations. Moreover, it makes it possible to demonstrate that these numbers, increasing by the square with disconcerting rapidity in proportion to the operations which will actually appear in the subjects' performance in an experimental situation, are not actually independent of each other; thus it is possible to reduce the 256 ternary operations to pairs or trios of binary operations or unitary-binary operations.

But especially, and this is most important for the problem at

---

[7] But topological structures do correspond to what we call **infralogical** operations (see Chap. 17, p. 273, note 2).

[8] See Piaget, *Essai sur les transformations des opérations logiques.*

hand, using symbolic logic as a means of analysis makes it possible to show that these sets of operations do not consist of simple series of juxtaposed elements; on the contrary, these collections of elements have structure as wholes. These wholes are the integrated structures of formal propositional operations (equivalent to the concrete *groupings* of classes and relations) which it would be instructive to analyze in order to resolve the psychological problems posed in the present work. In this respect, symbolic logic allows for an analysis which goes more deeply into the heart of intelligence than arithmetical or statistical calculations. Whereas the latter bear either on the results of operations or on "factors" which do not directly furnish meaning, the qualitative analysis available to symbolic logic reaches the structures themselves—*i.e.*, the operational mechanism as such and not simply its results or its more or less general conditions.

Of course, such a use of structural analysis implies that experimental results and theoretical analysis will be continually compared with each other. From the experimental standpoint, intelligence is first a coordination of acts—then operations which orient themselves step by step toward certain forms of equilibrium. But these equilibrium forms, which are of vital importance in the explanation of development, can be analyzed from both perspectives, the genetic or experimental and the theoretical. In genetic logic, empirical research consists of determining, by tests appropriate to the various levels of evolution, which operations are involved in the subject's cognitive acts and how these operations are gradually organized into structures to the point where certain empirically verifiable forms of equilibrium are reached. On the other hand, theoretical analysis describes these same structures in their general or abstract aspects so as to show how the most complex can be derived from the simplest, and it determines the system of possible operations which would permit the utilization of this or that actually performed operation.

It is clear from the start that these two types of research can be mutually reinforcing in furthering our understanding of cognitive equilibrium states. We have tried to understand the transition from the concrete to the formal level in the thinking of childhood and adolescence by proceeding from this continual comparison of their results, after having sought in vain for the criteria of formal thought in verbal primacy, in second-degree operations, etc.

Moreover, it is clear that such recourse to the theoretical analysis of structures in no sense constitutes logical investigation. Throughout, the question is a psychological one; the calculus of symbolic logic is brought in as an analytical instrument insofar as it is a more general algebra than the elementary algebra founded on numerical operations. If we may compare those sciences in the process of growth with the sciences which have attained full control of their methods, experimental psychology can be seen as corresponding, on the mental level, to experimental physics in the study of matter, with pure symbolic (or axiomatic) logic corresponding to mathematics. As for the discipline which deals with constructing a theory of mental operations by means of symbolic calculus, its relationship to experimental psychology would be comparable to the present day relationship between laboratory and mathematical physics. It would remain a branch of psychology, as mathematical physics is a branch not of mathematics but of physics. But it would utilize the algebra of symbolic logic as an analytic instrument in the same way that mathematical physics now makes use of the techniques and notation of mathematics.

17

Concrete and
Formal Structures

IN THE LAST CHAPTER we distinguished between concrete and formal equilibrium in that the second entails a gradual extension of the totality of potential or possible operations. In the present chapter, we will go on to compare the total structures which fit each of the two equilibrium states.

§ *The Groupings of Concrete Thought*

As we have just seen (Chap. 16), any state of equilibrium can be recognized by a characteristic form of reversibility. Consequently, in those cases where the equilibrium reached by thought processes is of an operational nature (as at stages II and III), this reversibility acts as the vital mechanism for the structural integration which each particular form of equilibrium calls up.

Reversibility is defined as the permanent possibility of returning to the starting point of the operation in question. From a structural standpoint, it can appear in either of two distinct and complementary forms. First, one can return to the starting point by canceling an operation which has already been performed— *i.e.*, by inversion or negation. In this case, the product of the direct operation and its inverse is the null or identical operation.

Secondly, one can return to the starting point by compensating a difference (in the logical sense of the term)—*i.e.*, by *reciprocity*. In this case, the product of the two reciprocal operations is not a null operation but an equivalence.

Inversion and reciprocity are necessary conditions of equilibrium for the most elementary behavior as well as for the most highly organized operations. Thus, with variations in form they are found at all developmental stages. Even at the perceptual level (although there is as yet no complete reversibility), we find *inversion* operative when elements are added or taken away and *reciprocity* in symmetries and similitudes. But since we are not directly concerned with the preoperational forms of inversion and reciprocity in the present work, we shall only analyze the forms found in concrete and formal operations.

But when we compare the structural integrations of concrete and formal thought, we immediately encounter important differences between them. Whereas formal structures bring together both inversions and reciprocities in a single system of transformations (in the I N R C group), concrete structures or "groupings" of classes and relations derive from either inversion (classes) or reciprocity (relations), but entail no general synthesis of the two forms of reversibility.[1]

To summarize our descriptions of the respective structures, we may remind the reader that in the most general sense concrete logical operations are actions performed on objects to bring them together into classes of various orders or to establish relations between them. One can also distinguish infralogical operations from concrete classifications or relational operations; their function is to integrate the parts of objects into a spatio-temporal whole—*i.e.*, a permanent object—and to place or displace these parts in continuous configurations.[2] However, from the standpoint of the present work, infra-logical operations can be reduced to the logical operations of class inclusion or relations.

However, the integrated structures of classes and relations

---

[1] Of course, there are many systems of classes and relations which do synthesize the two, but these systems are based on the "structured whole" and not on simple groupings. Moreover, since they are isomorphic with propositional systems, they do not appear before the formal level.

[2] Moreover, these operations can be described in terms of topological structures (see Chap. 16, p. 269, note 7).

which govern concrete operations are limited to bringing classes or relations together by a class inclusion or contiguous linkage which moves from one element to the next. Since they involve only contiguous class inclusions and fail to organize the combinatorial system essential to the "structured whole," which alone makes possible the synthesis of inversions and reciprocities in a single system, they do not develop beyond incomplete groups or semi-lattices.

It is in this sense that concrete class structures consist exclusively of simple (additive) or multiple (based on multiplicative tables or matrices) classifications. Since the concrete level subject has no combinatorial system to furnish all of the sub-sets of classes which potentially can be formulated within a given system, for him each class depends once and for all on those with which it is included at a given moment (A is included in B, B in C, etc.) and is formulated by opposition to its complement in relation to the class immediately above it ($A' = B - A$; $B' = C - B$; etc.). These structures are based exclusively on reversibility by inversion. They permit the subject to add two contiguous classes to form a single one ($A + A' = B$) or to subtract a class from the whole formed by addition ($A = B - A'$); he can also multiply two classes by each other ($A_1 \times A_2 = A_1 A_2$) or abstract one from the whole formed in this way ($A_1 A_2 : A_2 = A_1$); but the general form of reciprocity fails to appear in these systems.

Concrete structures of relations coordinate complete equivalences (equalities) or partial equivalences (alterites) in the case of symmetrical relations or serially ordered differences in the case of asymmetrical transitive relations (ordered series or serial linkages). In addition, they enable the subject to handle multiplicative systems (correspondences, etc.). Thus the reversibility characteristic of concrete systems of relations consists of reciprocity. For example, a symmetrical relation such as $A = B$ is identical with its reciprocal $B = A$. For asymmetrical relations, if $A < B$ is true, its reciprocal $B < A$ is false; but if they are both true ($A \gtrless B$) they can be reduced to $A = B$—i.e., to an equivalence. An asymmetrical relationship such as $A < B$ expresses the existence of a difference between the terms A and B; if this difference is canceled or if it is expressed in the opposite direction in the form $B > A$, the equivalence $A = B$, or $A = A$ is again encountered without the terms themselves being canceled. Thus we see that such systems cannot handle reversibility by inversion (negation), for inversion bears on the terms of the relations involved—*i.e.*, on classes, and not on the relations themselves.

In sum, the elementary "groupings" which constitute the only integrated structures accessible at the concrete operational level can be distinguished from formal structures on two counts:

1. The groupings are systems of simple or multiple class inclusion or linkage, but they do not include a combinatorial system linking the various given elements *n*-by-*n*. As a result they do not reach the level of a fully developed lattice structure, which would imply such a combinatorial system ("structured whole"); instead, they remain in the state of semi-lattices;

2. The mechanism of reversibility consists either of inversion (for classes) or reciprocity (for relations), but the two are not integrated into a single system. Consequently, they do not coincide with the group structure of inversions and reciprocities (the I N R C group which will be discussed more fully below) and remain in the state of incomplete groups.

To clarify the point, let us return to the multiplicative groupings or several-entry tables (matrices), the most complex groupings brought into play in concrete thinking. In the flexibility problem (Chap. 3) we saw how the stage II subject constructs tables for class multiplication. Here we must limit ourselves, for purposes of simplification, to the classes of brass rods (A_1) and nonbrass rods (A'_1), and rods which are sufficiently flexible (A_2) or not sufficiently flexible (A'_2) to reach the desired level. We see that the stage II subject can easily construct the four following multiplicative classes:

$$A_1 A_2 + A_1 A'_2 + A'_1 A_2 + A'_1 A'_2 .$$

Now let us suppose, as is actually the case, that all four possibilities occur in fact. The subject cannot conclude from his table alone either that brass rods tend to be more flexible or the contrary. How will he go about trying to solve the problem after he realizes this?

At the concrete level (stage II), the subject will simply continue to classify or to order the empirical data in series, for he does not know how to set up more than multiplicative associations and correspondences. For example, he might ascertain the occurrence of classes $A_1A_2A_3$ (where A_3 stands for circular cross-section bars) and $A_1A'_2A'_3$, from which he will conclude that inflexible brass rods $(A_1A'_2)$ do not bend because they do not have a circular section form (A'_3); etc. With this method he may come to isolate a certain number of factors by successive associations (between classes) and correspondences (between relations) (Chap. 3), but he will never be able to enumerate the factors exhaustively or to demonstrate conclusively the effects of those factors which he does discover, since he lacks a systematic method, notably the procedure of varying a single factor at a time while holding the others constant.

What then is the structural basis of this systematic method? If we turn by way of illustration to the behavior of the stage III subjects in the same experimental situation, the difference between simple multiplicative associations and n-by-n combinations becomes apparent. The formal level subject, when he encounters the four initial classes $A_1 A_2 + A_1 A'_2 + A'_1 A_2 + A'_1 A'_2$ and after he has discovered that factors other than A_1 and A'_1 must be called upon to explain the result of A_2 or A'_2, will immediately ask himself how many combinations are possible within the framework of these four classes. Thus he will formulate hypotheses of this type:

1. If no disturbing factors are present, only cases $A_1 A_2 + A'_1 A_2 + A'_1 A'_2$ actually occur (case $A_1 A'_2$ does not occur if A_1 is not accompanied by a factor excluding A_2). Once these three associations are set up, the subject is led to assume that factor A_1, once isolated, always entails result A_2, but that it is not the only factor which can produce this result (implication);

2. But it could also happen that only cases $A_1 A'_2 + A'_1 A_2 + A'_1 A'_2$ occur (in this case $A_1 A_2$ being the result of the intervention of a factor producing A_2 in spite of the presence of A_1). The occurrence of these three associations would signify that there is a relation of incompatibility between A_1 and A_2.

In short, instead of simply multiplying classes by successive associations, the stage III method is based primarily on sorting the four classes $A_1 A_2 + A_1 A'_2 + A'_1 A_2 + A'_1 A'_2$ within the set of the combinations possible when they are taken n-by-n. There are sixteen possible combinations, which can be enumerated as follows:

(1) o   (6) $A_1 A_2 + A_1 A'_2$ (11) $A'_1 A_2 + A'_1 A'_2$
(2) $A_1 A_2$ (7) $A_1 A_2 + A'_1 A_2$ (12) $A_1 A_2 + A_1 A'_2 + A'_1 A_2$
(3) $A_1 A'_2$ (8) $A_1 A_2 + A'_1 A'_2$ (13) $A_1 A_2 + A_1 A'_2 + A'_1 A'_2$
(4) $A'_1 A_2$ (9) $A_1 A'_2 + A'_1 A_2$ (14) $A_1 A_2 + A'_1 A_2 + A'_1 A'_2$
(5) $A'_1 A'_2$ (10) $A_1 A'_2 + A'_1 A'_2$ (15) $A'_1 A'_2 + A'_1 A_2 + A'_1 A'_2$
            (16) $A_1 A_2 + A_1 A'_2 + A'_1 A_2 + A'_1 A'_2$

Obviously, it would be useless for the subject to construct such a table in its complete form. The main point is simply that when he encounters the four cases $A_1 A_2 + A_1 A'_2 + A'_1 A_2 + A'_1 A'_2$, the potential relationships between A_1 and A_2 are seen as multiple. (And the experimental behavior of the adolescent demonstrates conclusively that he actually proceeds in this way.) In fact, the expectation on which the reasoning of the adolescent is based from the moment he first observes the associations of variables and results which actually occur is that there is a multiplicity of possible links; he assumes that only certain associations will remain true once the other factors are eliminated. In other words, the initial choice of the true relationships necessarily implies a combinatorial system. Hence (as we have repeatedly emphasized), however simple it may appear on the surface, the utilization of proof based on the schema "all other things being equal" is in itself a reliable sign of the presence of the combinatorial structures. For unless he tries to select the combinations which occur from the total number of possibilities, the subject would not feel the need to go beyond the empirically occurring associations to separate out the variables. Rather, he would be limited to accumulating new associations. On the other hand, if he is to isolate the combinations which actually occur, he has to begin by conceptualizing at least some of the possible combinations; thus he must make use of a combinatorial system.

In conclusion, the concrete grouping structures lack the combinatorial system inherent in the organization of the "structured whole." Said differently, propositional operations (implication, etc.) are not utilized at the concrete level, since these interpropositional operations are based on the same combinatorial structure.

So the first difference between elementary groupings of classes and relations (classifications, serial orders, and multiplicative correspondences) and the combinatorial structure is that the former are only semi-lattices while the combinatorial structure is a complete lattice. What is more, each of the operations comprised in the structured whole has both an inverse and a reciprocal. Hence the second essential difference between concrete groupings and the system of formal operations is that the former does not have any single group linking inversions and reciprocities in the midst of elementary groupings; the structured whole can be said to be present as soon as this group comes into play.

However, it is not enough to point out the differences between concrete and formal operations or to bring them all together in the development of a "structured whole." We must go on to examine how the transition from concrete to formal propositional operations actually comes about. In other words, we must find out what leads the 11–12-year-old child to organize "structured wholes."

§ *The Transition from Concrete Operations of Classes and Relations to Interpropositional Operations*

As a methodological experiment, we took the successive assertions of some stage II and stage III subjects and tried to reduce them entirely to logical formulae; our purpose was to see in which cases the subjects reasoned through arrangements of classes and relations and in which cases they used propositional operations. The attempt was very instructive. It was nearly impossible to decide between the two when we took any particular isolated statement as a point of reference. The question is psychologically meaningful only when the subjects' entire reasoning or a sufficiently systematic series of inferences is taken as a context. Consider, for example, the proposition: "This rod bends because it is made of steel." Can we say that it involves a class operation, in this case one which would be based on the transitiveness of inclusions and belongingness: "This bar is steel" and "All steel bars are flexible"? Or is it a true implication: "If a bar is steel, then it is flexible"? The second alternative would presuppose consideration of the three conjunctions $(p.q) \vee (\bar{p}.q) \vee (\bar{p}.\bar{q})$. Similarly, does a proposition which contains the word "or" translate a simple partial disjunction into classes or the propositional disjunction $(p \vee q)$? Etc.

But it is fruitless to look for an exclusively verbal or linguistic criterion—e.g., considering all statements containing the words "if . . . then" as implications while regarding the statements which do not contain them as inclusions or correspondences, etc. Rather, only certain expressions are conclusive indices. When a subject says, "If it were (such and such a factor) that played a part, then you should find (such and such an unobserved consequence)," one can be sure of the hypothetico-deductive and consequently propositional nature of the operations involved. But in the majority of cases language remains implicit and the subject does not disentangle the details of his inferences. Moreover, the details of verbal expression vary between subjects and sometimes even from moment to moment for a single subject. All in all, the subjects' language expresses their thoughts only in a rough way.

A second and more adequate method is to compare all of the statements and particularly the actions of a single subject. It then becomes clear whether he is limited to a simple registration of the raw experimental results, forming only the classifications, serial orders, and correspondences he sees as sufficient for solving the problem, or whether he tries to separate out the variables. The latter implies both hypothetico-deductive reasoning and a combinatorial system; when they appear, we have to interpret the stated judgments as propositional expressions, since the links between the successive statements (whether explicit or implicit) consist in interpropositional operations.

But the third and surest method of differentiation (which is actually a simple specialization of the second) is to analyze the proofs employed by the subject. If they do not go beyond observation of empirical correspondences, they can be fully explained in terms of concrete operations, and nothing would warrant our assuming that more complex thought mechanisms are operating. If, on the other hand, the subject interprets a given correspondence as the result of any one of several possible combinations, and this leads him to verify his hypotheses by observing their consequences, we know that propositional operations are involved.

However, in using such a procedure, are we merely begging

the question? Let us go over our method. We begin by defining concrete operations as independent of any combinatorial system, whereas formal operations are said to require a combinatorial system. Now, we readily admit that on the basis of language alone we cannot tell whether an isolated statement is concrete or formal. To do so we must examine the subject's train of reasoning; this examination enables us to distinguish between the presence and absence of a combinatorial system, but it subordinates the distinction between formal and concrete to a preconceived interpretation. The reader may object that this allows us to take any series of statements of a subject and translate it according to whim into *either* propositional language (even statements made at the earlier stages) or class and relational language (up to and including the later stages). There is no doubt that it is always possible to express propositional operations in the language of classes at the formal level. *But,* this can be done only on condition that a "structured whole" is introduced—*i.e.,* a combinatorial system as distinct from the elementary groupings which can alone be observed in the 7–8- to 11–12-year stages. Conversely, the operations inherent to these groupings (classifications, etc.) could be expressed in propositional language. But, if this were done, our analysis of the interpropositional combinatorial system would not be exhaustive, and we would be using a complex language for describing phenomena which do not go beyond much simpler structures in the subject's mind. This would be a mistake in psychology analogous to explaining the first "natural" numbers (which the child constructs on his own) in terms of a more general mathematical structure applicable to all "real" numbers.

If we accept the task of describing the structures which actually operate in the subjects' minds, we have to use the criteria furnished by the combinatorial system in distinguishing between concrete ("elementary groupings" of classes and relations) and formal ("lattice" and "group" of propositional operations) operations. But it is also clear that the transition is not completely discontinuous and that there must be many intermediate steps. We must now give an account of continuity of steps between these two clearly distinguished types of operations.

Moreover, both the continuity in development shown by the

existence of transitional phases and the difficulty of determining the exact boundary line between concrete and formal thinking described above are highly significant; they show us the degree to which more sophisticated forms of thinking, rather than being immanent from the start, derive from earlier forms. If someone wanted to say that an *a priori* form of reasoning accounts for the development of formal structures, he would have to accept the burden of proof for the fact that this *a priori* form emerges so late. Of course, he could always call on the effect of a late-maturing nervous structure, and such a structure is probably a necessary condition of the development of combinatorial operations. But the neurological explanation cannot in itself be sufficient because the occurrence of transitional phases shows that the new operations derive from earlier ones. Given this fact, it must be that a continuously operating equilibration factor plays a role beyond that of purely internal conditions of maturation, and the problem is to understand how a tendency toward equilibrium or its results can lead the subject to organize a formal combinatorial system.

Of course, it is important to attribute part of the developmental process to practice and to acquired experience. But it is equally important not to overemphasize their roles, for if experience were the only determining factor in learning, both the correct registration of the empirical data and their translation into the language of propositional logic would be present at the earliest levels (here we see the theses of "logical empiricism"). If this were so, we should not understand why data reading remains inadequate for such a long period or why propositional logic appears so late. But if operations are conceived of as a series of increasingly more complex coordinations deriving from one another, with the simplest proceeding from irreversible action, it becomes clear why the final equilibrium, based on combinatorial operations, is established so late. The subject has to separate means from goals before he can render his behavior reversible. The first reversible coordinations are elementary groupings that are still very close to object manipulation. Then, since he does not begin *immediately* to dissociate form from content in any complete manner, he must reconstruct these various groupings in the heterogeneous empirical factors which experience presents one after the other. It is only after the different contents of experience are structured by degrees that a general formal mechanism begins to emerge through coordination of heterogeneous factors into a whole.

Thus, the circumstances in which the transition from concrete to formal logic occurs are these: after the subjects have structured a number of qualitatively heterogeneous factors (linear lengths, surfaces, weights, rates of speed, time, etc.) at various stages (or at the same stage but without perceiving the interrelationships), they discover that in many real situations variables are interdependent—*i.e.*, that the independently structured factors may overlap. For example, one effect may be the result of several concomitant causes, or a single causal factor may be masked by several noncausal but concomitantly varying factors. Whereas concrete operations proceed from one dimension to the next, sooner or later reality will present a multifactor situation in which variables are mixed.[3] For such complex situations, new operational instruments have to be forged.

When this happens, the subject may employ either one of two methods, depending on whether he tries to coordinate the results of concrete operations—*i.e.*, to resolve the apparent contradictions due to the interaction of factors in a complex situation—or whether he tries to coordinate several of the concrete grouping operations among themselves. Both of these methods lead the subjects to discover formal propositional logic. They entail: (I) separating out the raw empirical data—*i.e.*, the variables structured through concrete operations alone—in such a way as to coordinate the results of concrete operations as a function of the various possible combinations; and (II) coordinating the various groupings of classes and relations into a single systematic whole. Furthermore, however different these two procedures may appear on the surface, it can be shown that they can be reduced to a single mechanism, for both are based on the combinatorial system. Thus the real problem is to understand the origins of this system.

I. Our observations show that at the level of concrete operations the subject tries to structure reality as completely as possible but that he remains close to reality in its raw form—*i.e.*, as it

---

[3] If, in fact, the causal series are completely independent, such a mixture of variables becomes a random distribution, but, to the extent that it is not random, some mutual dependence is present.

appears without isolation of variables. When the subject classifies, orders, formulates correspondences, etc., he registers the facts directly without adopting a critical attitude toward the empirical world or adopting systematic methodological precautions. This is due to the fact that as concrete operations develop in the 7 to 12-year-old child their function is to structure reality factor-by-factor; there may be a time lag of several years between dimensions (lengths are structured at 7–8 years, weights at 9–10 years, etc.).

When a problem involves several independently structured factors (thus several heterogeneous variables) which interfere with each other, sooner or later the subject comes up against results which are somewhat inconsistent or even contradictory. For example, he may ascertain that event x is nearly always associated with event y (thus most brass bars have a high flexibility), but he also finds cases in which $\bar{x}.y$ and even $x.\bar{y}$ appear—*i.e.*, there are exceptions to the rule. Or the reverse may occur: the subject may observe that in general there is no correspondence between two variables ($x.\bar{y}$ and $\bar{x}.y$) and yet find certain cases in which $x.y$ and $\bar{x}.\bar{y}$ occur. In short, the more accurately he analyzes reality at the concrete level (*i.e.*, by simple correspondences between distinct variables), the more likely he is to discover the mixture of partial regularities and exceptions that it contains which he cannot explain with any degree of certainty. Of course, the subject at first neglects both partial regularities and exceptions, but when he begins to take the demands of the experiment seriously he finds himself blocked in his attempt to describe the raw data in concrete terms.

Hence, sooner or later a new attitude toward the experiment emerges. Our observations show that in a number of subjects it appears in rough form during the concrete stage but it is not generalized until the formal level. This attitude is revealed in the act of separating out variables, a process which we must now try to understand in formulating our conclusions on the transition from child to adolescent thinking. We can ask the following questions: (A) In what forms does this attitude originate?; (B) In what forms does it generalize, and why does it generalize at such a late date?; (C) Finally, how and why is it necessarily linked with the structuring of a combinatorial system?

A. Even at the concrete level there are certain forms of isolation of variables, and they deserve careful examination. When he wants to know if a given factor influences a given result, even the stage II subject can carry out certain verification procedures; some of them are exclusively *observational*, others are *experimental*. To illustrate the first type, we can cite procedures which lead the subject to exclude a factor by attributing it to chance. For example, in the magnet experiment (Chap. 6), when a subject sees the needle stop on a particular color, he may at first be tempted to assign a causal role to the color. But after observing the random distribution of stops, he rejects his hypothesis.

In talking about this case, we must be sure to make the distinction between separating out a factor and excluding it. Separating out a factor can be imposed by simply observing what occurs in the experiment, for observation reveals cases where the color in question is sometimes present and sometimes absent. In some situations the subject may proceed from this separation of variables by presence or absence of the factor in the experimental results to an active and experimental separation and in this way controls the occurrence or nonoccurrence of the factor in question. For example, in the pendulum experiment (Chap. 4), he might verify the hypothesis that impetus plays a causal role by actually giving a push to the pendulum in certain cases and not pushing it in others.

However, in neither of these cases (observation or experimentation) do we have true isolation of variables; when isolation appears at the concrete level, it is isolation by negation—*i.e.*, the factor whose role is to be evaluated is seen by the subject to be present in the experimental results in some cases and absent in others (observation) or introduced by the subject in some cases and eliminated in others (experimentation). Thus, transformations by inversion or negation appear, but reciprocities do not—*i.e.*, the first, but not both forms of reversibility. The limitations of this method are clear in cases where a factor cannot be physically eliminated (*e.g.*, the weight of an object, the length of a rod, etc.). The concrete level subjects do not succeed in neutralizing it, whereas the stage III subjects are able to do so (mentally) and thus to calculate its effects.

A second difference between the elementary separations of

stage II and the true formal isolation of variables is apparent in cases involving two or more variables—*x, y,* etc. The concrete level subject is only able to introduce or eliminate variable *x* in order to see if *x* itself plays an active role and not as a means of studying variations in *y.* For example, in the pendulum problem he pushes or does not push the pendulum only when he wants to know whether or not impetus is causally effective. He does not do so with a view toward studying the effects of length of the wire, etc. On the other hand, the formal level subjects may eliminate factor *x* not only in order to control its own influence but also when they wish to analyze the variations of *y* without perturbations due to the presence of *x.* Furthermore, this second difference can explain the first; thus it is the more important of the two.

B. Thus the two discoveries found at the beginning of the formal level are (1) that factors can be separated out by neutralization as well as by exclusion and (2) that a factor can be eliminated not only for the purpose of analyzing its own role but, even more important, with a view toward analyzing the variations of other associated factors. Once these discoveries are made, isolation can be generalized to all cases, as happens during stage III (and completed during substage III-B). Consider, for example, a long brass bar whose flexibility is greater than that of a short steel bar. If the variable *type of metal* is not isolated from the variable *length,* the subject does not know whether the first bends more than the second because it is brass or because it is longer. If the concrete level subjects solve the problem at all they solve it by multiplying correspondences and ascertaining that sometimes type of metal and sometimes the length plays the causal role. But, since any real bar has both metal and length, not only do these subjects fail to exclude either factor in order to analyze the role of the other, but the idea of doing so does not even occur to them. On the other hand, the formal level subjects know (frequently during substage III-A and always at III-B) that, in order to determine the role of the metal, length must be eliminated (and vice versa) while the other factor is varied simultaneously by adding or taking away. Moreover, they know that the variable not being analyzed at a given moment may be neutralized by simple equalization as well as actual removal when its nature prevents the latter. Thus, they will hold the length constant in order to analyze the role of the metal and vice versa.

Given that these two discoveries are made, the crucial question from the genetic standpoint is: how can they arise from the limited capacities of the concrete level?

The origins of these new attitudes must doubtlessly be sought in the reversal of direction which results from the increasing complexity of the solutions obtained by concrete methods. The stage II subjects are not blocked when they meet a difficulty; they continue to move forward. But they simply multiply correspondences and make new attempts to find relationships, hoping that something meaningful will automatically emerge from the abundance of data. Sooner or later they have to retrace their steps, for if too complicated linkages are built up, the variables left unanalyzed at one moment will later reappear as disturbing influences. At this point the behavioral innovation unique to stage III appears. It consists of setting aside y in order to analyze x free from disturbing interference and vice versa. Thus the need to exclude one factor so as to vary another results from a reversal of direction in structuring correspondences; it involves an attitude toward abstraction or separating out variables instead of toward multiplication or association of empirical correspondences. Furthermore, it appears when the subject is faced with excessive complexity and too many contradictions in the raw empirical situation.

Given this complexity, certain variables can be eliminated by inversion or negation, but this is not true for all of them. Thus the second innovation of stage III is that of generalizing the method of excluding perturbating factors in cases where these factors do not involve negation. In this case it is no longer one of the terms (a property or an event) that is negated but the difference between the terms. In other words, the variable to be excluded is neutralized by the simple procedure of equalizing the expressions of that variable encountered in the experimental situation. Now this is no longer *inversion* but *reciprocity*. As an example of this procedure, if a subject wants to determine the role of metallic composition (variable x) in the flexibility of a pair of rods, he holds length (variable y) constant. This equivalence results from the elimination of the differences between the lengths of the two rods, just as negation results from the elimination of the factor itself in the case of impetus.

The result is that the separation of variables found at the formal level ushers in reversibility by reciprocity at the same time as reciprocity by inversion. Thus, it involves a parallel use of the two forms of reversibility which renders them functionally equivalent. The importance of this fact for the structural integration of stage III is clear.

C. We still do not know why these two new attitudes necessarily carry with them the structuring of a combinatorial system. But from the moment he is committed to isolating variables, the subject finds himself faced with new possibilities. For example, after having isolated *length* from *metal* in the flexibility problem, he must neutralize variables such as thickness and cross-section form, and, as an immediate consequence of the method adopted, he must continually ask himself whether he has left out a variable. How can he be sure he is on the safe side in such a situation? Naturally the first method would be to find all possible associations between variables, using 2-by-2, 3-by-3, etc., comparisons. But this form of thinking does not go beyond the simple multiplicative operations of the concrete level (double-, triple-entry tables, etc.), and we have shown that these associations or multiplicative correspondences are inadequate for the solution of the problems we posed, although of course they must be established before further reasoning is possible. Only afterwards, when these *base associations* have been structured, can the subject choose the crucial combinations from the total number of possible ones. Thus it must be at this point that the combinatorial system makes its appearance. It is conceivable that each one of the variables that appear together in the experimental situation plays an independent causal role, in which case the crucial combination is the one in which the factors are varied one at a time, the others being held constant (*e.g.*, the flexibility problem, Chap. 3). But it is also possible that two or three variables have to work together to produce the observed effect (as in the case of the colored liquids, Chap. 8). Or two variables might be mutually exclusive in the production of the effect, or one might work in the opposite direction from the other (as the decolorant in Chap. 7), etc. We see that separating out variables may lead to any one of a number of distinct *possibilities* which can be formulated by means of implications, equivalences, disjunctions, conjunctions, exclusions, incompatibilities, etc., depending upon the particular case. When these *possibilities* already exist in the subject's mind as expectations—*i.e.*, when as a result of previous acquisitions he possesses a wide enough range of possible combinations—it is evident that they can guide the separation of variables by a feedback effect which we can understand easily. However, if the subject is not aware of these possibilities but still manages to discover how to separate out variables toward the end of the period of concrete operations, the experimental isolation of variables itself leads him to the combinatorial system, because the crucial combinations vary as a function of the individual situation, and because in each situation the discovery of the crucial combinations (a discovery which is itself due to the isolation of variables) presupposes a choice among the possible associations. For example, for two factors x and y, the fact of varying x alone while leaving y constant (thus $x.y + \bar{x}.y$) constitutes a choice of the crucial combinations $(x.y + \bar{x}.y)$ among the four possible associations $(x.y + x.\bar{y} + \bar{x}.y + \bar{x}.\bar{y})$.

In sum, the isolation of variables necessarily leads the subject to combine the base associations among themselves n-by-n and thus to substitute for the simple multiplication and correspondence operations which gave rise to the base associations the combinatorial system which characterizes the "structured whole."

II. Now that we understand why isolation of variables (itself the result of a need to coordinate the increasingly entangled results of concrete operations of formulating relations and correspondence) necessarily ends up at a combinatorial system, we still have to show how the combinatorial system is structured and gives rise to formal thinking. This can be accomplished by analyzing the ways in which the *concrete operations*—rather than their results—are coordinated. It is this coordination which gives rise to the combinatorial system inherent to propositional logic and consequently to formal thinking.

After the factors in the experimental situation relevant to the solution of a given problem have been classified, ordered serially, and equalized, and correspondences set up, etc., it may still be necessary to integrate the operations performed up to this point into a single system before the problem can be solved. Thus, the need for integration results from the need to coordinate the results of concrete operations when they do not have sufficient internal consistency.

However, the subject does not possess any operations which allow him to integrate the various groupings of classes and relations into a single system directly unless he goes beyond simple additive or multiplicative class inclusions. In other words, he has to integrate them into a "structured whole"—*i.e.*, that very combinatorial system whose organization in the subject's thought we are now trying to understand. By contrast, either for classes or for relations (but not for both at the same time), there is a grouping more general than the others in the sense that it contains them or that they derive from it by successive specifications. The most general concrete grouping is the multiplicative grouping (of classes or relations) consisting of double- (or triple-, etc.) entry tables. For two events or properties, x and y, this method of grouping the data involves structuring the elementary associations $x.y + x.\bar{y} + \bar{x}.y + \bar{x}.\bar{y}$. However, as consideration of the isolation of variables and the resultant combinatorial system has just demonstrated, new problems arise for the subject as soon as he has to decide which of these associations are true and how to select sub-sets. What does he do in this situation? It is very important that we see that the choice or determination of sub-sets of true associations from among the possibilities $x.y + x.\bar{y} + \bar{x}.y + \bar{x}.\bar{y}$ results from the use of simple classification operations. However, at this stage they are applied to the associations ($x.y$, etc.) themselves and are generalized to all possible cases. The subject groups case $x.y$ with case $x.\bar{y}$, or case $x.y$ with $\bar{x}.\bar{y}$, etc., as if he were grouping objects characterized by common properties, whereas in fact the problem is to group associations—*i.e.*, situations in which two properties appear together (or one without the other, etc.) or where two events appear together (or one without the other, etc.).[4] In other words, the subject organizes the "struc-

[4] For example, in the flexibility problem (to which we have already referred in section I of this chapter), he will classify the brass rod with a circular cross-section which reaches a given degree of flexibility (xyz) with the steel rod of the same form which does not attain the same degree ($\bar{x}y\bar{z}$): from whence the class $xyz + \bar{x}y\bar{z}$; etc.

tured whole" by means of a new classification procedure in which the multiplicative system $x.y + x.\bar{y} + \bar{x}.y + \bar{x}.\bar{y}$ serves as a base; by this means he applies the simplest of groupings (classification) to the most general (the table of logical multiplications) and ends up with a sort of second-degree grouping which coordinates all of the groupings in a higher order system, since he cannot integrate them directly. Furthermore, the second-degree grouping formulated by application of the generalized classification [5] to multiplicative associations is none other than an n-by-n combinatorial system. The consequences of this fact for the development of formal thinking are the following:

1) In the first place, the generalized classification of associations $x.y$, etc., finally develops into a new mode of composition. Up to this point, the classifications utilized by the subjects are based essentially on simple class inclusions (for example, sparrow < bird < animal < living creature) congruent with the most elementary groupings: $A + A' = B$, $B + B' = C$, etc., or other classifications made according to two possible criteria (the Genevans plus the other Swiss = the Vaudois plus the other Swiss) *i.e.*, $A_1 + A'_1 = A_2 + A'_2$, etc. (*vicariance*). However, when the subsets of associations within a multiplicative grouping must be integrated by class inclusion in taking the different possibilities into account, the mode of composition will be completely different and will lead to an n-by-n combination by generalization of *vicariance*. In order better to understand this development, let us denote the four base associations derived from x and y by the numbers 1 to 4:

$$1 = x.y \; ; \; 2 = x.\bar{y} \; ; \; 3 = \bar{x}.y \text{ and } 4 = \bar{x}.\bar{y} \, .$$

Sixteen classes result from the various possible inclusions (*cf.* the table on page 277):

(0);
(1), (2), (3), (4);
(1 + 2), (1 + 3), (1 + 4), (2 + 3), (2 + 4), (3 + 4);
(1 + 2 + 3), (1 + 2 + 4), (1 + 3 + 4), (2 + 3 + 4);
(1 + 2 + 3 + 4).

[5] This generalized classification is what we have called elsewhere *vicariance*. In itself, it constitutes a distinct "grouping" of classes. *Translators' note:* since the term *vicariance* is specific to the work of Piaget, we have left it in the original. See *Traité de logique*, pp. 113–17, for definitions and examples.

Thus the structure of these classes, which the subject gradually builds up in his mind either by structuring empirical situations or by deducing from the possible combinations, differs radically from the structure of concrete groupings. In groupings of concrete operations, a given elementary class (A) is included exclusively in the higher classes of which it is a part (B, C, etc.) once and for all, even though it may be momentarily excluded (for example, birds belong to the class of vertebrates, and the vertebrates minus birds give fish, amphibians, etc.). For associations 1 to 4, on the other hand, each base association 1 ($x.y$) or 2 ($x.\bar{y}$), etc., reappears in seven higher-order classes (for example, 1 reappears in $1 + 2$, $1 + 3$, $1 + 4$, $1 + 2 + 3$, $1 + 2 + 4$, $1 + 3 + 4$, and in $1 + 2 + 3 + 4$); each double link ($1 + 2$, etc.) appears in three higher-order classes; and each triple link reappears in the same whole $1 + 2 + 3 + 4$. In other words, the new system which results from this combinatorial operation is no longer a simple classification (nor even a *vicariance* between classes of the same ranks): it is a generalized classification or a set of all possible classifications compatible with the given base associations. But this is exactly the same as the lattice structure, based on the "structured whole" of n-by-n combinations, in contrast to the structure of elementary groupings.

2) As a result, the negation of a given combination (for example, the conjunction $x.y$, association 1) no longer has to be built up from element to element in a series of complementary classes each included under the next largest class, as happens in the concrete groupings (where A is the class of sparrows and B that of birds, the complement selected by the subject for A is not the absolute negation $\bar{A}$—*i.e.*, all of reality with the exception of sparrows—but A'—*i.e.*, birds other than sparrows; where C is the class of animals, B' will be that of animals other than birds, etc.). Rather, the negation of a combination will be the conjunction of other elements—i.e., its complement with regard to the whole; thus the conjunction $x.y$ (association 1 will have for negation $x.\bar{y} + \bar{x}.y + \bar{x}.\bar{y}$ (*i.e.*, $2 + 3 + 4$)—*i.e.*, incompatibility (x/y).

3) As a result, the system built up in this way includes both inversions (the negation discussed above) and reciprocities ($x \supset y$ and $y \supset x$, etc.); thus inversions and reciprocities are integrated in a group of four transformations. Naturally, the subject

does not become conscious of the group in its abstract form, but, as we shall see presently, it has a number of repercussions in his thinking.

4) In a general sense, instead of dealing with objects directly—*i.e.*, classes of objects or relations between them—the combinatorial composition comes to handle more and more complex assemblages and their transformations. For example, the subject will eventually want to know whether two properties x and y are mutually exclusive (from whence $x.\bar{y} + \bar{x}.y$) or whether they are simply disjunctive although they may appear together (from whence $x.y + x.\bar{y} + \bar{x}.y$). When he asks such a question, the subject's reasoning deals not with reality directly but rather with reality as a function of possibility. Here addition ($+$) is no longer an addition of real cases but an addition of possibilities, for the real cases cannot always occur simultaneously. This is why the fundamental operation of propositional logic is noted v in the sense of "or": thus x v y signifies "either $x.y$ is true, or $x.\bar{y}$, or $\bar{x}.y$, or two of these cases out of three, or all three." The formula is equivalent to saying that the expression x v y is an addition of seven possibilities. Moreover, this is certainly what the integration of the possible associations means to the subject.

5) Finally, and by the same token, the combinatorial composition deals with *propositions*. Even during the concrete stage (and moreover, in preoperational thinking), reasoning is obviously based on propositions, with or without perceptual presence of the objects described. But the concrete operation consists in *de*composing and *re*composing the content of the propositions—*i.e.*, classes and relations as constituents of the proposition. Thus, at the concrete level a proposition is still linked to another not by virtue of its being a proposition but exclusively on account of its logical content, consisting of structures of classes and relations corresponding to actual objects. On the other hand, as soon as the proposition states simple possibilities and its composition consists of bringing together or separating out these possibilities as such, this composition deals no longer with objects but rather with the truth values of the combinations. The result is the transition from the logic of classes or relations to propositional logic. We will return to this point in the following pages.

In sum, thinking becomes formal as soon as it undertakes the coordination of concrete groupings into a single system (of the second degree) because it deals with possible combinations and no longer with objects directly. However hesitating, however incomplete the first trials of formal thinking at the beginnings of stage III may be, we can nevertheless see a tendency toward a new form of equilibrium which is characterized by a new type of structural integration deriving from both the lattice and the group of inversions and reciprocities. Although equilibrium is not achieved immediately, it will be achieved at substage III-B because operations cannot be coordinated without such processes resulting sooner or later in a system of necessary compensations, as the whole history of earlier operational thinking at the concrete level and even that of the intuitive regulations of the preoperational level goes to show.

§ The System of Sixteen Binary Operations

Now that we have described in a general way the transition from concrete to formal or propositional operations, we have to see whether the subject at substages III-A and III-B is actually able to use the sixteen binary operations of propositional logic, distinguishing each one from the others and putting them together as he makes inferences.

1 and 2. *Complete affirmation* $(p * q)$ *and negation* (o).[6]

The formation of the operation $p * q$ and its negation is easily explained. The first, which is the assertion of the possible occurrence of any one of the four base associations $(p.q \text{ v } p.\bar{q} \text{ v } \bar{p}.q \text{ v } \bar{p}.\bar{q})$ is only the translation of the four associations already understood in a concrete double-entry table in terms of propositions p and q. In the flexibility experiment (Chap. 3), when the subject classifies the rods into brass or non-brass and circular or noncircular sections, he discovers that all four possible associations occur. As a result, even at the concrete level he is able to conclude that, in the apparatus furnished, these two properties are in part linked together and in part disjunctive. But the operation does not take on propositional meaning until it is placed in opposition to the other possible combinations. When he tries to find out whether

[6] These two operations correspond to expressions (16) and (1) of the table on p. 277.

property x carries with it or results from property y, the subject will conclude from the occurrence of the four associations that the two properties can be assumed to be independent of each other but mutually compatible. (These two assertions remain subject to the qualification of later proofs if the variables have not yet been separated out.) For example, in the pendulum problem (Chap. 4), if p states an increase in weight and q an increase in acceleration, it is possible to have $p.q$ v $p.\bar{q}$ v $\bar{p}.\bar{q}$ (if in $p.q$ and $p.\bar{q}$, the length of the pendulum has been increased at the same time, and if, in $p.q$ only the weight has been changed). Observation of these empirical associations (without isolating the other variables) will furnish sufficient demonstration that weight is not decisive. But for an accurate exclusion, the subject has to use other operations.

3 and 4. *Conjunction* $(p.q)$ *and incompatibility* (p/q).[7]

Conjunction $p.q$ may have either one of two meanings; the general one holds when the association $p.q$ is linked to others, the particular one when association $p.q$ is the only true combination (by exclusion of $p.\bar{q}$, etc.).

In the first sense, conjunction $p.q$ is only the propositional expression of the multiplicative operation already known at the concrete level for the case in which two classes A_1 and A_2 are associated with each other: $A_1 \times A_2 = A_1A_2$; or if $A + A' = B$, the association $A \times B$ gives AB, which is identical to A itself (example: vertebrate birds constitute the same class as birds). But on the concrete level, the multiplicative class AB is included only in B (and in C, etc., if C includes B), and class A_1B_2 is included only in B_1B_2; etc. On the other hand, conjunction $p.q$ can be integrated with seven other formal combinations $(p \supset q, p$ v $q, p = q, q \supset p, p * q$, etc.) which cannot be included in each other. Thus, when confronted with the conjunctive association $p.q$, the formal level subjects will not draw conclusions before they have determined the other associations $(p.\bar{q}$, etc.) with which it integrates. This reaction is uncertain at substage III-A; it is systematic at III-B. In particular, the stage III subjects struggle against the temptation to conclude too quickly from the occurrence of $p.q$ the assertion of an equivalence $p \gtreqless q$ or of a simple implication

[7] These two operations correspond to combinations (2) and (15) of the table on p. 277.

$p \supset q$. The contrast becomes evident when the subjects of sub-stage III-A and III-B are compared: at III-A, the subject gives in too rapidly to this temptation without taking care to verify whether or not $p.q$ is accompanied by $\bar{p}.q$ and $p.\bar{q}$.

On the other hand, in its more restricted sense the conjunction $p.q$ signifies that the single association $p.q$ is true, with the three others excluded. From the standpoint of experimental methodology, the subject has to test the three other possible associations if he is to assume $p.q$ in this particular sense. But once established, the conjunction $p.q$ acquires the strong sense of the necessary union of two properties in a definition (static) or two factors which must come together for a given result to be obtained. In the color experiment (Chap. 7), the substage III-A and III-B subjects manage to discover a ternary conjunction by this procedure (sulphuric acid, oxygenated water, and potassium iodide have to be brought together for the yellow color to be produced). However, it is interesting to see that this discovery presupposes the utilization of a combinatorial system: each one of the variables is tried out with all of the others—2-by-2, 3-by-3, and the 4 together. Only the elimination of some variables calculated from the false combinations makes it possible for the subject to discover the conjunction $p.q.r$ (where p, q, and r state the respective intervention of the three causal factors) and the law $(p.q.r.) \supset s$ (where s stands for the color). Doubtless, in this case, the combinatorial system deals with objects before it deals with propositions, but we have seen that the two systems are isomorphic and that the subject translates his physical combinatorial system into a propositional combinatorial system.

The inverse of the operation of conjunction $p.q$ is incompatibility $(p/q = p.\bar{q} \vee \bar{p}.q \vee \bar{p}.\bar{q})$, which means that the characteristics denoted by p and q are never conjunctive—that is to say, that where one is present the other or both are absent. Thus, in the color experiment the subject discovers as follows that one of the liquids bleaches the mixture (expressed by $p.q.r$) referred to above. If s states the presence of color and $\bar{s}$ states its absence and t and $\bar{t}$ the presence or absence of the fourth liquid, the subject ascertains that only the combination $s.\bar{t} \vee \bar{s}.t \vee \bar{s}.\bar{t}$ is verified. But does he need formal operations to make this discovery? On the one hand, beginning with substage II-B (9–11 years), the subject discovers that thiosulphate bleaches the liquid when it is colored and has no effect when it is not. On the other hand, if younger subjects are given a set of geometric figures comprising black squares, white circles and rectangles, etc., and various other shapes of different colors, they see right away that the group can be divided into (non-white squares) $+$ (white non-squares) $+$ (neither square nor white figures); this constitutes a distribution of multiplicative associations of isomorphic classes involving an incompatibility. But, as we said earlier in this chapter, we cannot tell whether or not the subject's reasoning derives from propositional logic until the total set of his experimental procedures and verbal statements have been examined. The substage II-B subject discovers the bleaching influence of thiosulfate only in mixing the liquids at random, without systematic combinatorial procedures, and the younger subject who classifies shapes and colors notes the absence of white squares only when he perceives a gap in his multiplicative table (which consists of simple 2-by-2 associations and not n-by-n combinations). Thus we would not want to speak of incompatibility p/q except for those subjects capable of placing this combination in opposition to the set of fifteen other combinations in the system or to the principal ones among these ($p.q$, $p \vee q$, $p \supset q$, etc.). For to do this would be to commit the "fallacy of the implicit." However, the stage III subjects, who utilize a complete combinatorial system in their experimental trials on the colorants and who also express the results by means of adequate statements, are able to place their combination in opposition to the fifteen others.

5 and 6. *Disjunction* ($p \vee q$) *and conjunctive negation* ($\bar{p}.\bar{q}$).[8]

Disjunction $p \vee q$ signifies that p is true or q is true or both are true. Thus it serves to express the case where an effect may be due to two causes acting either independently of each other or conjointly: $p \vee q = p.q \vee p.\bar{q} \vee \bar{p}.q$. Thus its negation $\bar{p}.\bar{q}$ expresses the simultaneous absence of both causes. We have seen repeated examples of this operation, even in its ternary form $p \vee q \vee r$, as for example in the case of the dumping apparatus (Chap. 13): "To make it go down, you could either pull up the line or take off some weight (on the counterweight) or add some

8 These operations correspond to combinations (12) and (5) of the table on p. 277.

in the wagon" (CLAU; 11 years). Comprehension of the inverse follows immediately: $(\bar{p}.\bar{q}.\bar{r}) \supset \bar{s}$ (where $\bar{s} =$ no change).

Here also, it is clear that, beginning at the concrete level, we find cases in which the subject ascertains the absence of $A'_1A'_2$ after having multiplied two classes B_1 and B_2; thus, the remaining associations $A_1A_2 + A_1A'_2 + A'_1A_2$ will constitute the equivalent of a disjunction. But here also, the propositional disjunction must be distinguished by the fact that it is placed in opposition to the fifteen other possible combinations.

7 and 8. *Implication* $(p \supset q)$ *and nonimplication* $(p.\bar{q})$.[9]

Implication $p \supset q$ which expresses the combination $(p.q) \vee (\bar{p}.q) \vee (\bar{p}.\bar{q})$ is employed by the subject every time a cause, expressed by proposition p, produces an effect, expressed by q, but is not the only cause which can produce the same effect. In the preceding example, where the descent of a wagon (expressed by s) can be related either to an increase in the incline of the track (p) or the diminution of the counterweight (q) or an increase in the load carried by the wagon (r), the subject knows that $p \supset s$, $q \supset s$, and $r \supset s$; in the case $p \supset s$, one actually has $p.s$ (when the first factor is in actual operation) but also $\bar{p}.s$ (when it is the two others) and, naturally, $\bar{p}.\bar{s}$ (when none are operating).

In the case of implication, more than in that of any other propositional operation, we can get the illusion that it is found even at the concrete level. From the class standpoint, it corresponds to the inclusion $A < B$ (where $B = A + A'$). If we call S the class including the set of events described by proposition s (the wagon's descents) and P the class of descents linked to a greater inclination of the track (*cf.* prop. p), the inclusion $P < S$ will signify $P + P' = S$–*i.e.*, that all P's are S's, but that there are some S's which are not P's (which then correspond to the descents due to changes in weight: *cf.* propositions q and r). Thus, the implication $p \supset s$ could be expressed by the inclusion $P < S$. From the standpoint of relations, the same implication corresponds, on the other hand, to a multifactor correspondence: to the effects S correspond the causes P, Q, R, whence the correspondence $S \leftarrow Q$

$$S \begin{array}{c} \nearrow P \\ \leftarrow Q \\ \nwarrow R \end{array}$$

---

[9] Corresponding to combinations (14) and (3) in the table on p. 277.

in which P corresponds unequivocally to S, but this unequivocal relationship is not reciprocal (one cause corresponds only to its effect, but the same effect may correspond to several causes).

However, here again, the difference between a true implication and a many-one correspondence can be recognized psychologically by the progression of the totality of the subject's reasoning. As long as he proceeds by inclusions or correspondences (stage II), the subject is limited to classifying and ordering serially the raw experimental data, whereas the discovery of implication as such consists of differentiating it from the other possible combinations (p v q, $p = q$, etc.). Moreover, this discovery is distinguished by the fact that the subject begins to separate out the potential factors; his goal is to verify exactly which combinations occur among the possible combinations compatible with the given situation.

But, although by this criterion the propositional implication is clearly psychologically distinct from class inclusion, it is still worth while to try to find out how the transition between them is accomplished. Moreover, it is important to remember that even at stage III the subject begins by classifying the data, by relating them, etc.; in other words, one must keep in mind that a concrete structuring of the data is an indispensable prerequisite of the propositional structure. Thus, we must ask: what are the most spontaneous forms of implication and what are their relations to the corresponding concrete linkages?

But, implication can be expressed in at least three equivalent ways: $p \supset q$; $\bar{p}$ v q and $p = p.q$, expressions which can be transformed to give the same product $p.q$ v $\bar{p}.q$ v $\bar{p}.\bar{q}$. For example, if p expresses the fact that a rod is thin and q the fact that it is flexible, it does not matter whether the proposition is stated: "If it is thin, then it is flexible" ($p \supset q$); "Either it is not thin, or it is flexible" ($\bar{p}$ v q); or "To say that it is thin is equivalent to saying it is both thin and flexible" ($p = p.q$). It should also be noted that, according to a well-known law of lattices, given $p = p.q$, it follows that $q = p$ v q, an equivalence which is itself equal to $p \supset q$ (for its transformation also gives $p.q$ v $\bar{p}.q$ v $\bar{p}.\bar{q}$). This said, it seems clear that the simplest psychological form of implication must be $p = p.q$. For, before he can maintain "If this rod is thin, then it is flexible," the subject must assure himself that thin always means thin and flexible. Moreover, $p = p.q$ and $q = p \vee q$ are the most direct translations of the product $A \times B = A$ and the sum $A + B = B$, foundations of the inclusion $A < B$.

The negation of implication is nonimplication $\overline{p \supset q} = p.\bar{q}$. This operation, like all those which are formed from a single pair (we have seen this in reference to $p.q$) can appear either in an isolated state or integrated with others in the following seven different forms: p/q, $p \vee q$, $q \supset p$, $p \vee\vee q$, $p[q]$, $\bar{q}[p]$, and, of course, $p * q$.

In its isolated form, the negation of implication is frequently employed by the subjects to prove nonintervention of a possible factor. For example, in the case of the pendulum (Chap. 1), increasing the weight (expressed by p), leaving all other factors equal, and ascertaining that the acceleration does not increase ($\bar{q}$), will convince the stage III subjects that the weight plays no role: $p.\bar{q} = (p \supset q)$. But, here as always, the formal character of this operation (by contrast with the statements of noncorrespondence characteristic of the concrete level) can be seen in its integration with the totality of the combinatorial system—before concluding on the basis of observation of $p.\bar{q}$ that it must be a case of nonimplication, the stage III subject assures himself that he has eliminated all other combinations which include not only $p.\bar{q}$ but also $p.q$. For this reason when he studies the effects of the variable under consideration, he holds the other variables constant in such a way as to isolate the occurring combinations from the possible ones.

9 and 10. *Reciprocal implication $(q \supset p)$ and its negation $(p.q)$.*[10]

Naturally, implication $q \supset p$ alone does not have a meaning which differs from that of $p \supset q$, since it is always possible to call q proposition p and vice versa. Thus the operation $q \supset p$ has no distinct consequences except as compared to an operation $p \supset q$ already determined or in the course of being determined. In this case the two problems which arise for the subjects are the following:

If the link $p \supset q$ is true (for example, $p =$ the statement of an increase in the angle of incidence in the billiard experiment (Chap. 1) and $q =$ an increase in the angle of reflection) is the link $q \supset p$

[10] Corresponding to combinations (13) and (4) of the table on p. 277.

also true? This would mean that, in ascertaining an increase in the angle of reflection, the subject will conclude that the angle of reflection also increases even if he has not been able to observe the firing conditions). If it is, we then have $(p \supset q).(q \supset p) = (p = q)$ —i.e., there is propositional equivalence between p and q ($= p$ and q are always both true or both false).

If the linkage $p \supset q$ is not yet established but remains doubtful throughout the experimental trials, the subject may ask himself whether it is $p \supset q$ or $q \supset p$ which is the true implication. For example, in the case of indicators and colorants (Chap. 7), the subject may ask himself whether it is the last liquid added that colors the mixture (the dominant hypothesis for the youngest subjects: the last flask used contains the coloring potentiality) or whether it is the mixture which colors each component, including the last. Thus the problem is, does $p \supset q$ or $q \supset p$. Moreover, it is immediately obvious that the fact of raising or even of grasping such a question is itself an index of formal fluidity. Even more than the others, the operation $q \supset p$ does not acquire operational meaning until it is related to the system as a whole.

As for the negation of $q \supset p$—i.e., $\bar{p}.q$—it warrants the same comments as the operation $p.\bar{q}$, discussed under (8). Let us note only that the integration of the two nonimplications $p.\bar{q}$ and $\bar{p}.q$ is a reciprocal exclusion—i.e., $p.\bar{q}$ vv $\bar{p}.q$—which will be discussed further below.

11 and 12. *Equivalence ($p = q$) and its negation, reciprocal exclusion.*[11]

Propositional equivalence is neither an identity nor an equality but simply the assertion that two propositions are always both true or both false. From the standpoint of classes—i.e., of the set of objects to which the propositions apply—equivalence corresponds to the identity of a single class, but a class which is multiplicative as well as additive. For example, the equivalence between "This animal is a protozoon" (A) and "This animal is a unicellular invertebrate" (B) corresponds to the same additive class, for if $A' =$ pluricellular invertebrates, the equivalence corresponds simply to $A = B - A'$. But the equivalence between "This animal has a vertebral column" and "This animal has a spinal marrow" corresponds to the product of two classes,

[11] These operations correspond to combinations (8) and (9) on p. 277.

A_1 and A_2. Thus, from the standpoint of relations, equivalence means that there is a one-to-one reciprocal correspondence which is always reflexive but not necessarily reflected. In the pendulum experiment (Chap. 4), the increase in length (stated by p) corresponds to an increase in oscillation rate (stated by q), and the converse is also true; whence $(p \supset q).(q \supset p) = (p = q)$. The subject understands this equivalence when, after having constructed the implication $p \supset q$, he succeeds in discovering that the effect (q) has no other possible cause except (p) and the two variables always have to correspond.

Reciprocal exclusion $p \vee\vee q$ is the negation of the equivalence and describes the integration of the two nonimplications $p.\bar{q} \vee \bar{p}.q$, just as equivalence states the product of the two implications $p \supset q$ and $q \supset p$. From the class standpoint, exclusion describes mutually-exclusive membership in each of two entirely disjoint classes. From the standpoint of relations, it describes an inverse correspondence. For example, in the balance experiment (Chap. 11), if the subject considers all of the equilibrium states in which the balance is horizontal, he finds that for any increase in weight (stated by p), there is a corresponding decrease in distance ($\bar{q}$), and for any increase in distance (q), there is a corresponding decrease in weight ($\bar{p}$): whence $p.\bar{q} \vee \bar{p}.q$. In this case the reciprocal exclusion describes two serial orders in inverse correspondence. As for correspondence with classes, we can refer back to the example of the colorants cited above in discussing incompatibility. The only difference between incompatibility and reciprocal exclusion relates to the presence or absence of association $\bar{p}.\bar{q}$; this may be important when the problem is to know whether either p or q is always true, but usually neither is found, since in this case the association $\bar{p}.\bar{q}$ describes a situation in which nothing happens.

13 and 14. *The affirmation and negation of p—i.e., p[q] and $\bar{p}[q]$.*[12]

Operation $p[q]$ is equivalent to $p.q \vee p.\bar{q}$, and its negation $\bar{p}[q]$ to $\bar{p}.q \vee \bar{p}.\bar{q}$. Thus, both operations amount to affirming (or denying) that p is true when q is either true or false. In other words, both operations affirm (or deny) p independently of q.

Furthermore, this particular relationship of relative independ-

12 These two operations correspond to combinations (6) and (11) on p. 277.

ence but not exclusion is extremely important for the functioning of formal thought; ordinarily, it is because they possess operations $p[q]$ and $\bar{p}[q]$ that the stage III subjects are able to discover that a hypothetical variable does not actually contribute to or determine the production of a given phenomenon. In other words, if the variable stated by p appears both in the case in which the event stated by q does not occur $(\bar{q})$ and the case in which it does (q), it must be that the variable stated by p is not the cause of the event described by q. Thus, it is because he possesses operation $p[q]$ that the subject is able to exclude the hypothesis that a given variable is influential. Conversely, if $\bar{p}[q]$ occurs—*i.e.*, $\bar{p}.q$ and $\bar{p}.\bar{q}$ —the absence of the variable stated by p does not prevent the event described by q from occurring or not occurring; this confirms the necessity of excluding the hypothesis that the variable stated by p influences the result. But operations $p[q]$ and $\bar{p}[q]$ entail only simple exclusion and not reciprocal exclusion as in $p \text{ vv } q$, for although both $p[q]$ and $p \text{ vv } q$ contain $p.\bar{q}$, operation $p[q]$ also contains the association $p.q$, absent in $p \text{ vv } q$ and p/q.

For example, in the magnet experiment (Chap. 4), the weight of a box can be increased (an increase which can be stated by p). Sometimes the needle will stop in front of this box on the next trial (thus q), and in other cases it will not $(\bar{q})$. Consequently, the subject concludes from the combination $p.q \text{ v } p.\bar{q}$ that increasing the weight has no effect and that the stopping or nonstopping of the needle $(q$ or $\bar{q})$ is due to other variables. Conversely, non-increase in weight (stated $\bar{p}$) is also accompanied by both the events described in q and $\bar{q}$—*i.e.*, $\bar{p}.q \text{ v } \bar{p}.\bar{q} = \bar{p}[q]$. This counterproof confirms the earlier conclusion.

15 and 16. *The affimation of q and its negation—i.e., q[p] and $\bar{q}[p]$.*[13]

The two operations $q[p]$ and $\bar{q}[p]$ have the same structure as the above except for permutation of p and q. Actually $q[p]$ is equivalent to $p.q \text{ v } \bar{p}.q$ and $\bar{q}[p]$ to $p.\bar{q} \text{ v } \bar{p}.\bar{q}$. Thus considered in isolation, they add nothing to the system, but, compared to the previously described operations, they make new combinations possible.

For example, if $p[q]$ and $q[p]$ are both true at the same time, and they alone are true, $p.q \text{ v } p.\bar{q} \text{ v } \bar{p}.q$ is obtained, thus $p \text{ v } q$. If

[13] These two operations correspond to combinations (7) and (10) on p. 277.

pq and $\bar{q}[p]$ are both and alone true, one obtains $p.q \vee p.\bar{q} \vee \bar{p}.\bar{q}$, thus $q \supset p$. If $\bar{p}[q]$ and $q[p]$ are both and alone true, one obtains $p.q \vee \bar{p}.q \vee \bar{p}.\bar{q}$, thus $p \supset q$. And if $\bar{p}[q]$ and $\bar{q}[p]$ are both and alone true, one obtains $p.\bar{q} \vee \bar{p}.q \vee \bar{p}.\bar{q}$, thus p/q. Moreover, each one of these compositions may describe an effective structuring of the data at one point during the subject's run of trials. Thus operations $q[p]$ and $\bar{q}[p]$ are not simple replicas of $p[q]$ and $\bar{p}[q]$ but play a real role in discovery.

§ *The Application of the System of Sixteen Binary Operations and the Process of Formal Reasoning*

We have just seen that each of the binary propositional operations describes the particular partial structure of one or several of the eight elementary "groupings" of classes and relations on the plane of concrete operations. Once the eight "groupings" (which we know are worked out simultaneously, although they are applied to different dimensions in succession) are available, one might say that concrete level subjects have an operational range equivalent to that of the sixteen fundamental propositional operations and functionally equivalent to a combinatorial system. However, we do not think that this is the case, for there are no general transformations by which the concrete level subject can pass from one of the groupings of classes or relations to another. But the propositional operations, on the other hand, form a single system such that it is possible to move with accuracy from any one of its sixteen elements to each of the others. For this reason, as we have seen, the coordination of the groupings of classes and relations into a single system requires the introduction of a new structure, that of the "structured whole" with its n-by-n combinatorial system. The starting point in the formation of the latter is the multiplicative groupings; but in addition it involves a generalization of the classification applied to elementary associations (the second-degree groupings described earlier). Once organized, this single system constitutes the system of sixteen binary propositional operations.

But the question which arises next is: is the subject aware of the fact that the system of propositional operations exists as a system? If we mean, by "aware," does he think about the system as a system, it is obvious that he is not, because such a conscious logic never existed prior to the work of logicians. However, as Brunschvicg said, logic is like literary criticism, which codifies the laws of poetry which has already been written but is not present at its creation. In the realm of the adolescent's action logic, which as yet has nothing to do with the formulated logic of the logician, we can only mean by "awareness of the system" that a motivated attempt to look for relationships between the possible operations within the set is made for its own sake. In this sense, the substage III-B subjects seem to be aware of the system,[14] although a phase of gradual organization where little coordination is found takes place during substage III-A. They know when they see an elementary association $p.q$ or $p.\bar{q}$, etc., that it may be included in any one of several combinations ($p * q$, $p[q]$, or $p \supset q$, for example), and they can verify its truth or falsehood more or less systematically. Conversely, when they assume a complex combination such as $p \supset q$ as a hypothesis, they know how to verify it by going back to its elements $p.q$ and $\bar{p}.q$ or by looking for a counterproof in the falsehood of $p.\bar{q}$.

This continuous linking of various operations or possible combinations within a system has two further consequences; the first relates to the subject's train of inferences or reasoning, and the second to the working of integrated operational structures as such.

We have just examined the role of the sixteen binary operations in the thinking of the formal level subjects. However, they do not in themselves constitute reasoning or inferences. They describe only more or less complex judgments but not actual series of inferences. Even the implication $p \supset q$ is limited to asserting a link which can be reduced to a single judgment $p = p.q$. Given this, two questions arise: (1) What is the nature of the actual reasoning of our subjects? (2) Do we have to formulate a new set of links over and above those which we have already described in order to account for the essentially deductive aspect of spontaneous formal thinking?

Logicians are careful to distinguish the two realms of elemen-

[14] Of course it is possible to be aware of an operational system without knowing how to translate it into algebraic symbols; for example, this is true for serial ordering operations.

tary operations from the mechanisms of inference. First, they make the distinction on the basis of the operations themselves: for example, in a deduction such that $[(p \supset q).(q \supset r)] \supset [p \supset r]$ ("if p implies q and p implies r, then p implies r"), the three implications $p \supset q$, $q \supset r$, and $p \supset r$ will be considered as simple or "material" implications, and the implication linking the first two to the third as "inferential" implication. The latter can then be considered an operation of a new and original type. Secondly, from a logical standpoint, it goes without saying that the solution of the problem of deduction or "decision" requires the use of new axioms and a much more complicated formal construction (with distinction between logical syntax and semantics, etc.) that is not necessary in the theory of elementary operations.

However, none of these problems need concern us here, as it may be that the distinction between the two domains is only relevant to the needs of axiomatization or normative logic.[15] From the psychological standpoint, which alone is relevant here, we need to know only whether stage III subjects use operations in their reasoning other than those described above and, if not, how they go about reasoning with these same operational mechanisms. The answer which the facts force on us is very simple. Insofar as the judgments stated by the subjects correspond to operations in the propositional system, and insofar as these operations can be formulated by means of algebraic symbols (as we have formulated them here), the reasoning of these subjects corresponds to the transformations which link these operations together. No further operations need be introduced since these transformations correspond to the calculus inherent to the algebra of propositional logic. In short, reasoning is nothing more than the propositional calculus itself. Although, in the subjects' thought, this calculus is linked to current speech patterns, it can be expressed symbolically in terms of the algebra of propositional logic.

For example: no subject, after he has established any implication $p \supset q$ by observing three associations $p.q$, $\bar{p}.q$, and $\bar{p}.\bar{q}$ and after he has done the same with the implication $q \supset r (= q.r \,\mathrm{v}\, \bar{q}.r \,\mathrm{v}\, \bar{q}.\bar{r})$, doubts that the result is the implication $p \supset r (= p.r \,\mathrm{v}$

---

15 In particular, we do not need to distinguish syntax from semantics, for the operations which our subjects use have meaning from the start.

$\bar{p}.r$ v $\bar{p}.\bar{r}$). Does he have to bring in at a later point, in addition to the elementary implications, the inferential implication discussed above $[(p \supset q).(q \supset r)] \supset [p \supset r]$, or is the calculation sufficient to impose $(p \supset r)$? If we multiply $(p \supset q)$ by $(q \supset r)$ by means of operation (.), we obtain the following ternary composition:

$$(p.q \text{ v } \bar{p}.q \text{ v } \bar{p}.\bar{q}).(q.r \text{ v } \bar{q}.r \text{ v } \bar{q}.\bar{r}) =$$
$$p.q.r \text{ v } \bar{p}.q.r \text{ v } \bar{p}.\bar{q}.r \text{ v } \bar{p}.\bar{q}.\bar{r}$$

However, one can see immediately that these four trios contain $(p.r \text{ v } \bar{p}.r \text{ v } \bar{p}.\bar{r})$, therefore $(p \supset r)$. Thus they furnish the sought-after conclusion, which, moreover, can be calculated in an even more direct way.[16]

But to what do these calculations refer in the subject's mind, since he possesses neither symbols nor symbolic logic? In this particular case, they merely correspond to the subject's awareness that implications are transitive. First, the subject has an idea of implication which he may express in the words: "If x, then y" (without the reciprocal's being true) or "if y, then z" (without reciprocity). In addition, if the subject really has this idea (based on the composition of cases such as $x.y \text{ v } \bar{x}.y$, etc., and on a purely verbal expression of these combinations) we can say with some justification that he has the ability to group these implications into a single sequence: "If x implies y and if y implies z, then x implies z," since the same conceptions and the same combinations corresponding to simple verbal expressions are sufficient for the organization of such a transitive series. (Moreover, they are acquired during the concrete stage for serial ordering of asymmetrical transitive relations: $A < B$, $B < C$, thus $A < C$.) Thus, to say that this reasoning amounts to an operational calculus does

---

[16] The reader should remember that $p \supset q$ may also be written $p = p.q$. Secondly, q in these cases is equivalent to $q = p \text{ v } \bar{p}.q$. We then have the series of equivalences (where $\equiv$ stands for the identity of two kinds of symbolism):

$$(p \supset q) \equiv (p = p.q). \tag{1}$$
$$(q \supset r) \equiv (q = q.r). \tag{2}$$
$$q = p \text{ v } \bar{p}.q \text{ (this follows from 1)}. \tag{3}$$
$$p = p.q.r \text{ (product of 1 and 2)}. \tag{4}$$
$$p = [(p.r).(p \text{ v } \bar{p}.q)] = p.r \text{ (product of 3 and 4)}. \tag{5}$$
$$(p = p.r) \equiv (p \supset r). \tag{6}$$

Thus property (6)–i.e., $p \supset r$, results from (1) and (2).

not imply any mysterious power of intelligence on the subject's part, if one assumes that the act of intelligence is the grouping of operations among themselves. The only mystery (but it subsists in all interpretations of logical reasoning) is that after a certain stage of development, language is adequate to allow for an approximate expression of the combinations that symbolic logic translates into abstract symbols in another context.

Now, if reasoning consists of an operational calculus which the *subject* performs continuously,[17] it follows that the entire set of operations that is available to him constitutes an algebraic structure which can operate as much as a totality (by its structural laws) as in terms of the respective effects of its particular operations. We have seen that the structuring of a combinatorial system presupposes the elaboration of a "structured whole" and consequently of a *lattice* structure with the general laws of reciprocity which characterize it. As for the group structure, it appears in the use of the sixteen binary operations in the systematic utilization of negation or inversion (for example, in the fact that an association $p.\bar{q}$ is understood as contradicting the link $p \supset q$, etc.).

But must we look even further and ask whether we can see that these fundamental structures play a role in the psychological functioning of propositional operations and, especially, in the constructive reasoning to which they lead? By fundamental structures, we mean, of course, the group of four transformations: direct, inverse, reciprocal, and inverse of the reciprocal.[18] We firmly believe that this is the structure actually present, but in order to demonstrate this we must examine what may be called the *operational schemata,* as opposed to the particular operations that constitute these *schemata.*

§ The Formal Structured Operational Schemata

It is clear that, from the start, the psychological functioning of the sixteen binary operations requires structural organization or

---

[17] and having the potentiality of reduction to operations (v), (.), (=), and N.

[18] The I N R C group.

integration, since these operations constitute a system even in the mind of the subject. (This is true whether or not this system immediately acquires the properties of the I N R C group, which will be given further consideration in the following discussion.) Secondly, this structure entails certain forms of reversibility which assure a continuous compensation among the operant transformations, and by this very fact it defines a certain equilibrium state. Moreover, an equilibrium state is a system which entails potential transformations, since their algebraic sum expresses exactly the compensation of possible modifications in both positive and negative directions. Psychologically, this means that, alongside the operations actually performed by the subject, the system itself implies a set of potential transformations which may become manifest or remain latent depending on particular conditions (see Chap. 16).

The problem of operational schemata has to be raised precisely because of the existence of such potentialities; operational schemata are defined as the concepts which the subject potentially can organize from the beginning of the formal level when faced with certain kinds of data, but which are not manifest outside these conditions. From the point of its appearance during substage III-B, and often even during III-A, formal thinking makes its presence known not only by the constant utilization of the sixteen binary propositional operations and some ternary or superior combinations which derive from them but also by the sporadic elaboration of some concepts or schemata which are inaccessible at the concrete level because their development presupposes the earlier operations. These operational schemata consist of concepts or special operations (mathematical and not exclusively logical), the need for which may be felt by the subject when he tries to solve certain problems. When the need is felt, he manages to work them out spontaneously (or simply to understand—*i.e.*, to rework them in cases when academic instruction has already dealt with the relevant concepts). Before the formal level he is not able to do this.

So we have a very important psychological phenomenon here: the synchronized emergence of a set of concepts or operations whose interrelationships are not clear at first (because they are not interdependent in the way the sixteen binary operations are), but which are nevertheless bound up with each other as a result of deeper links. The roots of this relationship must be sought in the combined structure of the lattice and the group, which *is* the structure of formal thought.

If the interconnectedness of these operational schemata is not immediately discernible, it is because the subject constructs them in a way which on the surface is analogous to the way he works out a number of concepts that are useful to him—*i.e.*, by the intermediary of propositional operations, but in a sequence which is determined by unpredictable needs resulting from experimental conditions and the nature of the particular problems. For example, in the course of the experiments described in chaps. 1 to 6, the stage III subjects have elaborated concepts such as the laws of flexibility, the oscillations of the pendulum, the equality between the angles relevant to the rebounds of a ball, the law of floating bodies, etc., by means of formal operations. However, none of these concepts is itself formal from a structural standpoint. Although they do not appear before the formal level because use of formal operations is a prerequisite to their discovery or construction, they are not formal as concepts because they relate to particular experimental situations or to the geometry of physical bodies. But among the concepts or operations that the subject apparently organizes in the same way—*i.e.*, by deduction or invention in the experimental situation after he has observed empirical combinations—there are some which reappear in the most varied situations or in reference to the most diverse problems. Our analysis shows that these concepts have somewhat different properties from the others.

In this category, for example, we would include the notion of proportions. It is applied to several problems whose content is unrelated: equilibrium between action and reaction, combinatorial probability, or even mathematical operations of combinations, etc. Such notions have three common features: (1) They are more general than the others and thus constitute operational schemata susceptible of varied applications rather than concepts in the narrow sense of the term; (2) From the standpoint of psychological development, they are less discovered in objects than deduced or abstracted with the subject's own operational structures serving as the starting point; (3) They all show some relationship to the lattice or group structures, and several of them, to the I N R C group of inversions and reciprocities.

The simultaneous appearance of these schemata at the beginning of the formal stage (III-A) is so important to us because of these three features. Insofar as the operational schemata are bound up with the structural integration which defines the type of equilibrium found in formal thought, one may ask whether the capacity to organize such schemata (a capacity which, may we repeat, can remain latent but becomes manifest when required by the nature of the problems to be solved) does not constitute one of the manifestations of the potential transformations inherent in any form of equilibrium, especially when the equilibrium has a structure as complex as the integrated group and lattice found in the system of formal operations.

I. *The combinatorial operations.* The "structured whole," which can be drawn from a set of 4 elements $p.q$, $p.\bar{q}$, $\bar{p}.q$, and $\bar{p}.\bar{q}$ considered 1-by-1, 2-by-2, 3-by-3, all 4, or 0, constitutes a system $= 16$ combinations; this combinatorial system gives rise to 2^{2^2} of the 16 binary operations (likewise the $16 \times 16 = 256$ ternary operations, etc.). However, we have never encountered a stage III subject or an adult (logicians excluded) who has successfully calculated these 16 possible combinations or who has even become aware of the existence of such a combinatorial system in any explicit form. The deliberate and reasoned use of these combinations is as foreign to the subject who begins to reason formally as are the laws of harmony to the child or to the popular singer who retains a melody or whistles an improvised tune.

But it happens that in the color experiment (Chap. 7), where the subjects are given four flasks containing various liquids and a dropper containing a fifth but are not given any instructions about the combinations, it is precisely at substage III-A that they begin spontaneously to make systematic 1-by-1, 2-by-2, 3-by-3, and 4-by-4 combinations.

Secondly, as we have shown elsewhere,[19] if children are given 5 or 6 cups containing various colored tokens with explicit instructions to make up all possible pairs with tokens taken from the cups, again it is only at substage III-A that a systematic method

---

[19] *La Genèse de l'idée de hasard chez l'enfant,* Chap. 8.

is used to solve the problem. At the concrete level, the combinations remain incomplete and are obtained by simple trial-and-error. As for the permutations [20] (6 permutations between 3 tokens of different colors and 24 permutations between tokens of 4 colors) and arrangements [21] (find all the numbers which can be constructed with 2, 3, 4, . . . figures), it is only at substage III-B that these operations are acquired systematically—*i.e.*, without explicit formulation of mathematical expressions, but with a performance based on an exhaustive method.

There are four ways of explaining the convergence between the spontaneous appearance of mathematical combinatorial operations (which are not taught in school at the ages considered) and the equally spontaneous organization of the propositional combinatorial system. First, one could say that it is nothing more than a coincidence. Secondly, one could consider the (mathematical) calculation of the combinations as the primitive acquisition and the propositional combinatorial system as a derivative application. Third, the propositional combinatorial system could be considered the primitive acquisition and the mathematical combinations the derivative applications. Finally, one could assume that, from the standpoint of the psychological functioning of operations in the subject's thinking, the two kinds of combinatorial systems constitute a single unified mechanism.

The first possibility seems to us untenable on the basis of probability. Of the hundreds of subjects examined (more than 1500 for the problems of experimental induction and at least 300 for the mathematical combinatorial operations), the synchronism of processes described could not be attributed to chance. Secondly it is difficult to hold to the view that two related kinds of acquisitions, one logical and the other mathematical, could be observed at the same ages without any communication being established between the corresponding pigeon-holes in the brain or the mind, even if such pigeon-holes can be said to exist.

The second possibility is improbable, first from the developmental standpoint; and secondly, because its meaning only *seems* to be clear. Against it, it can be said first that it contradicts the developmental data usually found because a mathematical opera-

20 *Ibid.*, Chap. 8.
21 *Ibid.*, Chap. 9.

tion is generally more complicated than the corresponding logical operation. In fact, the difference between the mathematical combination operations and a logical combinatorial system is that the first relate to simple units, whereas the second relates to qualities of objects of which no one is exactly equivalent to another. But it can be said that the unit is more abstract than the qualified object because it presupposes a preliminary elimination of all qualities.[22] Thus it is doubtful that the formation of mathematical combination operations precedes that of the logical combinatorial system. But even if this were the case, what would be the meaning of the hypothesis according to which the latter results from the application of mathematical operations formed earlier? It is important for the reader to understand that our subjects did not discover and do not know any mathematical formulas for the calculation of combinations. What they do discover is an operational procedure which is actually only a simple method of action and not knowledge consciously thought over and formulated. Thus, to say that this procedure first applies to any objects whatever and is finally generalized to logical thought and its deductive operations has no clear meaning, since logical thinking is already effective in the operations applied to objects. Hence, the only question is whether the nascent combinatorial system deals first with objects as units or first with qualities of objects; but, may we repeat, numerous

[22] In the case of the propositional combinatorial system, the elements are the associations $p.q$, $p.\bar{q}$, $\bar{p}.q$, and $\bar{p}.\bar{q}$, which are not mutually equivalent and are even qualitatively quite distinct. It is true that we are dealing here with stated associations relating to objects or qualified events, which implies a greater degree of abstraction. However, the reader will remember that this propositional combinatorial system arises from the general classification of elements in a double-entry table found in concrete multiplicative groupings. Thus at its source, this combinatorial system deals with simple associations of descriptively qualified objects. In the case of colorants, the given elements are *conceived* of as potentially qualitatively different, but they are in fact *perceived* as indistinguishable except by their rank order numbers. In part, this case involves a combination of units. Finally, in the token example, the situation is equally mixed. Each element is a unit, but selected in a qualitatively defined set (by color). Thus it is hardly possible, if analysis is restricted to the several examples of combinatorial systems cited above, to decide which are the simplest and which are the most complex. Although the propositional combinatorial system is purely logical, it makes for variable difficulty (in actual application) according to the given data. As for the other two cases, both involve a mixture of logical and mathematical structures and thus cannot be compared to the first without ambiguity.

analogies (formations of numbers compared to that of classes and relations, etc.) favor the latter solution.

Does this mean that we have to reverse the terms of the second argument and adopt the third? In one sense, yes, but since the logical combinatorial system is also structured in action and not in pure reflection, this third argument does not differ essentially from the fourth. After all, what does the use of the propositional combinatorial system mean for a subject in the process of working out formal structures? Simply this, that when confronted with facts to be explained, he combines their qualitative links according to all the combinations accessible to him (and formulates a complete system without knowing it, in the case of the n-by-n combinations of binary associations). Faced with objects to combine, he does nothing else, save that he may add an enumeration of units to the qualitative combination; for permutations (serial ordering of series), he does the same, etc.

Thus it looks as if the combinatorial operations constitute an operational *schema* that is quite general beginning with a particular stage in development (III-A): in other words, a method or a way of proceeding which on some occasions is adopted spontaneously without conscious or explicit decision and on others used intentionally when the subject is faced with problems whose solution requires a systematic table of combinations. This schema is formal and not concrete, since we have demonstrated that the essential difference between the concrete groupings and the logic of stage III derives from the absence or presence of a systematic table. Finally, we do not have to remind the reader at this point how this table is linked to the lattice structure and the construction of the "structured whole." Nevertheless, we have kept our promise to return to the combinatorial operations so as to make them the first of the "operational schemata" considered in this chapter. The combinatorial operations do not actually belong to the set of propositional operators and do not derive from them; on the contrary, they are the prerequisite condition of their development (and as such they are quite different). Secondly, they can be generalized to new situations as soon as they serve in this development. Thus, these psychological properties show how the combinatorial operations plunge their roots into a deeper reality, the reality of the integrated structure in which the propositional operators originate—and they express some of its laws of totality.

II. *Proportions.* If the combinatorial operations express some of the laws of totality found in the lattice constituted by the "structured whole," in contrast, the proportional schema effects the transition between the schemata originating in the lattice [23] and those which are integral with the group structure, more particularly, the group of inversions and reciprocities (I N R C).

Mathematical proportions consist simply of the equality of two ratios $x/y = x'/y'$. Their formation raises a psychological problem only because it does not take place during the concrete operational stage. The subject at this level can already construct fractions or numerical ratios as naturally as he equalizes differentially distributed quantities. Moreover, from the qualitative standpoint, beginning with the concrete level, we see evidence of an operation that Spearman has called the "eduction of correlates" in which the subject formulates the links in a double-entry table in such a way as to forecast proportions: for example, "Rome is to Italy as Paris is to France." This is why we wonder why the 8- to 11-year-old subjects are not able to discover the equality of two ratios which form a proportion, and why the discovery is not made before the formal level. But in the course of the above experiments (Chaps. 11 to 14), we have been able to observe repeatedly that proportions are not acquired before substage III-A; this has been shown in the most diverse areas (space, speed, probabilities, etc.). It is not enough to explain it as a result of timing in the academic curriculum. In the first place, we have seen subjects construct the notion in the experimental situation before they have learned it in school. Secondly, if it could be understood earlier, we can be sure that its presentation in the academic program would accelerate the timing of acquisition! Thus we have to look for the explanation of its late comprehension in the actual structure of the operations available to the subjects at the different levels.

But, like all the other operational schemata considered here, the proportional schema has two aspects, one logical and the other mathematical. In its general logical form, a proportion is the equivalence of the relations connecting two terms a and β to the

[23] See the *Comment* at the end of Chap. 13.

relations connecting two other terms γ and δ. Thus, by definition we have:

$$\frac{a}{\beta} = \frac{\gamma}{\delta} \text{ if } \begin{array}{l} (1) \quad a.\delta = \beta.\gamma \\ (2) \quad a \vee \delta = \beta \vee \gamma \end{array} \quad \begin{array}{l} (3) \ a.\bar{\beta} = \gamma.\bar{\delta} \text{ and } \bar{a}.\beta = \bar{\gamma}.\delta \\ (4) \ a.\bar{\gamma} = \beta.\bar{\delta} \text{ and } \bar{a}.\gamma = \bar{\beta}.\delta \,. \end{array}$$

For example:

$$\frac{p}{q} = \frac{\bar{q}}{\bar{p}} \text{ because } \begin{array}{l} (1) \ p.\bar{p} = q.\bar{q} \ (=0) \quad (2) \ p \vee \bar{p} = q \vee \bar{q} \ (= p * q) \\ (3) \ p.\bar{q} = \bar{q}.\bar{\bar{p}} \text{ and } p.\bar{\bar{q}} = q.\bar{p} \,. \end{array}$$

Defined in this way, the logical proportions derive from both the group structures of inversions and reciprocities (I N R C) and lattice structures. A given propositional expression (a) is to its reciprocal (β) as the negation of the latter (γ) is to the negation of the first (δ):

$$\frac{a}{\mathrm{R}a} = \frac{\mathrm{C}a}{\mathrm{N}a} \cdot$$

For example:

$$\frac{p \vee q}{p \mid q} = \frac{p.q}{\bar{p}.\bar{q}} \text{, or } \frac{p \supset q}{q \supset p} = \frac{\bar{p}.q}{p.\bar{q}} \text{,}$$

etc., which verifies properties (1) to (4).

On the other hand, the conjunction (the logical product or lower bound) of any two elements of a propositional lattice, for example p and q, is to one of them as the other is to their disjunction (sum, or upper bound):

$$\frac{p.q}{p} = \frac{q}{p \vee q} \text{ whose complete form is } \frac{p.q}{p[q]} = \frac{q[p]}{p \vee q} \,.$$

Thus, the notion of logical proportions is inherent in the integrated structure which seems to dominate the acquisitions specific to the level of formal operations (stage III). For this reason, one can ask whether the elaboration of the operational schema of proportions does not derive from these logical proportions, whose late appearance would be explained by their necessary relationship to the structural integration of the formal stage. Actually, whenever a system of proportions comes into play, before the subject arrives at the calculation of numerical relations, he isolates an anticipatory schema for qualitative proportionality. Second, this schema, simply a logical one at first, leads at a later point to the discovery of metrical proportions. For example, in the balance problem the subject first discovers that a given increase in weight can be compensated by a proportional increase in the distance from the central axis: in placing a light weight at a great distance and a heavy weight at a small distance, he reaches equilibrium and concludes that the four values have a proportional relationship. But, at first, the compensation as well as the proportion are exclusively qualitative. If we let p stand for the statement of an increase in weight and q an increase in distance, and $\bar{p}$ and $\bar{q}$ the corresponding decreases, the subject begins by conceptualizing simply the following links:

$$p = \mathrm{R}\bar{q} \text{ and } \frac{p}{\bar{q}} = \frac{q}{\bar{p}}, \text{ from whence } p.\bar{p} = q.\bar{q} \text{ and } p \vee \bar{p} = q \vee \bar{q} \, ;$$

i.e., that the increase in weight is to the corresponding decrease in distance as an increase in distance is to the corresponding decrease in weight (still independently of any measurement).

The qualitative schema which serves as a starting point for the subject in the discovery of proportionality seems to be of this type. Essentially, it is based on the reciprocity of weights and distances which the experiment suggests to him: the increase in weight can be canceled by taking away the added weight ($\mathrm{N}p = \bar{p}$), but it is equally possible to compensate this increase by reducing the distance ($p = \mathrm{R}\bar{q}$ or $\mathrm{R}p = \bar{q}$). When this is done, the preceding proportion can also be written: [24]

$$\frac{p}{q} = \mathrm{R}\,\frac{\bar{p}}{\bar{q}}, \text{ from whence } p.\bar{q} = \mathrm{R}(\bar{p}.q).$$

The subject's reasoning usually appears in this latter form: increasing the weight and reducing the distance ($p.\bar{q}$) is equivalent to ($\mathrm{R} =$ compensates) decreasing the weight and increasing the distance.

Once these two schemata have been acquired, the subject can

---

[24] In other words, if $p.\bar{q} = \mathrm{R}(\bar{p}.q)$ we have also $p.q = \mathrm{R}(\bar{p}.\bar{q})$ and $p.\bar{p} = \mathrm{R}(q.\bar{q})$—*i.e.*, $(\mathrm{o} = \mathrm{Ro})$ for $p.\mathrm{C}\bar{p} = p.\bar{p}$ and $q.\mathrm{C}\bar{q} = \bar{q}.\bar{q}$ since $\mathrm{C}p = p$ and $\mathrm{C}\bar{p} = \bar{p}$.

at a later point insert the numerical values which are furnished by his measurements, for a metrical proportion corresponds to each one of the two (see proportions [13] and [14] in Chap. 11):

$$\text{to } \frac{p}{q} = \frac{\bar{q}}{\bar{p}} \text{ corresponds } \frac{nx}{ny} = \frac{n : y}{n : x},$$

$$\text{and to } \frac{p}{q} = R \frac{\bar{p}}{\bar{q}} \text{ corresponds } \frac{nx}{ny} = \frac{x : n}{y : n}.$$

Thus, the acquisition of the operational schema of numerical or metrical proportions presupposes qualitative expectations in the form of compensations by equivalence and in the form of logical proportions. The latter are part of the integrated structure from which propositional operations are derived. Such would thus be the explanation of the late appearance of the concept of proportions and the synchronism which is observed between its appearance and the emergence of the combinatorial operations and the other operational schemata which we will take up presently. No doubt, it would seem that one could obtain the proportional schema by numerical quantification with a system of class products as a starting point. Certainly it is prepared for by the use of "eductions of correlates" mentioned above. But the analysis of this elementary structure shows that it does not yet have the general properties found in propositional proportions.[25]

III. *Coordination of two systems of reference and the relativity of motion or acceleration.* The third operational schema relates to our earlier research and not to experiments included in this work.[26] Nevertheless, we think it important to discuss them here, both to demonstrate the general nature of the phenomenon of

[25] For example, in the correlate "hair is to mammals as feathers are to birds," let us call A_1 epidermic organs including both hair and feathers and A'_1 other organs and call A_2 the organs of mammals and A'_2 those of birds (assuming that $A_1 + A'_1 = B_1$ and $A_2 + A'_2 = B_2$). We then have the correlate $A_1 A_2 / B_1 A_2 = A_1 A'_2 / B_1 A'_2$. But it is not possible to conclude that $(A_1 A'_2 + B_1 A'_2) = (B_1 A_2 + A_1 A'_2)$ (*i.e.*, the hair of mammals plus the organs of birds = the organs of mammals plus the feathers of birds), etc. Rather, the similarities in relationships are closer to propositional proportions, but on the condition that they do not deal with mutually identical relations.

[26] See J. Piaget, *Les Notions de mouvement et de vitesse chez l'enfant,* Paris, (Presses Universitaires de France), 1946, Chaps. 5 and 8.

formal operational schemata and to prepare the reader for an understanding of the example to follow; this example concerns mechanical equilibrium, which is dependent upon the same structure.

In this experiment, a snail is set in motion on a plank which can be moved either in the same direction as the motion of the snail or in the opposite direction. The subjects at the concrete level know very well that the snail can move from left to right, then return from right to left by an inverse operation which cancels the preceding. Likewise, they know that if the snail is immobile on the plank, moving it from left to right will cause the snail to end up at the same point (in relation to an external frame of reference) and that the opposite motion would return him to his starting point. But it is not before the level of formal operations that predictions can be made for both sorts of motion simultaneously, for in this case two systems of reference must be coordinated, one of which is mobile and the other immobile. The difficulty lies, for example, in understanding the fact that a movement from left to right made by the snail can be compensated by a displacement of the plank from right to left; in this case the snail remains in the same place (in relation to the frame of reference) without any reverse movement.

Similarly, we set up a number of toy cyclists to move with uniform rhythm and rate in front of an observer. We ask whether the observer will see more cyclists in the same period of time if he remains stationary before his door, if he walks in the opposite direction from the cyclists, or if he walks in the same direction as they do. This problem too is solved only at stage III, and for the same reasons. But at this level, the subject shows a good understanding of relativity: one subject told us, for example, that if the observer marches against the cyclists, "it's as if he didn't move and as if the cyclists went faster."

Actually, the difficulty in these problems lies in distinguishing and combining two types of transformation: (1) *cancelation* (for example, when the snail returns from B to A after having moved from A to B); and (2) compensation (for example, when the snail goes from A to B while the plank is displaced from B to A). Thus the problem involves the coordination of two systems, each involving a direct and an inverse operation, but with one of the systems in a relation of compensation or symmetry with respect to the other.

Moreover, one can see immediately that this coordination is the same one that is attained by the I N R C group, since N is the inverse of I, and C of R, whereas R is symmetrical to or compensates I (reciprocity). So the problem is to distinguish inversion N from reciprocity R at the same time that one is coordinating them. This is why the problem cannot be solved before the formal level, when a schema based on the I N R C group is in operation.

In other words, if we call I the snail's motion from A to B, N will be his motion from B to A; R will be the plank's motion from B to A (thus R = C of N) and C will be the plank's motion from A to B (thus C of I = N of R). But there is no need to work out further this schema of reasoning, for we will now see its equivalent in the problem of mechanical equilibrium.

IV. *The concept of mechanical equilibrium.* When the processes operating in a system of mechanical equilibrium can be directly understood, thanks to perceptual configurations, the subject does not have much trouble in grasping the notion of the equality of opposite directions and he does so from the concrete stage on. For example, this would be true in the case of a balance with fixed weighing pans, where the distances from the central axis do not vary and where the elements in equilibrium are of the same nature (weight) and can be seen to act in opposite directions from each other. But, in this case, no general equilibrium schema is worked out, and difficulties appear the moment the subject is faced with an apparatus in which the elements in equilibrium are no longer of the same nature and, particularly, in which only the action is visible while the reaction remains invisible except for its results. The late appearance in the history of science of the principle of equality between action and reaction, which was unknown to the Greeks, is evidence that such a difficulty exists. Likewise, from the individual standpoint, the general equilibrium schema is not organized before the level of formal operations; and we shall now try to explain why this is the case.

The first reason is natural enough: since the notion of equilibrium requires compensation between the potential transformations of the system, it is a typical example of the notions which establish a connection between reality and possibility. As such, it requires the organization of the cognitive instrument which is especially adapted to possibility—*i.e.*, formal thought. However, if this were the only reason, the notion of equilibrium could be considered as one of those which, although requiring propositional operations as preconditions of discovery, would not necessarily be formal in *structure*.

A second reason immediately comes to mind which leads us to compare the equilibrium schema and the operational schemata of combination and proportion which, as we have just seen, are structurally formal as well as formal in terms of the means by which they are organized. In other words, the notion of equilibrium requires simultaneously the distinction and the intimate coordination of two complementary forms of reversibility—inversion and reciprocity. Inversion takes place in an equilibrium state whenever the elements of the system are modified by adding or taking away an element and, thus, when transformations are involved whose compensation form is the null operation. On the other hand, there is reciprocity when transformations performed in opposite directions do not annul each other (the null operation signifying the absence of actions) but only compensate each other in terms of a form of compensation we call equivalence; the equality of action and reaction is the general form of this compensation by reciprocity. Furthermore, inversion and reciprocity are always integrated here. In adding or in taking away some elements of a part of a system (inversion), one modifies with the same stroke the relations of reciprocity between this part of the system and those parts which equilibrate it. In reverse, to modify the equivalence (reciprocity) between elements acting in opposite directions, obviously one must take away or add something; this is an inversion transformation.

Moreover, this interdependence of transformations by inversion and by reciprocity implies a formal structure. As we have seen, it is only on the level of formal operations that these two forms of reversibility can be integrated into a single system. As opposed to elementary groupings which depend either on inversion (classes) or reciprocity (relations), the propositional operations all have an inverse (N), a reciprocal (R), and a correlative (C)—*i.e.*, the inverse of the reciprocal. With the identical transformation (I), these transformations constitute a commutative group such that

$$NR = C, \ CR = N, \ CN = R \text{ and } NRC = I .$$

But these are exactly the four transformations of the group used by the subjects in explaining equilibrium in a mechanical system. As we have just seen, in order to understand such a system a subject must be able to simultaneously differentiate and coordinate its modifications, using both inversion and reciprocity. We have here a situation which is highly instructive from the psychological standpoint. Naturally, the subject knows nothing about the "group" as a matter for theoretical reflection, but he uses elementary group structures operationally even at the concrete level (for example in the composition of arithmetical additions and subtractions, since they constitute a group of two transformations). When he begins, at the formal level, to analyze mechanical systems in equilibrium, he has to differentiate and coordinate the operant transformations as objective modifications of physical reality; he also structures them according to a group model as a result of the laws of his thought processes. Consequently, this operational model, which happens to be that of any equilibrium, corresponds to the internal equilibrium of his own logical operations without his being aware of it. This occurs in such a way that, in the explanation of a mechanical system in equilibrium, the group of inversions and reciprocities (I N R C group) comes into play on two completely different levels at the same time. First, it governs the propositional operations which the subject uses to describe and explain reality; as such it constitutes an integrated structure at the interior of his thought, a structure of which he is naturally not aware. But second, as a direct result of the first function, it is projected outside into the phenomena under analysis (since, in the given data, these consist of a physical system whose equilibrium represents the very problem to be resolved). Thus the group gives rise to the operational schemata which the subject uses in this and similar situations to account for the physical modifications he finds and their coordination.

For example, in the experiment in which the piston exerts pressure on a liquid (Chap. 10), the four transformations to be distinguished are the following: (1) first, there is an action of the weight of the piston and the additional weights which can be placed on it (direct transformation); (2) inversely, one can take off either the weights placed on the piston or the piston itself (inverse transformation); (3) furthermore, there is reaction of the liquid in the form of a resistance which is a function of its quantity and density (reciprocal transformation); and (4) finally, one can diminish either the quantity or density of the liquid in replacing it with a smaller amount or a less dense liquid (inverse of the reciprocal). Moreover, in examining our subjects, we find that their principal difficulty is understanding that the mass of the liquid exerts pressure, a pressure that acts in the opposite and not in the same direction as that of the piston. Thus, understanding the equilibrium requires not only understanding of the transformations which play the direct and inverse roles (1 and 2) but also an understanding of the specific modifications constituted by cases (3) and (4). The liquid's resistance (3) is also a pressure, but one which is both equivalent to that of the piston and oriented in the opposite direction in such a way that it is a reciprocal and not an inverse transformation, although at the same time it acts in the same direction as the inverse transformation. As for the inverse transformation of the reciprocal (4)—i.e., the diminution of the liquid's resistance—it acts in the same direction as the direct transformation (1) but is not identical to it.

Thus we are not exaggerating when we say that understanding such a system entails both the differentiation and coordination (equally difficult and interdependent) of the four transformations isomorphic to the I N R C group. In this and similar examples, the notion of mechanical equilibrium certainly corresponds to an operational schema structured at the formal level. Its structuring can be explained by the now constant use of inversions and reciprocities in carrying out propositional operations (for example $p \supset q$ transformed into $p.\bar{q}$, or into $q \supset p$). Just as any casual explanation consists of deducing modifications in the empirical world by assimilating them to cognitive operations,[27] the interpretation of mechanical systems by means of the operational schema of equilibrium means (as described above) assimilating the distinct and integrated modifications of the physical sys-

---

[27] Cf. J. Piaget, *Introduction à l'épistémologie génétique*, vol. II, Chap. 8, #10. To appear in English translation.

tem to the fundamental I N R C transformations on which the structural integration of the formal operations used mentally to understand the physical system is based.

V. *The notion of probability.* We now turn to a group of concepts deriving simultaneously from one or two of the above operational schemata which play a sufficiently general role to permit us to rank them with the others in the set of formal operational schemata.

The first is the notion of probability in the form which reaches the formal level of development at stage III—*i.e.*, when it acquires a general combinatorial structure. In our study of the development of conceptions of chance,[28] we learned that the discovery of (spatio-temporal and physical or logico-arithmetical) indeterminacy, as opposed to what is operationally determined, entails even for the concrete level subject a preliminary separating out among what is possible, what is real, and what is deductively necessary (which remain undifferentiated at the preoperational level); whence stems the appearance of an elementary notion of probability by relating favorable and possible cases.[29] However, at the concrete level, the determination of what is possible (and consequently the determination of what is probable) is limited to those cases in which an operational composition is accessible to the subject—*i.e.*, to the cases of additive composition in contrast to combinatorial compositions: for example, when elements A are drawn from a collection B formed of two parts $A + A'$ (thus $A = B - A'$). Thus concrete possibility is nothing more than a prolongation of reality, in situations in which variables are mixed randomly and are indeterminate; the concrete probable still relates to elementary operational groupings. But at the formal level, as we have seen (Chap. 16), possibility is extended to the point where stage III subjects' deduction begins with possibility (hypothesis), to end up at reality conceived of as a realized sector of the total number of possible combinations. It follows that the notion of probability acquires an inherently broader extension and greater accuracy. It is still a relationship between the confirming and the possible cases, but both begin to be calculated as a function of the combinations, permutations, or arrangements com-

[28] *La Genèse de l'idée de hasard chez l'enfant.*
[29] For example, when drawing from a collection of specified sub-sets.

patible with the given elements. The difference is clearly revealed, for example, when the problem is to draw pairs or trios of elements from collections containing several varieties of objects according to a specified numerical distribution. Whereas the concrete level subjects fail because they lack a combinatorial system, those at stage III succeed without any regular difficulty. Thus the combinatorial conception of probability can be considered a formal operational schema, a schema whose formation is easily explained in the light of the general operations examined above. We need not refer back to the examples of probabilistic attitudes noted in the present research in regard to partial chance distributions of results (see Chap. 15).

VI. *The notion of correlation.* Correlation is a notion which derives simultaneously from that of probability and from a structure close to the one governing proportions. Let us take the set of four possible base associations which can be formulated between two propositional affirmations and negations: $p.q \vee p.\bar{q} \vee \bar{p}.q \vee \bar{p}.\bar{q}$. Like many others discussed in this work, this set corresponds to the four compartments of a double-entry table referring to the product of two classes $(A_1 A_2 + A_1 A'_2 + A'_1 A_2 + A'_1 A'_2)$ available from the initial stages of the concrete level. But, two of these possible associations $p.q \vee \bar{p}.\bar{q}$ express the equivalence between p and q, thus a term-by-term correspondence between the values involved in case of serial ordering. For this case, we can speak of perfect positive correlation. But the two other associations, $p.\bar{q} \vee \bar{p}.q$, if they are taken alone, express reciprocal exclusion between p and q, thus an inverse correspondence or a perfect negative correlation. On the other hand, if all four associations occur and they correspond to an equal numerical distribution of the events stated by the propositional conjunctions, there is a zero correlation.

But when the stage III subject wants to find out whether there is a relationship between the facts described by p and q, when the empirical distribution is irregular (thus involving a mixture of chance interferences and underlying causality), he uses a method based on these operations. Since the subjects do not know any metrical formula for correlations (in fact know nothing about mathematical probability), they are limited to the use of a logical schema. They have to guess at the corresponding numerical frequencies and simply compare the number of confirming cases, corresponding to $p.q \vee \bar{p}.\bar{q}$, to the number of nonconfirming cases, corresponding to $p.\bar{q} \vee \bar{p}.q$. If one of the sets is numerically superior to the other with a sufficiently observable margin, they conclude that there is a causal relationship, whether positive or negative, and explain the minority cases by the interference of chance variables.

The problem is to understand why this conception of correlation does not appear during the concrete stage. First, one could answer that we are dealing with a probabilistic notion which is of great complexity even at this point. A correlation is sought only when a causal relationship remains in part veiled by partial chance distribution, and in order to think of separating out these two sorts of elements one has to distinguish the probable from the determinate—i.e., the set of confirming cases within the four possible associations. However, we have just seen that the schema of combinatorial probability is a formal one. But, is it really a question of combinatorial probability, or is the additive mode of composition of the concrete level adequate to solve the problem? The search for correlation does in fact require a combinatorial system, since the subject's problem is not simply to classify the four possible cases but to distinguish the various realized and realizable combinations among them. In order to place $p.q \vee \bar{p}.\bar{q}$ in opposition to $p.\bar{q} \vee \bar{p}.q$, the subject must separate out these two combinations (equivalent to $p = q$ and to $p \vee\vee q$) from the 16 possible combinations such that $p.q \vee p.\bar{q}$—i.e., $p[q]$ or $p.q \vee \bar{p}.q \vee \bar{p}.\bar{q}$—i.e., $p \supset q$, etc. In this sense the correlation schema depends on the propositional combinatorial system. That is why it appears late and cannot be observed before the level of formal operations (Chap. 15).

But the notion of correlation is also related to the concept of proportions, in that the associations $p.q$, $p.\bar{q}$, etc., which appear in the logical structure of correlations (as well as the corresponding numerical values) can be put in the form of proportions. Thus, one of these possible propositions has the special importance of being a negative proportion [30] from the logical standpoint, and

[30] For a definition of the concept of negative logical proportion, see J. Piaget, *Essai sur les transformations des opérations logiques*, Appendix, pp. 226–227.

of corresponding to a null correlation—*i.e.*, the proportion:

$$\frac{p.q}{p.\bar{q}} \equiv N \frac{\bar{p}.q}{\bar{p}.\bar{q}},$$

where

$$(p.q).(\bar{p}.\bar{q}) = (p.\bar{q}).(\bar{p}.q) = 0,$$
$$(p.q) \text{ v } (p.\bar{q}) = N[(\bar{p}.q) \text{ v } (\bar{p}.\bar{q})],$$
$$(p.q) \text{ v } (\bar{p}.\bar{q}) = N[(p.\bar{q}) \text{ v } (\bar{p}.q)], \quad \text{and}$$
$$(p.q) \text{ v } (\bar{p}.q) = N[(p.\bar{q}) \text{ v } (\bar{p}.\bar{q})].$$

From the numerical standpoint, if p corresponds to n, q to m, $\bar{p}$ to n', and $\bar{q}$ to m'; we have:

$$\frac{n.m}{n.m'} = \frac{n'.m}{n'.m'}, \text{ from whence } n.m.n'.m' = n.m'.n'.m.$$

Thus, this proportion is always true, whatever the numbers n, m, n', and m' (and not only if n' is the inverse of n and m' of m). Secondly it corresponds to a correlation which is always zero, since $n.m.n'.m' - n.m'.n'.m = 0$.

VII. *Multiplicative compensations.* The multiplicative compensations are directly related to the notion of proportions, since, if one has $x.y = x'.y'$, one by definition also has $x/x' = y'/y$. But from the psychological standpoint, although there is no doubt that the organization of proportions always begins with the discovery of compensations (see II below), the latter does not always imply the former. Second, there are additive compensations (for example, when one element simply gains what another has lost) whose comprehension naturally begins at the concrete level and, for this reason, comes long before understanding of multiplicative compensations.

We have studied these cases of multiplicative compensations with regard to the potential canceling out of effects between certain factors in flexibility (Chap. 3), etc. But we had already encountered a noteworthy case in another study relative to conservation of volume. If a subject wants to justify the fact that a given volume is conserved even though its form is changed, he needs to understand that what the volume gains or loses in one dimension is compensated by what it loses or gains in the other two. It is clear that a multiplicative compensation, which as always implies the organization of possible proportions, is involved. Take, for example, a parallelepiped of $3 \times 4 \times 5$ units which is transformed into another of $6 \times 3 \times 3.33$. . . . It is evident that if volume is conserved by multiplicative compensation of dimensions, it is because when one of the sides is doubled (from 3 to 6) the product of the other two is reduced by one half ($4 \times 5 = 20$ and 3×3.33 . . . $= 10$). Thus the compensation is based on a proportion such that:

$$\frac{a_1}{a_2} = \frac{b_2 \times c_2}{b_1 \times c_1},$$

or any other analogue.

But the important fact for the theory of operational equilibrium is that the child does not come to control the notions of multiplicative compensation and conservation of volume until the age when he discovers proportions in other areas. But he makes the discovery without metrical calculation and without realizing that the numerical calculation of the compensations that he conceptualizes qualitatively implies the use of proportions.[31] Thus, everything appears to indicate that when an operational schema is organized, the subject discovers its various consequences simultaneously, even without explicitly bringing together the various aspects of the schema. In this case, on the one hand he discovers certain multiplicative compensations, but without realizing that they imply proportions, and on the other he discovers certain proportions, but without extracting multiplicative compensations from them. In a manner which is even more general, he discovers simultaneously notions of proportions, equilibrium, correlations, multiplicative compensations, etc., without realizing that they stem from a common operational base and without knowing anything about the nature of the group (the I N R C group) from which they derive.

For multiplicative compensation, this situation is especially striking. The subject comes to the idea of compensation qualita-

[31] To find the stage when conservation of volume develops, we simply immersed a small ball or a cylinder of modeling clay in one glass, and a second cylinder which is lengthened or shortened, etc., in a second glass; conservation of volume is revealed by the equality of the water levels. In this situation, conservation of volume is understood at 11–12 years. See *Développement des quantités*, Chap. 3.

tively, without using any calculation. Thus he acts in conformity with a sort of schema of expectations, consisting of operations which he could perform to demonstrate the compensation, which is taken for granted. In other words, the compensation is in this way recognized as possible and often as necessary before the operational procedures which could justify it are made explicit. But it is evident that this does not mean that the subject does not employ any operations when he encounters a volume whose form has been changed. His operational schema appears as an expectation only in relation to metrical operations which he does not actually perform (or in relation to allied schemata with which he does not establish any links), but it does require qualitative operations. They enable him to understand that what is taken away on one dimension is not simply put back qualitatively but is conserved as part of a product. We have here something analogous to what can be observed when the subject has the feeling that a proportionality exists before calculating it. In this second case, he understands in advance that the problem involves the identical transformation of the same relationship, while in the first case, multiplicative compensation, he expects the return of the same product. In the second case, he does not establish a link between the notion of proportions and that of multiplicative compensation because he fails to make explicit the metrical operations anticipated by the qualitative logical schema.

VIII. *The forms of conservation which go beyond direct empirical verification.* We have summarized the development of the notion of conservation of volume from the schema of multiplicative compensation. However, although the conservation of volume is acquired only at the beginning of the formal stage, it has this in common with the conservation concepts structured during the concrete stage—although empirical evidence is not enough for their discovery (since they require an operational composition), it does suffice for complete verification. In contrast, there are other conservation concepts that empirical evidence verifies only in the negative sense that it never contradicts them. But it cannot verify them completely in a positive way, because this verification could not take place within the given limits of space and time or would contradict physical conditions which the experimenter has to accept. An important example is the principle of inertia. If the subject wants to demonstrate the conservation of uniform rectilinear motion in a controlled experiment, he has to face the fundamental difficulties that any motion which is created experimentally is eventually slowed down by external obstacles and that his observation is limited in space and time. Thus, the inertia principle has to be deduced and verified from its implied consequences. Strictly speaking, it does not give rise to observable empirical evidence.

However, in relation to the structure of the I N R C group of four transformations to which we have already referred in explanation of the formation of the notion of equilibrium, the substage III-B subjects do come to discover an elementary process whose starting point is the obstacles which would stand in the way of verification—*i.e.*, the causes of loss of motion. The reasoning which follows is extremely simple but that much more significant. From the fact that when any object loses motion (stated by p) the intervention of observable variables is implied (stated by $q \vee r \vee s$ . . .), they come to the hypothesis that in eliminating all of these variables (*i.e.*, $\bar{q}.\bar{r}.\bar{s}$ . . .), all loss of motion would be eliminated at the same time; the result would be conservation of motion (m) with its rate of acceleration (see Chap. 8):

If $p \supset (q \vee r \vee s$ . . .), then $\bar{q}.\bar{r}.\bar{s}$ . . . $\supset \bar{p}$, where $\bar{p} \supset m$.

We can see how this reasoning uses simultaneously negation or inversion (N) and contraposition which is a form of reciprocity (R).

§ The Integrated Structure of Formal Operations as a Final Form of Equilibrium in Mental Operations

The conclusion we may draw from the foregoing analysis is that the various operational possibilities implied by the integrated structure comprising the lattice and group found in formal thinking give rise to the organization of operational schemata. These schemata often differ greatly from each other and are not linked together by the *subject*, but they appear in more or less integrated form and even synchronously during stage III.

Therefore, the next problem is to understand the psychological correlates of this structural integration. Since it is the source not only of the propositional operations first utilized at stage III but also of the operational schemata whose vital role in the intellectual behavior of this level has just been summarized, it appears to play a continuous role in the causal structuring of the subject's behavior and thinking. Moreover, he does not become aware of it in spite of its general importance. Given this fact, it could not be a product of acquired experience. And it could not be the expression of an *a priori* form of the mind, since it does not appear until the last stage of mental growth—*i.e.*, at the level which defines the end of childhood and points the way across adolescence to adult achievement. One could assume that it is linked to late-maturing nervous coordinations, but this, however, still leaves us with the problem of why it does not appear in its totality as soon as it is organized and put into operation in certain determinate organs but rather is limited to partial manifestations such as those discussed above. On this count, an argument which locates such a structure in the nervous system rather than in the mind or in the intellectual unconscious does not further solution of the problem. The real question is whether or not this integrated structure exists "somewhere" ready-made or whether its existence is of another nature. At this point we find that if we want to give it a place in the causal mechanism of intelligence and at the same time give due consideration to the specific characteristics of its mode of action, we have to conceive of it as a type of equilibrium. (Moreover, this leads us back to the considerations set forth in the last chapter and the beginnings of the present one.)

In other words, if the integrated group and lattice structure found in formal thinking is a type of equilibrium, it must exist as a set of possibilities among which only the operations and operational schemata actually organized in performance are manifest. The others must exist only as latent transformations which may appear in performance in the appropriate situation.

Given this, we can understand why the subject would not be aware of the general structure as a totality; this totality is formed in part of simple possibilities. But we also can understand why this totality plays a causal role, since, as we have seen (Chap. 16), psychological possibility can orient manifest mental productions. The set of latent transformations actually constitutes a system in the strict sense of the term; its structure obeys group and lattice laws in such a way that the new organizations of the cognitive field as a whole, made after some parts have already been organized, do not occur at random but are directed toward filling in the totality according to laws of composition dominated by the relationships of inversion and reciprocity. (This is particularly true of the operational schemata discussed in the present chapter.)

We have yet to understand why this form of equilibrium is the inevitable result of all of the earlier mental development and why for this very reason it can be considered final in relation to later stages (adult thinking). First, we have to remember that, from stage to stage in mental development, the equilibrium of actions or operations at each new plateau is both more stable and covers a more extensive field than the preceding ones. Although even perception obeys an equilibrium law, as Gestalt theory has demonstrated, this equilibrium covers nothing more than momentary states whose conditions are continuously modified ("displacements of equilibrium," as physicists say). Sensori-motor actions encompass perceptions and movements in systems which are somewhat more extensive, but which are bound to present situations and subject to the same displacements. With the use of representations, the equilibrium field is widened still more in preoperational behavior, but the coordinations effected still depend on immediately present configurations and are thus always dominated by the succession of displacements of equilibrium. With the structuring of concrete operations, on the other hand, the complete reversibility of the nascent operational systems assures that the equilibrium will maintain a limited stability within the limits of the various fields corresponding to the "groupings" which have been structured. But since the form of these operational systems has not yet been entirely disentangled from the empirical content, we still encounter successive levels of equilibrium, defined in function of the heterogeneous factors to be structured; as yet there is no general form of equilibrium which cuts across these various operations independently of content. But finally, with the appearance of formal thinking, such a form is organized; its necessity is related to the dual requirements that coordination be achieved between a set of operations of diverse kinds and that the form be liberated from particular contents. Thus this general form of equilibrium can be conceived of as final in the sense that it is not modified during the life span of the individual (although it may be integrated into larger systems [polyvalent logics]) and in the sense that it integrates into a single system the groupings which were without operational linkages among each other up to this point. Moreover, it necessarily results from previous development, because structural evolution must be conceived of as a growth of equilibrium, thus as a directed evolution. This does not mean that it is determined by a final cause. The increase of entropy in physics is a good example of a progression toward equilibrium or of directed evolution which is independent of any finalism; the evolution of operations obeys an analogous law, although there are these two differences: (1) operational equilibrium increases in mobility as it increases in stability; and (2) potential or possible transformations play the kind of causal role specific to mental reality that we have analyzed above.

Thus, to consider the integrated structure which directs formal thinking as a form of equilibrium is not the same as to satisfy a simple need for symmetry in relations to physics and even less to support a philosophical thesis. On the contrary, it is a simple consequence of the fact that psychology is not logic. Since it is not logic, it cannot be satisfied with explaining facts by means of an abstract structure. The facts with which its investigations deal are actions, and notably the internalized or mental actions which we call operations. These actions and operations act and react on each other according to causal laws, whereas consciousness translates them in the form of implications in the largest sense of the term (i.e., links between concepts and between values). From the causal standpoint, they may give rise to more or less equilibrated confusion or end up with the organization of equilibrated structures. Empirical observations show the importance of the second potentiality. Where structure is present, psychology must make use of a tool which permits deduction of its possibilities and prediction of effects so that equilibrium forms can be defined. Such a tool is found in symbolic logic; but for psychology its algebra is nothing more than a symbolic translation whose principal utility is that of a means of analysis. The empirical reality behind this symbolic translation is the field of coordinated behavior. The concept of equilibrium proves indispensable to causal explanation from this standpoint; it makes it possible for us to understand how at a given level of development intelligence takes up simultaneously all of the directions opened up in this field as a function of the potential transformations which characterize it as well as of the portions already structured. If neurological considerations come to round out our explanation at some later date, the structures of groupings, lattices, and groups will reappear in this new perspective, and, as a result, these laws of equilibrium will prove to be more general than when linked to behavior patterns alone. Already cybernetics permits us to understand how this linkage is possible, since the solution of a problem in a *homeostat* also proceeds by successive equilibrations within a system with a combinatorial structure (lattices) and essential laws of reversibility (regulations and groups).

18

Adolescent Thinking

It is surprising that in spite of the large number of excellent works which have been published on the affective and social life of the adolescent—we hardly need remind the reader of the studies of Stanley Hall, Compayré, Mendousse, Spranger, Charlotte Bühler, Landis, Wayne Dennis, Brooks, Fleming, or Debesse, or those by psychoanalysts such as Anna Freud and Helene Deutsch, and by sociologists and anthropologists such as Malinowski and Margaret Mead, not to mention others—so little work has appeared on the adolescent's *thinking*.

The few detailed studies of adolescent thinking which do exist are all the more valuable because of their scarcity. But, until now, there have not been enough to approximate a coherent outline of the whole. On the one hand, intelligence tests such as Terman's, Burt's, and especially Ballard's nonsense phrases have furnished information on the hypothetico-deductive nature of formal thought. With a different emphasis a number of works on adolescent mathematical and physical thought—Johannot, Michaud, etc. —have brought out the residues of infantile thinking found throughout adolescence; they result from a sort of overflow of concrete level problems onto a more abstract plane.

In the light of this deficit, in this final chapter, we should like to see whether the results of the earlier chapters—on the experimental thinking of adolescents in situations which impel them toward both action and thought at the same time—enable us to set down the broad lines of this picture which neither tests nor the study of verbal (or even mathematical) thought have outlined before.

From the standpoint of logical structures, this work seems to imply that the thinking of the adolescent differs radically from that of the child. The child develops concrete operations and carries them out on classes, relations, or numbers. But their structure never goes beyond the level of elementary logical "groupings" or additive and multiplicative numerical groups. During the concrete stage, he comes to utilize both of the complementary forms of reversibility (inversion for classes and numbers and reciprocity for relations), but he never integrates them into the single total system found in formal logic. In contrast, the adolescent superimposes propositional logic on the logic of classes and relations. Thus, he gradually structures a formal mechanism (reaching an equilibrium point at about 14–15 years) which is based on both the lattice structure and the group of four transformations. This new integration allows him to bring inversion and reciprocity together into a single whole. As a result, he comes to control not only hypothetico-deductive reasoning and experimental proof based on the variation of a single factor with the others held constant (all other things being equal) but also a number of operational schemata which he will use repeatedly in experimental and logico-mathematical thinking.

But there is more to thinking than logic. Our problem now is to see whether logical transformations fit the general modifications of thinking which are generally agreed—sometimes explicitly but often implicitly—to typify adolescence. We should like to show briefly not only that they do but also that the structural transformation is like a center from which radiate the various more visible modifications of thinking which take place in adolescence.

However, we must begin by eliminating a possible source of ambiguity. We take as the fundamental problem of adolescence the fact that the individual begins to take up adult roles. From such a standpoint, puberty cannot be considered the distinctive feature of adolescence. On the average, puberty appears at about the same ages in all races and in all societies, although there is widespread opinion to the contrary. (In fact, a short delay has been verified in Canada and in Scandinavia, but not the wide gap between north and south, etc., that legend would have us believe.) But the age at which adult roles are taken up varies considerably among societies and even among social milieus. For our purposes, however, the essential fact is this fundamental social transition (and not physiological growth alone).

Thus we will not attempt to relate formal thinking to puberty. There are, of course, a number of links between the rise of formal structures and transformations of affective life which we shall consider in greater detail presently. But these relations are complex and are not one-way affairs. Even at this point, our thinking would be muddled before we started if we wished to reduce adolescence to the manifestations of puberty. For example, one would then have to say that love appears only in adolescence; but there are children who fall in love and, in our societies, what distinguishes an adolescent in love from a child in love is that the former generally complicates his feelings by constructing a romance or by referring to social or even literary ideals of all sorts. But the fabrication of a romance or the appeal to various collective role models is neither the direct product of the neuro-physiological transformations of puberty nor the exclusive product of affectivity. Both are also indirect and specific reflections of the general tendency of adolescents to construct theories and make use of the ideologies that surround them. And this general tendency can only be explained by taking into account the two factors which we will find in association over and over again—the transformations of thought and the assumption of adult roles. The latter involves a total restructuring of the personality in which the intellectual transformations are parallel or complementary to the affective transformations.

However, even though the appearance of formal thought is not a direct consequence of puberty, could we not say that it is a manifestation of cerebral transformations due to the maturation of the nervous system and that these changes do have a relation, direct or indirect, with puberty? Given that in our society the 7–8-year-old child (with very rare exceptions) cannot handle the structures which the 14–15-year-old adolescent can handle easily, the reason must be that the child does not possess a certain number of coordinations whose dates of development are determined by stages of maturation. In a slightly different perspective, the lattice and group structures are probably isomorphic with neurological structures [1] and are certainly isomorphic with the structures of the mechanical models devised by cybernetics in imitation of the brain.[2] For these reasons, it seems clear that the development of formal structures in adolescence is linked to maturation of cerebral structures. However, the exact form of linkage is far from simple, since the organization of formal structures must depend on the social milieu as well. The age of about 11–12 years, which in our society we found to mark the beginning of formal thinking, must be extremely relative, since the logic of the so-called primitive societies appears to be without such structures. Moreover, the history of formal structures is linked to the evolution of culture and collective representations as well as their ontogenetic history. Since Greek adults became aware of some of these structures only in their logical and mathematical reflection, it is probable that the Greek children were behind our own. Thus the age of 11–12 years may be, beyond the neurological factors, a product of a progressive acceleration of individual development under the influence of education, and perhaps nothing stands in the way of a further reduction of the average age in a more or less distant future.

In sum, far from being a source of fully elaborated "innate ideas," the maturation of the nervous system can do no more than determine the totality of possibilities and impossibilities at a given stage. A particular social environment remains indispensable for the realization of these possibilities. It follows that their realization can be accelerated or retarded as a function of cultural and educational conditions. This is why the growth of formal thinking as well as the age at which adolescence itself occurs—i.e., the age at which the individual starts to assume adult roles—remain dependent on social as much as and more than on neurological factors.

[1] We know that W. McCulloch and W. Pitts (*Bull. Math. Biophys.* [Chicago, 1943] Vol. V, pp. 115–133), have applied the schemata of propositional logic to neuronal connections.

[2] See J. Piaget, "Structures opérationnelles et cybernétique," *Année Psychologique*, Vol. 33 (1953), pp. 379–388.

As far as formal structures are concerned, we have often taken special note of the convergence between some of our subjects' responses and certain aspects of instruction in school. The convergence is so striking that we wonder whether the individual manifestations of formal thinking are not simply imposed by the social groups as a result of home and school education. But the psychological facts allow us to reject this hypothesis of complete social determinism. Society does not act on growing individuals simply by external pressure, and the individual is not, in relation to the social any more than to the physical environment, a simple *tabula rasa* on which social constraint imprints ready-made knowledge. For, if the social milieu is really to influence individual brains, they have to be in a state of readiness to assimilate its contributions. So we come back to the need for some degree of maturation of individual cerebral mechanisms.

Two observations arise out of this circular process which characterizes all exchanges between the nervous system and society. The first is that the formal structures are neither innate *a priori* forms of intelligence which are inscribed in advance in the nervous system, nor are they collective representations which exist ready-made outside and above the individual. Instead, they are forms of equilibrium which gradually settle on the system of exchanges between individuals and the physical milieu and on the system of exchanges between individuals themselves. Moreover, in the final analysis the two systems can be reduced to a single system seen from two different perspectives. And this comes back to what we have said many times before.

The second observation is that between the nervous system and society there is individual activity—*i.e.*, the sum of the experience of an individual in learning to adapt to both physical and social worlds. If formal structures are laws of equilibrium and if there is really a functional activity specific to the individual, we would expect adolescent thinking to show a series of spontaneous manifestations expressing the organization of formal structures as it is actually experienced—if adolescence is really the age at which growing individuals enter adult society. In other words, formal development should take place in a way that furthers the growth of the adolescent in his daily life as he learns to fill adult roles.

But first we must ask what it means to fill adult roles? As opposed to the child who feels inferior and subordinate to the adult, the adolescent is an individual who begins to consider himself as the equal of adults and to judge them, with complete reciprocity, on the same plane as himself. But to this first trait, two others are indissolubly related. The adolescent is an individual who is still growing, but one who begins to think of the future—*i.e.*, of his present or future work in society. Thus, to his current activities he adds a life program for later "adult" activities. Further, in most cases in our societies, the adolescent is the individual who in attempting to plan his present or future work in adult society also has the idea (from his point of view, it is directly related to his plans) of changing this society, whether in some limited area or completely. Thus it is impossible to fill an adult role without conflicts, and whereas the child looks for resolution of his conflicts in present-day compensations (real or imaginary), the adolescent adds to these limited compensations the more general compensation of a motivation for change, or even specific planning for change.

Furthermore, seen in the light of these three interrelated features, the adolescent's adoption of adult roles certainly presupposes those affective and intellectual tools whose spontaneous development is exactly what distinguishes adolescence from childhood. If we take these new tools as a starting point, we have to ask: what is their nature and how do they relate to formal thinking?

On a naïve global level, without trying to distinguish between the student, the apprentice, the young worker, or the young peasant in terms of how their social attitudes may vary, the adolescent differs from the child above all in that he thinks beyond the present. The adolescent is the individual who commits himself to possibilities—although we certainly do not mean to deny that his commitment begins in real-life situations. In other words, the adolescent is the individual who begins to build "systems" or "theories," in the largest sense of the term.

The child does not build systems. His spontaneous thinking may be more or less systematic (at first only to a small degree, later, much more so); but it is the observer who sees the system from outside, while the child is not aware of it since he never thinks about his own thought. For example, in an earlier work on the child's representation of the world, we were able to report on a number of systematic responses. Later we were able to construct the systems characterizing various genetic stages. But *we* constructed the system; the *child* does not try to systematize his ideas, although he may often spontaneously return to the same preoccupations and unconsciously give analogous answers.[3] In other words, the child has no powers of reflection—*i.e.*, no second-order thoughts which deal critically with his own thinking. No theory can be built without such reflection.

In contrast, the adolescent is able to analyze his own thinking and construct theories. The fact that these theories are oversimplified, awkward, and usually contain very little originality is beside the point. From the functional standpoint, his systems are significant in that they furnish the cognitive and evaluative bases for the assumption of adult roles, without mentioning a life program and projects for change. They are vital in the assimilation of the values which delineate societies or social classes as entities in contrast to simple interindividual relations.

Consider a group of students between 14–15 years and the *baccalaureat*.[4] Most of them have political or social theories and want to reform the world; they have their own ways of explaining all of the present-day turmoil in collective life. Others have literary or aesthetic theories and place their reading or their experiences of beauty on a scale of values which is projected into a system. Some go through religious crises and reflect on the problem of faith, thus moving toward a universal system—a system valid for all. Philosophical speculation carries away a minority, and for any true intellectual, adolescence is the metaphysical age *par excellence*, an age whose dangerous seduction is forgotten only with difficulty at the adult level. A still smaller minority turns from the start toward scientific or pseudo-scientific theories. But

---

[3] For an example, see *Play, Dreams and Imitation in Childhood*, Chap. IX.

[4] *Translators' note: baccalaureat*—a French examination taken at the end of secondary school or about 18–19 years of age. Although, in its details, the analysis of the adolescent presented below fits the European better than the American pattern, one might suggest that even if metaphysical and political theories are less prominent, the American dating pattern and other phenomena typical of youth culture are a comparable "theoretical" or "as if" working out of types of interpersonal relations which become serious at a later point; thus the difference is one of content but not of structure.

whatever the variation in content, each one has his theory or theories, although they may be more or less explicit and verbalized or even implicit. Some write down their ideas, and it is extremely interesting to see the outlines which are taken up and filled in in later life. Others are limited to talking and ruminating, but each one has his own ideas (and usually he believes they are his own) which liberate him from childhood and allow him to place himself as the equal of adults.[5]

If we now step outside the student range and the intellectual classes to look at the reactions of the adolescent worker, apprentice, or peasant, we can recognize the same phenomenon in other forms. Instead of working out personal "theories," we would find him subscribing to ideas passed on by comrades, developed in meetings, or provoked by reading. We would find fewer family and still fewer religious crises, and especially a lower degree of abstraction. But under different and varied exteriors the same core process can easily be discerned—the adolescent is no longer content to live the interindividual relations offered by his immediate surroundings or to use his intelligence to solve the problems of the moment. Rather, he is motivated also to take his place in the adult social framework, and with this aim he tends to participate in the ideas, ideals, and ideologies of a wider group through the medium of a number of verbal symbols to which he was indifferent as a child.

But how can we explain the adolescent's new capacity to orient himself toward what is abstract and not immediately present (seen from the outside by the observer comparing him to the child), but which (seen from within) is an indispensable instrument in his adaptation to the adult social framework, and as a result his most immediate and most deeply experienced concern? There is no doubt that this is the most direct and, moreover, the simplest manifestation of formal thinking. Formal thinking is both thinking about thought (propositional logic is a second-order operational system which operates on propositions whose truth, in turn, depend on class, relational, and numerical operations) and a reversal of relations between what is real and what is possible (the empiri-

---

[5] Of course, the girls are more interested in marriage, but the husband they dream of is most often "theoretical," and their thoughts about married life as well often take on the characteristics of "theories."

cally given comes to be inserted as a particular sector of the total set of possible combinations). These are the two characteristics— which up to this point we have tried to describe in the abstract language appropriate to the analysis of reasoning—which are the source of the living responses, always so full of emotion, which the adolescent uses to build his ideals in adapting to society. The adolescent's theory construction shows both that he has become capable of reflective thinking and that his thought makes it possible for him to escape the concrete present toward the realm of the abstract and the possible. Obviously, this does not mean that formal structures are first organized by themselves and are later applied as adaptive instruments where they prove individually or socially useful. The two processes—structural development and everyday application—both belong to the same reality, and it is *because* formal thinking plays a fundamental role from the functional standpoint that it can attain its general and logical structure. Once more, logic is not isolated from life; it is no more than the expression of operational coordinations essential to action.

But this does not mean that the adolescent takes his place in adult society merely in terms of general theories and without personal involvement. Two other aspects of his entrance into adult society have to be considered—his life program, and his plans for changing the society he sees. The adolescent not only builds new theories or rehabilitates old ones; he also feels he has to work out a conception of life which gives him an opportunity to assert himself and to create something new (thus the close relationship between his system and his life program). Secondly, he wants a guarantee that he will be more successful than his predecessors (thus the need for change in which altruistic concern and youthful ambitions are inseparably blended).

In other words, the process which we have followed through the different stages of the child's development is recapitulated on the planes of thought and reality new to formal operations. An initial failure to distinguish between objects or the actions of others and one's own actions gives way to an enlargement of perspective toward objectivity and reciprocity. Even at the sensorimotor level, the infant does not at first know how to separate the effects of his own actions from the qualities of external objects or persons. At first he lives in a world without permanent objects and without awareness of the self or of any internal subjective life. Later he differentiates his own ego and situates his body in a spatially and causally organized field composed of permanent objects and other persons similar to himself. This is the first decentering process; its result is the gradual coordination of sensori-motor behavior. But when symbolic functioning appears, language, representation, and communication with others expand this field to unheard-of proportions and a new type of structure is required. For a second time egocentrism appears, but this time on another plane. It still takes the form of an initial relative lack of differentiation both between ego's and alter's points of view, between subjective and objective, but this time the lack of differentiation is representational rather than sensori-motor. When the child reaches the stage of concrete operations (7–8 years), the decentering process has gone far enough for him to be able to structure relationships between classes, relations, and numbers objectively. At the same stage, he acquires skill in interindividual relations in a cooperative framework. Furthermore, the acquisition of social cooperation and the structuring of cognitive operations can be seen as two aspects of the same developmental process. But when the cognitive field is again enlarged by the structuring of formal thought, a third form of egocentrism comes into view. This egocentrism is one of the most enduring features of adolescence; it persists until the new and later decentering which makes possible the true beginnings of adult work.

Moreover, the adolescent manifestation of egocentrism stems directly from the adoption of adult roles, since (as Charlotte Bühler has so well stated) the adolescent not only tries to adapt his ego to the social environment but, just as emphatically, tries to adjust the environment to his ego. In other words, when he begins to think about the society in which he is looking for a place, he has to think about his own future activity and about how he himself might transform this society. The result is a relative failure to distinguish between his own point of view as an individual called upon to organize a life program and the point of view of the group which he hopes to reform.

In more concrete terms, the adolescent's egocentrism comes out in a sort of Messianic form such that the theories used to represent the world center on the role of reformer that the adolescent feels himself called upon to play in the future. To fully understand the adolescent's feelings, we have to go beyond simple observation and look at intimate documents such as essays not written for immediate public consumption, diaries, or simply the disclosures some adolescents may make of their personal fantasies. For example, in the recitations obtained by G. Dumas from a high-school class on their evening reveries, the most normal students—the most retiring, the most amiable—calmly confessed to fantasies and fabulations which several years later would have appeared in their own eyes as signs of pathological megalomania. Without going into the details of this group, we see that the universal aspect of the phenomenon must be sought in the relationship between the adolescent's apparently abstract theories and the life program which he sets up for himself. Then we see that behind impersonal and general exteriors these systems conceal programs of action whose ambitiousness and naïveté are usually immoderate. We could also consider the following sample taken from the dozen or so ex-pupils of a small-town school in Rumansch Switzerland. One of them, who has since become a shopkeeper, astonished his friends with his literary doctrines and wrote a novel in secret. Another, who has since become the director of an insurance company, was interested among other things in the future of the theater and showed some close friends the first scene of the first act of a tragedy—and got no further. A third, taken up with philosophy, dedicated himself to no less a task than the reconciliation of science and religion. We do not even have to enumerate the social and political reformers found on both right and left. There were only two members of the class who did not reveal any astounding life plans. Both were more or less crushed under strong "superegos" of parental origin, and we do not know what their secret daydreams might have been.

Sometimes this sort of life program has a real influence on the individual's later growth, and it may even happen that a person rediscovers in his adolescent jottings an outline of some ideas which he has really fulfilled since. But in the large majority of cases, adolescent projects are more like a sort of sophisticated game of compensation functions whose goals are self-assertion, imitation of adult models, participation in circles which are actually closed, etc. Thus the adolescent takes up paths which satisfy him for a time but are soon abandoned. M. Debesse has discussed this subject of egotism and the crisis of juvenile originality. But we believe that, in the egocentrism found in the adolescent, there is more than a simple desire to deviate; rather, it is a manifestation of the phenomenon of lack of differentiation which is worth a further brief discussion.

Essentially, the process, which at any one of the developmental stages moves from egocentrism toward decentering, constantly subjects increases in knowledge to a refocusing of perspective. Everyone has observed that the child mixes up subjective and objective facts, but if the hypothesis of egocentrism did nothing more than restate this truism it would be worth next to nothing.[6] Actually, it means that learning is not a purely additive process and that to pile one new learned piece of behavior or information on top of another is not in itself adequate to structure an objective attitude. In fact, objectivity presupposes a decentering—*i.e.*, a continual refocusing of perspective. Egocentrism, on the other hand, is the undifferentiated state prior to multiple perspectives, whereas objectivity implies both differentiation and coordination of the points of view which have been differentiated.

But the process found in adolescence on the more sophisticated plane of formal structures is analogous. The indefinite extension of powers of thought made possible by the new instruments of propositional logic at first is conducive to a failure to distinguish between the ego's new and unpredicted capacities and the social or cosmic universe to which they are applied. In other words, the adolescent goes through a phase in which he attributes an unlimited power to his own thoughts so that the dream of a glorious future or of transforming the world through Ideas (even if this

[6] *Translators' note:* This passage refers to an opinion more prevalent in Europe than in America, namely that the authors' work simply demonstrates a normative view of the child as an irrational creature. In the United States, where problems of motivation are more often given precedence over purely intellectual functions both from the normative standpoint and in psychological research, another but parallel misinterpretation has sometimes been made; namely, that in maintaining that the child is egocentric, the authors have neglected the fact that he is capable of love. It should be made clear in this section that egocentrism, best understood from its root meaning—that the child's perception is cognitively "centered on his own ego" and thus lacks a certain type of fluidity and ability to handle a variety of perspectives— is not to be confused with "selfish" or "egoistic."

idealism takes a materialistic form) seems to be not only fantasy but also an effective action which in itself modifies the empirical world. This is obviously a form of cognitive egocentrism. Although it differs sharply from the child's egocentrism (which is either sensori-motor or simply representational without introspective "reflection"), it results, nevertheless, from the same mechanism and appears as a function of the new conditions created by the structuring of formal thought.

There is a way of verifying this view; namely, to study the decentering process which later makes it possible for the adolescent to get beyond the early relative lack of differentiation and to cure himself of his idealistic crisis—in other words, the return to reality which is the path from adolescence to the true beginnings of adulthood. But, as at the level of concrete operations, we find that decentering takes place simultaneously in thought processes and in social relationships.

From the standpoint of social relationships, the tendency of adolescents to congregate in peer groups has been well documented—discussion or action groups, political groups, youth movements, summer camps, etc. Charlotte Bühler defines an expansive phase followed by a withdrawal phase, although the two do not always seem clearly distinguishable. Certainly this type of social life is not merely the effect of pressures towards conformity but also a source of intellectual decentering. It is most often in discussions between friends, when the promoter of a theory has to test it against the theories of the others, that he discovers its fragility.

But the focal point of the decentering process is the entrance into the occupational world or the beginning of serious professional training. The adolescent becomes an adult when he undertakes a real job. It is then that he is transformed from an idealistic reformer into an achiever. In other words, the job leads thinking away from the dangers of formalism back into reality. Yet observation shows how laborious and slow this reconciliation of thought and experience can be. One has only to look at the behavior of beginning students in an experimental discipline to see how long the adolescent's belief in the power of thinking endures and how little inclined is the mind to subjugate its ideas to the analysis of facts. (This does not mean that facts are accessible without theory, but rather that a theoretical construction has value only in relation to empirical verification.)

From this standpoint, the results of chaps. 1–15 of this work raise a problem of general significance. The subjects' reactions to a wide range of experimental situations demonstrate that after a phase of development (11–12 to 13–14 years) the preadolescent comes to handle certain formal operations (implication, exclusion, etc.) successfully, but he is not able to set up an exhaustive method of proof. But the 14–15-year-old adolescent does succeed in setting up proofs (moreover, spontaneously, for it is in this area that academic verbalism is least evident). He systematically uses methods of control which require the combinatorial system—*i.e.*, he varies a single factor at a time and excludes the others ("all other things being equal"), etc. But, as we have often seen, this structuring of the tools of experimental verification is a direct consequence of the development of formal thought and propositional logic. Since the adolescent acquires the capacity to use both deduction and experimental induction at the same time, why does he use the first so effectively, and why is he so late in making use of the second in a productive and continuous task (for it is one thing to react experimentally to an apparatus prepared in advance and another to organize a research project by oneself)? Furthermore, the problem is not only ontogenetic but also historical. The same question can be asked in trying to understand why the Greeks were limited (with some exceptions) to pure deductive thought[7] and why modern science, centered on physics, has taken so many centuries to put itself together.

We have seen that the principal intellectual characteristics of adolescence stem directly or indirectly from the development of formal structures. Thus, the latter is the most important event in the thinking found in this period. As for the affective innovations found at the same age, there are two which merit consideration; as usual, we find that they are parallel to intellectual transformations, since affectivity can be considered as the energetic force of

[7] No one has yet given a serious explanation of this fact from the sociological standpoint. To attribute the formal structures made explicit by the Greeks to the contemplative nature of one social class or another does not explain why this contemplation was not confined to metaphysical ideologies and was able to create a mathematical system.

behavior whereas its structure defines cognitive functions. (This does not mean either that affectivity is determined by intellect or the contrary, but that both are indissociably united in the functioning of the personality.)

If adolescence is really the age at which growing individuals take their place in adult society (whether or not the role change always coincides with puberty), this crucial social adjustment must involve, in correlation with the development of the propositional or formal operations which assure intellectual structuring, two fundamental transformations that adult affective socialization requires. First, feelings relative to ideals are added to interindividual feelings. Secondly, personalities develop in relation to social roles and scales of values derived from social interaction (and no longer only by the coordination of exchanges which they maintain with the physical environment and other individuals).[8]

Naturally, this is not the place for an essay on the psychology of affects; still, it is important to see how closely these two essential affective aspects of adolescence are interwoven with the transformations of behavior brought on by the development of formal structures.

First, we are struck by the fact that feelings about ideals are practically nonexistent in the child. A study of the concept of nationality and the associated social attitudes [9] has shown us that the child is sensitive to his family, to his place of residence, to his native language, to certain customs, etc., but that he preserves both an astonishing degree of ignorance and a striking insensitivity not only to his own designation or that of his associates as Swiss, French, etc., but toward his own country as a collective reality. This is to be expected, since, in the 7–11-year-old child, logic is applied only to concrete or manipulable objects. There is no operation available at this level which would make it possible

---

[8] *Translators' note:* "Interindividual" and "social" are used as oppositional terms to a greater extent in French than in English. The first refers to face-to-face relationships between individuals with the implication of familiarity, and the second to the relationship of the individual to society as a whole, to formal institutional structures, to values, etc. Here the meaning is that the child relates only to small groups and specific individuals while the adolescent relates to institutional structures and to values as such.

[9] J. Piaget and A. M. Weil, "Le développement chez l'enfant de l'idée de patrie et des relations avec l'etranger," *Bulletin international des Sciences sociales* (UNESCO), Vol. III (1951), pp. 605–621.

for the child to elaborate an ideal which goes beyond the empirically given. This is only one among many examples. The notions of humanity, social justice (in contrast to interindividual justice which is deeply experienced at the concrete level), freedom of conscience, civic or intellectual courage, and so forth, like the idea of nationality, are ideals which profoundly influence the adolescent's affective life; but with the child's mentality, except for certain individual glimpses, they can neither be understood nor felt.

In other words, the child does not experience as social feelings anything more than interindividual affects. Even moral sentiments are felt only as a function of unilateral respect (authority) or mutual respect. But, beginning at 13–15 years, feelings about *ideals* or *ideas* are added to the earlier ones, although, of course, they too subsist in the adolescent as well as the adult. Of course, an ideal always exists in a person and it does not stop being an important interindividual element in the new class of feelings. The problem is to find out whether the idea is an object of affectivity because of the person or the person because of the idea. But, whereas the child never gets out of this circle because his only ideals are people who are actually part of his surroundings, during adolescence the circle is broken because ideals become autonomous. No commentary is needed to bring out the close kinship of this affective mechanism with formal thought.

As for personality, there is no more vaguely defined notion in psychological vocabulary, already so difficult to handle. The reason for this is that personality operates in a way opposite to that of the ego. Whereas the ego is naturally egocentric, personality is the decentered ego. The ego is detestable, even more so when it is strong, whereas a strong personality is the one which manages to discipline the ego. In other words, the personality is the submission of the ego to an ideal which it embodies but which goes beyond it and subordinates it; it is the adherence to a scale of values, not in the abstract but relative to a given task;[10] thus it is the eventual adoption of a social role, not ready-made in the sense of an administrative function but a role which the individual will create in filling it.

[10] For the relationship between personality and the task, see I. Myerson, *Les fonctions psychologiques et les oeuvres* (Vrin).

Thus, to say that adolescence is the age at which adolescents take their place in adult society is by definition to maintain that it is the age of formation of the personality, for the adoption of adult roles is from another and necessarily complementary standpoint the construction of a personality. Furthermore, the life program and the plans for change which we have just seen as one of the essential features of the adolescent's behavior are at the same time the changing emotional force in the formation of the personality. A life plan is above all a scale of values which puts some ideals above others and subordinates the middle-range values to goals thought of as permanent. But this scale of values is the affective organization corresponding to the cognitive organization of his work which the new member in the social body says he will undertake. A life plan is also an affirmation of autonomy, and the moral autonomy finally achieved by the adolescent who judges himself the equal of adults is another essential affective feature of the young personality preparing himself to plunge into life.

In conclusion, the fundamental affective acquisitions of adolescence parallel the intellectual acquisitions. To understand the role of formal structures of thought in the life of the adolescent, we found that in the last analysis we had to place them in his total personality. But, in return, we found that we could not completely understand the growth of his personality without including the transformations of his thinking; thus we had to come back to the development of formal structures.

Index